Progress in Ergometry: Quality Control and Test Criteria

Fifth International Seminar on Ergometry

Edited by
H. Löllgen and H. Mellerowicz

With 104 Figures and 88 Tables

Springer-Verlag
Berlin Heidelberg New York Tokyo 1984

Prof. Dr. Herbert Löllgen
Abteilung Innere Medizin, Kardiologie, Sportmedizin,
St. Vincenz-Krankenhaus, D-6250 Limburg (FRG)

Prof. Dr. Harald Mellerowicz
Institut für Leistungsmedizin, Forckenbeckstraße 20,
D-1000 Berlin 33 (FRG)

Technical assistance:
Dr. Inge Heidelbach, Scheffelstraße 27, D-7815 Kirchzarten (FRG)

5th International Seminar on Ergometry
Titisee/Freiburg (FRG), September 29 to October 1, 1983
Sponsored by International Council of Sport Sciences
and Physical Education (ICSSPE)

ISBN 3-540-13570-7 Springer-Verlag Berlin Heidelberg New York Tokyo
ISBN 0-387-13570-7 Springer-Verlag New York Heidelberg Berlin Tokyo

Library of Congress Cataloging in Publication Data. International Seminar on Ergometry
(5th : 1983 : Titisee, Germany and Freiburg im Breisgau, Germany)
Progress in ergometry. Bibliography: p. Includes index. 1. Function tests (Medicine) –
Congresses. 2. Work – Physiological aspects – Congresses. 3. Physical fitness –
Measurement – Congresses. 4. Work measurement – Congresses. I. Löllgen, H. (Herbert),
1943 –. II. Mellerowicz, Harald. III. Title. RC71.8.I58   1983   616.07'54   84-10659

The use of registered names, trademarks, etc. in the publication does not imply, even in the
absence of a specific statement, that such names are exempt from the relevant protective
laws and regulations and therefore free for general use.

Product Liability: The publisher can give no guarantee for information about drug dosage
and application thereof contained in this book. In every individual case the respective user
must check its accuracy by consulting other pharmaceutical literature.

Typesetting and bookbinding: G. Appl, Wemding. Printing: aprinta, Wemding
2119/3140–5 4 3 2 1 0

Table of Contents

List of Senior Authors

Amecke, F., Institut für Kreislaufforschung und Sportmedizin der Deutschen Sporthochschule Köln, Carl-Diem-Weg, D-5000 Köln 41 (FRG)

Bachl, N., Österreichisches Institut für Sportmedizin, Possingergasse 2, A-1150 Wien (Austria)

Borg, G., Department of Psychology, University of Stockholm, S-10691 Stockholm (Sweden)

Bunc, V., Physical Culture Research Institute, Újezd 450, CS-11807 Prague (Czechoslovakia)

Cramer, E., Physikalisch-Technische Bundesanstalt, Institut Berlin, Abbestraße 2–12, D-1000 Berlin 10 (FRG)

Eynde, B. van den, Instituut Voor Lichamelijke Opleiding Tervuurse Vest, B-3030 Herverlee (Belgium)

Franz, I.-W., Institut für Leistungsmedizin, Freie Universität Berlin, Forckenbeckstraße 20, D-1000 Berlin 33 (FRG)

Goffloo, K., c/o Seca, Hammer Steindamm 7–25, D-2000 Hamburg 76 (FRG)

Graham, T.E., School of Human Biology, College of Biological Sciences, University of Guelph, Guelph, Ontario N1E 2 Wl (Canada)

Heck, H., Institut für Kreislaufforschung und Sportmedizin der Deutschen Sporthochschule Köln, Carl-Diem-Weg, D-5000 Köln 41 (FRG)

Heller, J., Physical Culture Research Institute, Újezd 450, CS-11807 Prague (Czechoslovakia)

Heimer, S., Faculty of Physical Culture, Arnoldova 4, YU-41000 Zagreb (Yugoslavia)

Horák, J., Institute of Sportsmedicine, Charles University, Central Army Hospital, Salmovska 5, CS-Prague (Czechoslovakia)

Iliev, I., Department of Physiology, Higher Institute of Physical Culture "G. Dimitrov", T. Kirkova 1, BG-1000 Sofia (Bulgaria)

Jokl, E., University of Kentucky Medical Center, Lexington, Kentucky (USA)

Just, H., Medizinische Universitäts-Klinik, Abteilung Kardiologie, Albert Ludwig-Universität, Hugstetter Straße 55, D-7800 Freiburg (FRG)

Karlsson, J., Department of Clinical Physiology, Karolinska Hospital, S-10401 Stockholm (Sweden)

Ketusinh, O., Sports Science Centre, Sports Organisation of Thailand, Hua Mahk Sports Complex, Bangkok 10240 (Thailand)

Landry, F., Physical Activity Sciences Laboratory and Department of Mechanical Engineering, Laval University, Quebec G1K 7PA (Canada)

Löllgen, H. (formerly: Universität Freiburg), St. Vincenz-Krankenhaus, Auf dem Schafsberg, D-6250 Limburg (FRG)

Macková, E. V., Slezská 107, CS-13000 Praha 3 (Czechoslovakia)

Mellerowicz, H., Institut für Leistungsmedizin, Forckenbeckstraße 20, D-1000 Berlin 33 (FRG)

Novák, J., Physical Culture Research Institute, Újezd 450, CS-11807 Prague (Czechoslovakia)

Novak, L. P., Department of Anthropology, Southern Methodist University of Texas Health Science Center, Dallas, TX (USA)

Reinke, A., Institut für Kreislaufforschung und Sportmedizin der Deutschen Sporthochschule Köln, Carl-Diem-Weg, D-5000 Köln 41 (FRG)

Reißmann, H., Sportphysiologische Abteilung, FB Sport, Johannes Gutenberg-Universität, Saarstraße 21, D-6500 Mainz (FRG)

Rüddel, H., Medizinische Universitäts-Klinik, Sigmund-Freud-Straße 25, D-5300 Bonn (FRG)

Samek, L., Benedikt-Kreuz-Rehabilitationszentrum, Südring 15, D-7812 Bad Krozingen (FRG)

Semiginovský, B., Department of Physiology, Faculty of Physical Education and Sports, Charles University, Újezd 450, CS-11807 Prague (Czechoslovakia)

Simoons, M. L., Thoraxcenter, BD 322, Erasmus University, P. O. Box 1738, NL-3000 DR Rotterdam (The Netherlands)

Stegemann, J., Institut für Kreislaufforschung und Sportmedizin der Deutschen Sporthochschule Köln, Carl-Diem-Weg, D-5000 Köln 41 (FRG)

Stoeckle, B., Rektorat, Albert Ludwig-Universität, H.-v.-Stephan-Straße 25, D-7800 Freiburg (FRG)

Tahy, A., Cardiopulmonary Department, Institute for Pulmonology, H-7257 Mosdós (Hungary)

Thadani, U., Department of Medicine, Division of Cardiology, Oklahoma University Health Sciences Center, P.O. Box 26901, Oklahoma City, OK 73190 (USA)

Ulmer, H.-V., Sportphysiologische Abteilung, FB Sport, Johannes Gutenberg-Universität, Saarstraße 21, D-6500 Mainz (FRG)

Wollschläger, H., Medizinische Universitäts-Klinik, Abteilung Kardiologie, Albert Ludwig-Universität, Hugstetter Straße 55, D-7800 Freiburg (FRG)

List of Contributors

---

1 Page, on which contribution commences

Welcoming Remarks

B. Stoeckle

The 5th International Seminar on Ergometry of the International Council of Sports and Ergometry (ICSPE) continues the tradition of these Seminars, this being the first one held outside Berlin.

Papers will be presented from scientists from Westen and Eastern Europe, Scandinavia, the United States, and Canada. These seminars are significant activities under the leadership of the UNESCO. The main topics of this meeting are standardization and the quantification of ergometric results.

Contact between scientists from all over the world is extremely useful. This holds especially true for analysis of physical fitness and interpretation of ergometric data. Recent results on these topics are important steps towards comparing results and employing a methodology which has been generally agreed upon.

I will hope that the papers presented at this meeting will provide interesting results, and that all the guests from here and abroad will be welcomed in a friendly and hospitable manner. Personal and scientific exchange should take place at this meeting. I wish all the participants a successful meeting.

Welcoming Address

H. Just

On behalf of the Albert Ludwig University and our Medical Faculty I welcome you to the 5th International Seminar of the scientific committee of the ICSSPE, the International Council of Sport Sciences and Physical Education.

We are very pleased that you have chosen Freiburg as the site of your symposium. A short historical reminiscence may illuminate the background for your meeting. Five hundred and twenty-six years ago, that is in 1497, the Albert Ludwig University was founded. A medical faculty was already included at this time, but it was still very small.

During the following 300 years medical education was limited to hospital care for the poor. In those days, however, patient care was not the special duty of physicians; this responsibility was assumed by nurses and monks of the monasteries.

In 1767 a general hospital with 14 beds was instituted in Freiburg for clinical teaching. This facility was made possible through generous donations from Katharina Egg, daughter of a merchant, and from the famous sculptor and painter, Christian Wenzinger. It is worthy to mention that Wenzinger's donation was made on condition that professors be obliged to visit patients "daily or as often as necessary," an unusual request for its time. Wenzinger himself took a special interest in caring for the poor and the sick, and his example has connected medical care with medical teaching from the foundation of the Medical Faculty. The first clinical teaching took place in 1773 in the newly added Nosokomium-Klinikum.

The empress Maria Theresia – Freiburg being then property of the house of Habsburg and therefore belonging to Austria – realized the necessity of instituting clinical training, and established the basis for clinical scientific research by imposing regulations such as the law on autopsy. In 1781 she founded the "Institutum Klinikum," the University Clinic in Freiburg.

In the past 200 years medicine has seen remarkable progress, which paralleled, if not preceded, the explosion-like increase of knowledge in the natural sciences and technology. Today it is possible to comprehend, with detailed measurements, the most subtle and differentiated organ functions. The development of obligatory standards of measurement and quality control for the various tests of function is today the presumption for reliability and comparison of clinical testing and an important basis for all scientific research.

Today, even the capacity for physical exertion can be measured. We have, in Freiburg, a reputation in physical performance medicine. It was here in Freiburg, that Prof. Reindell developed his conception of the heart's adaptation to physical exertion. He pointed out that increased heart size is a physiological normality and may, or indeed must be, the presumption for special physical ability. His school of thought is carried on today at the Institute for Physical Performance Medicine through the work of Prof. Keul, under whose direction metabolic factors important for the physical adaptation of heart and circulation have been determined.

The chairman of our meeting has also contributed to the quantification methods of physical exertion capacity. His findings, which have recently been documented in an issue of the *Documenta Geigy,* are equally important for clinical practice and research. Moreover, he is a qualified clinician and cardiologist. It seems especially important to me that experienced clinicians such as he contribute their knowledge and expertise to the field of physical performance medicine.

This meeting occurs at a historical place. In the year 1770 the 15-year-old daughter of emperor Franz I of Austria and empress Maria Theresia travelled from Vienna to Paris to marry the Dauphin, later king Louis XIV. Accompanied by a great number of attendants, she crossed the road which passes the location of our meeting, full of hope for future glory and importance. Indeed, the intelligent and vivacious queen was received with great appreciation. Her destiny, to be executed in the year 1793, is common knowledge. She bore her fate with outstanding firmness.

I hope that your meeting will be a great success, and that this opportunity for international exchange will be creative and stimulating for all. The Albert Ludwig University, and especially the Medical Faculty, supports your intentions and wishes you the best.

Opening Remarks

E. Jokl

I consider it a privilege to have been asked to open the 5th International Seminar on Ergometry in Titisee. Having participated in most of the preceding seminars, I am in a position to evaluate the remarkable progress that has been made over the years in respect to the scientific role played by the subject. Perhaps it is of interest today to recall the early beginnings of the studies to which our meeting is going to devote its time.

The first ergometric device that was used for exercise studies was located in the laboratories of the Agricultural College *(Landwirtschaftliche Hochschule)* in Berlin more than a hundred years ago. It was constructed at the request of Professor Nathan Zuntz, the college's director. Zuntz had been chief assistant to Professor Pflüger in Bonn prior to his appointment in Berlin. The ergometric device was operated with the help of a steam engine; it served initially as a method of assessing the performance capacity of horses. That research was planned at the request of the Prussian military authorities, who desired information concerning the potential scope and limit of the physical performance of cavalry units under battle conditions. It was only later that tests were undertaken with young soldiers as experimental subjects. The first metabolic analyzers were constructed for the latter studies. The analyzers could be carried by the soldiers. Over the years they were improved and eventually led to the design of measuring units of the kind now in use. The pioneers in research using metabolic analyzers were Zuntz, Adolf Loewy, Geppert, and Caspari. The original models are depicted in the monograph *Höhenklima und Bergwanderungen,* published in 1906, a remarkable opus whose study can be recommended to all students of ergometry.

In the 1920s, Professor A. V. Hill and Professor Otto Meyerhof, both recipients to Nobel Prizes in 1923, introduced the terms "oxygen debt" and "steady state", now routinely used in the literature on exercise physiology. It is of interest to recall that both terms were originally used to describe experimental observations with isolated frog muscles. Their applicability in human biology and medicine came as a surprise.

The adjectives "aerobic" and "anaerobic", fundamental descriptive terms of biochemistry, were first used by Louis Pasteur. It was Otto Warburg who introduced them into German scientific literature 60 years ago. Warburg, whose 100th birthday was celebrated recently, made numerous discoveries of importance to exercise physiology, among them the clarification of the transfer of oxygen from blood into tissues through mitochondria. He identified the role of the latter during his stay at the Marine Biological Institute in Naples in 1913.

Quantitative evaluation and standardization of ergometry owes its advancement to the work of Professor Bruce Dill in Boston and Professor Harald Mellerowicz in

Berlin. The Berlin team was among the first to recognize the possibilities offered by the availability of electronic recording and computing units now ubiquitously attached to ergometric testing apparatus. The methodological advancement thus brought about led to a detailed understanding of the components involved in the oxygen transport from environmental air via respiration and circulation into the tissues and facilitated the introduction of ergometry into clinical cardiology. The "stress test" has become an irreplacable diagnostic device now in use in all departments of cardiology.

I wish the 5th International Seminar on Ergometry much success and close my address with an expression of thanks to Professor Löllgen, who has prepared our meeting with much care, and to Professor Mellerowicz, whose research over the years has decisively contributed to the advancement of ergometry.

Introductory Remarks

J. Horák

Ergometry has recently become an indispensable investigation method not only for sports medicine but also for industrial medicine, internal medicine, and rehabilitation. The continuously growing application of sports activity in primary and secondary prevention of the diseases of civilization gives ergometry a position of permanent importance with regard to noninvasive methods for the investigation of circulation and respiration.

It is necessary to stress the contributions of Knipping's school and its adherents, Bolt, Valentin, Venrath, and Hollmann. In the field of sports medicine and cardiology, much significant work has been done by Reindell and his group. Mellerowicz and his associates have done much highly meritorious work, and his book on ergometry (now in its third edition), published with Smodlaka and his team, represents a further enrichment of the literature and will help to improve functional diagnosis. The merits of the Scandinavian authors Christensen and Åstrand, as well as those of the Switzerland school headed by Fleisch, Rossier, Bühlmann, and Wiesinger, cannot be neglected.

Spiroergometry now represents a worldwide recognized method suitable for noninvasive investigation and examination and for objective assessment of cardiorespiratory system performance. It also assesses the metabolic adaptability of the cardiorespiratory system to standard submaximal or maximal work load. It makes possible an assessment of organism adaptability to physical load and an objective evaluation of sports training and/or physical rehabilitation. In this way, it enriches our knowledge about man and his functional abilities. It is an experimental method. Its accuracy depends on the quality of the instruments, on standardized examination, and on the methods used to evaluate results.

In this connection, attention should be given to the fact that an empirical description of organism activity based on experiment is closely related to the systematization of results. The ability to describe causal relationships between observed phenomena makes it possible to describe empirical regularities based only on the relations between the phenomena. Correlations between observed phenomena also represent one form of empirical regularity. With regard to the systematization of a great number of variables featuring the cardiorespiratory system, a method using a simple linear correlation and regression does not assure reliable interpretation. For such a system, a model of latent variables including factor analysis becomes very important. This model helps to find important relations among the many investigated variables and to distribute them into logically coherent groups. The structure and function of the investigated organism can then be more accurately classified. Circulatory and respiratory systems constitute the only functional complex which fully serves metabolism. The degree of adaptability to physical load as measured on the circulation, respiration, and metabolism – tested by means of ergometry – reflects man's adaptability.

Efforts toward achieving a more objective interpretation of ergometry results have led to the development of precision instruments and calibration equipment, as well as to development of adequate forms of load and load standardization in order to assure the compatibility of results from different investigators.

The four previous international ergometric seminars, which were held in West Berlin, presented the most up-to-date knowledge about ergometry – testing, methods, and new applications in sports medicine and other clinical disciplines. These seminars made possible personal contact among specialists and featured a creative working atmosphere that encouraged vivid discussion, while allowing the possibility to confront results and to interpret them. Therefore, we are grateful to Dr. Mellerowicz and Dr. Löllgen, who have arranged (under the sponsorship of the ICSSPE) the fifth international seminar on ergometry in this beautiful spa, Titisee.

I am sure that the proceedings of this seminar will yield new knowledge and experience, enrich us and our colleagues, and make possible further improvement in the quality of our work to benefit sportsmen, patients, and the older population (particularly regarding prevention of ischemic heart disease by means of physical education and sports). In the name of the ICSSPE, I wish good luck for these proceedings.

Preliminary Remarks on the Present State and Future Tasks of Ergometry

H. Mellerowicz

During the four international seminars on ergometry held in Berlin 1965, 1967, 1972, and 1981, there was an opportunity to exchange experiences and results and to contribute to the development of ergometry. Our efforts at standardization, which were difficult and controversial, led to initial success in 1981.

Standardization is of importance for scientific and practical comparison and reproducibility of ergometric measurements all over the world. This is especially true because ergometry is applied increasingly in nearly all fields of medicine: in internal medicine, cardiology, pneumonology, surgery and orthopedics; for pre- and postoperative diagnosis; in pediatrics and geriatrics; in insurance medicine, industrial medicine, and sports medicine; and in preventive and rehabilitative medicine. However, we are still far from having exactly comparable and standardized ergometric methods in all the fields of medicine and in all the countries in which ergometry is used. To achieve this we probably need several more decades of resolute effort.

Nevertheless, we have taken some steps in the right direction. Let us realize that after more than 100000 years of human history it was only with in the last 100 years that physical performance was measured for the first time on the meter-kilogram-second system of units, i. e., in watts. Physiological and pathological functions during a given ergometric load were defined. It may be said that this new field of measuring methodology proved very useful. Let me remind you that during several thousand years, physicians examined their patients in the supine position in the state of rest. Medical examination during a certain physical performance disclosed a new type of quantitative diagnosis in human medicine.

Ergometry found many applications, e. g., in preventive, curative, and rehabilitative cardiology. Especially the combination of ergometry with other medical methods proved very useful. For example, ECG during ergometry turned out to be a most reliable noninvasive method for early recognition and quantification of coronary artery disease. Many other examples could be given.

Ergometry is still a very young medical method. There is still much to be done to render it really reliable and scientifically well confirmed.

What is to be done in ergometry and what goals may be reached in the near future?

1. Standardized, comparable, and reproducible ergometry is to be applied in scientific and practical medicine.
2. For that we need ergometers providing precise physical and biological measurements. Their calibration is to be examined and certified systematically. Ergome-

ters should be constructed as simply, reliably, and inexpensively as possible and be no more complicated than necessary.

3. Quality control of ergometric measurements must be established. That is the main topic of this seminar. We should come to an agreement concerning practical control of ergometry.

4. Well-confirmed mean values and standard deviations of physiological functions in relation to sex and age during ergometric performance are urgently needed. They must be based on standardized ergometric measurements of a sufficiently large number of demographically representative people. Only a few of the so-called normal values presently used meet these conditions.

5. There is no doubt that a large number of doctors apply ergometry without sufficient knowledge of its fundamentals. Systematic and qualified teaching of ergometry is urgently required in order that ergometry not be discredited before it is well established.

Let us join together to achieve these five goals in the near future.

Introductory Comments on Quality Control and Test Criteria in Ergometry

H. Löllgen

Quality control is an old approach in medicine. In Greek medicine, quality was achieved by control in the teacher-pupil relationship. Tradition and observation contributed to quality control. In the 15th to 18th centuries, quality control was a right of the medical faculties. General practitioners became members of the faculty when starting to practice. At the same time, some British authors requested quality control of medical performance (John Gregory of Edinburgh and Thomas Percival of Manchester). Especially Gregory claimed a quality control of medical performance as the basis of consulting colleagues [1].

Quality control became important in the United States in the 19th century, when a large number of medical schools were founded. In an attempt to establish a comparable quality, the Flexner Report was written with support from the Carnegie Foundation for the Advancement of Teaching [2, 3].

Recently, proposals for some aspects of quality control in ECG and ergometry have been reported by committees of the American Heart Association. Similar recommendations are now being worked on in Germany by committees of the German Society for Cardiovascular Research.

Therefore, it is not surprising that the main topic of this 5th International Seminar on Ergometry was chosen to be quality control and test criteria. Recent investigations of calibration deviations in commercially available ergometers underline the need of such a quality control. Due to the theme, there will be some overlap between standardization and quality control in ergometry.

This meeting is being held for the first time outside of Berlin and has brought together experts from the United States, Canada, and many countries in Europe (from Scandinavia to Bulgaria). Therefore, this book presents papers related not only to the main topic but also to actual aspects of ergometry, thus contributing to progress in ergometry. We also hope that proposals presented in the papers and at the end of the book will contribute to discussions in national and international societies (like ICSSPE) to finally elaborate recommendations on quality control in ergometry.

References

1. Ackerknecht E.H. (1974) Zur Geschichte der medizinischen Ethik. Schweiz Rundsch Med 53: 578–581
2. Ackerknecht E.H. (1979) Der Flexner-Report. Schweiz Ärztezeitung 607–612
3. Rodegra H. (1981) Qualitätssicherung unter medizinhistorischen Aspekten. In: Qualitätssicherung in der Medizin. Selbmann H.K., Schwartz F.W., v. Eimeren W (eds) Springer, Berlin Heidelberg New York, pp 2–10

Quality Control and Test Criteria in Ergometry

H. Löllgen

Introduction

Quality control is an established requirement in medicine. It is strongly related to standardization on the one hand and to test criteria on the other. Quality assurance is a basic procedure in clinical chemistry, but it has been neglected in ergometry. Recent results on ergometer calibration tests yielded deviations between measured and indicated power output of about 50% [2, 11]. Obviously, there is a need for quality control in cardiovascular testing, especially in stress testing. This can be underlined by recommendations proposed by the American Heart Association (AHA) [5] and by recent efforts of the German Society for Cardiovascular Research to establish quality control criteria. Further, as has been shown by Philbrick et al. [15], exercise testing requires methodological improvements in patient selection, data collection and data analysis. The enthusiasm about the Bayes' theorem again highlights the need for an improved approach to interpretation of stress testing, another aspect of quality control [3, 14, 16, 20].

Quality control in ergometry is concerned with three aspects:
- control of instruments
- control of performance of the experimental procedure and stress test protocol, including patient (subject) and investigator bias
- control of the test-related knowledge of the physician and technician, including continuing education and operator training

Instruments and Quality Control

Quality control of instruments can easily be obtained in most cases, as calibration devices have been developed for most apparatus. However, the results mentioned above [11] demonstrate that this goal has not yet been reached. Table 1 gives a checklist for use with exercise testing equipment.

Ergometers

New ergometers should have a certification rendered by the vendor indicating correct calibration. Supervision of the calibration procedure by independent testing laboratories is inevitable. Calibration of ergometers requires special and reliable in-

Table 1. Checklist for exercise testing equipment

Treadmill: handrail assembly, emergency stop, calibration of speed and elevation, service after 1000 h of use

Bicycle ergometer: calibration (checked every ± yr); width of base is one-third its height; handlebars, seat, and/or pedals adjust for height of a patient; visible indications of speed and power output; accuracy 5% [2]

Recorders: ECG: see recommendations of the AHA, e.g., stability of gain and baseline, linearity, frequency limits, response time, chart speed

Receptors of air flow: range 0.03–15 l/s, accuracy better than 2%, small resistance, small dead space; response time: less than 10–30 ms during forced expiratory breathing, stability of gain and baseline, applicability (for details, see [7a, 13])

Receptors of gases: for details, see [17]

Receptors of pressures: recommendations of the AHA and the German Society for Cardiovascular Research

Physical environment: thermometer, hygrometer

Emergency equipment, drugs, and procedures

Pretest procedures: medical history, physical examination, ECG, current medication known by the tester; information and instructions to the patient; written and informed consent

Laboratory staffing: personnel are competent, continuing training and education, emergency training every ± weeks, equipment functions well, testing procedure is known to operator (protocol, end point, contraindications, and indications to stop the test), physician present

Documentation of results

Self-certification: special checklist to be filled out every 6 months and to be maintained in the files of the laboratory

---

struments, which have been developed recently [7, 11]. However, in most cases, the calibration device is not available to the individual purchaser. Therefore, repeated calibration should also be provided by the vendor or by independent testing laboratories. Regular calibration is recommended every 2 years.

A simple approach for a rough calibration is the physiological calibration test [21]. Volunteers ride different ergometers with a given power output and with one ergometer calibrated exactly as the "golden" standard. Physiological variables such as heart rate, ventilation, or oxygen uptake should be identical at the same work load [21].

Detailed recommendations on exercise testing devices, including treadmills and bicycle ergometers, have been presented in an AHA committee report [10], in other papers [2, 9], and in reports at this meeting [2, 9].

Analyser

Analysers are widely used in ergometry, especially when measuring oxygen uptake and ventilation. Analysers in this context are those for

- pressure
- flow
- volume

Manometers usually are calibrated by the manufacturer. In addition, calibration can be performed in the laboratory using standard mercury manometers. This holds

Table 2. Requirements of analysers in ergometry

Range: to be identified

Accuracy (precision): variance of a value after repeated measurements with defined samples (flow, pressure) to be given as coefficient of variance (%) or as standard deviation

Stability of gain and baseline: % of control value

Linearity: % of range, present or not

Response time: t_{90} or 63%

Frequency response (catheters, monitors, direct writers)

    Limits: number of cycles per second (Hz) at which amplification is reduced to 70.7% (3 dB) of its value at the midband frequency

    ECG: frequency limits from 0.05 to 100 Hz

Calibration (ECG): standardized voltage 1 mV $\pm$ 2% rise time of less than 1 ms, time constant not less than 100 s

Chart speed (direct writers): accuracy of paper speed better than 99.5%

Flow analyser: dead space, insensitivity to low flow, no cross sensitivity, independence of physical gas characteristics (gas mixtures)

true for sphygmomanometers or those for recording intravascular pressures. Some of the characteristics which should be fulfilled are shown in the checklist of Table 2. Pneumotachygraphs should be regularly calibrated by devices producing constant air streams with a defined flow. Such devices are commercially available. Similar criteria as presented in the table can be applied for gas analysers. Data such as sensitivity, stability of gain and baseline, linearity, and time response (t_{90}) are essentials in the cardiopulmonary or exercise testing laboratory. These parameters should be checked regularly. Mass spectrometers should be calibrated with at least three test gases of known concentration, as nonlinearity occurs when inert gases are measured over a wide range [17].

The problem posed by quality control in *blood gas analysis* has already been resolved, and there is a standard procedure for use in the laboratory. A number of test sera are commercially available, and progress has recently been made in assuring precision and reliability of sera [13]. The same holds true for lactate analysis, which can be checked for precision by test substances. However, there is a need for quality control in the collection and storage of the blood samples. Recommendations do exist in the literature [13] on how to handle the samples, but control is necessary in daily routine.

Recorders in biomedical investigations should be subject to quality control. This is mostly guaranteed by the manufacturer. However, paper speed, deflection time of the pens, damping, and stability should be checked at regular intervals. Detailed recommendations for ECG recording at rest and during exercise have been given by the above-mentioned AHA report [10].

Besides quality control of single instruments, control assurance is also necessary for the whole setup of the experimental apparatus. In determining $\dot{V}O_2$, for example, time response has to be evaluated for the system as a whole, including analyzers, mixing chamber (if present), and flow- or volume-measuring device. Detailed results of control in similar systems have been presented by Ulmer et al. [19]. Results of our own experiments are presented in Table 3. This implies that technical data of systems for analyzing physical fitness should be evaluated for each laboratory be-

Table 3. Some test criteria evaluated for spiroergometry in an open system

Response time (90%)	O_2 analyser:	0.2 s
	CO_2 analyser:	0.1 s
System (from mask to analyser)	O_2 (0–4.56 vol%)	30.9 s
	CO_2 0–4.2 vol%)	30.4 s
Reproducibility	$(n=4, \dot{V}O_{2max})$	
2.8 ± 70		
s_x in individual tests: ± 66		

fore performing large series of experiments or research studies. Regular check-ups are required to allow quick recognition of defects such as leakages of masks or valves or instability of analyzers.

When analysing quality control, biological variability has to be taken into consideration as a large source of error. Biological variability is composed of

- individual baseline
- individual and intraindividual variability
- random error (day-to-day variability) of a variable (e.g., precision) [22]
- systematic error

In addition, the variability is enhanced by observer bias due to different reasons (ignorance, outdated knowledge, careless interpretation of test results, education due to style of a "school", etc.). Allowable limits for the procedure precision have been presented by Tonks [18]:

$$ALE = \frac{1/4 \text{ (normal range)}}{\text{mean of normal range}} 100\%$$

with ALE as the allowable limit of error.

Control of Stress Testing Performance

Quality assurance is similarly concerned with organization, protocol, and documentation of stress testing. Important points of quality control are pretest information on the patient, pretest examination (physical examination, ECG, roentgenograms), and information and instructions to the patients (Table 4). It is self-evident that the exercise protocol should be standardized (fiction or fact?), even though a wide variety of "standardized" protocols have been reported. The stress testing protocol should be known to all operators and physicians working in the exercise performance laboratory. There is no doubt that regular checking of routine protocol adherence is necessary. This holds also true for the interpretation of the test results. Physicians new in the field of exercise testing need control and supervision, and examination of actual knowledge in this area should be performed by personnel with previous work experience. Quality control deals with continuing education and instruction of all the persons involved in stress testing.

Table 4. Checklist for operation of an exercise laboratory [5]

. . .	Physician present for exercise test
. . .	Laboratory well ventilated with comfortable temperature and humidity range
. . .	Calibrated bicycle ergometer or treadmill or step ergometer
. . .	Oscilloscope visible at working distance
. . .	ECG machine meeting AHA standards
. . .	Defibrillator present and tested before each day of use
. . .	Procedures outlined for management of cardiopulmonary complications
. . .	Medical kit fully stocked with drugs which are within the date for use
. . .	Procedure for accepting patients
. . .	Preexercise ECG
. . .	Preexercise procedures for history taking and examination by physician
. . .	Exercise protocol
. . .	Initial and subsequent loads identified
. . .	End point identified
. . .	Indications to stop test clearly understood by operator of test
. . .	ECG interpreted by physician
. . .	Exercise test interpretation adequate
. . .	Work capacity recorded in patient's chart
. . .	ST changes recorded in patient's chart
. . .	Dysrhythmias recorded in patient's chart
. . .	Format for report to referring physician
. . .	Dated and signed by physician

Continuing education also comprises training for emergency situations. Emergency equipment and drugs for basic and advanced life support should be present and should work. Complications are rather rare in stress testing but emergency preparation demands regular review nonetheless. A written plan for handling complications during stress testing is recommended. Phone numbers for emergency calls should be readily available.

Interpretation of Stress Testing

Wide ranges of intra- and interobserver variability have been reported for exercise electrocardiography [4, 8]. Even the golden rule in coronary artery disease is subject to considerable inter- and intraobserver variability [8]. This underlines the need for control in interpretation of exercise testing. Interpretation requires understanding of pathophysiology and training and experience.

Correct mathematical approaches for interpretation should be considered (Fig. 1).

Quality control in methodology requires a glossary of generally agreed upon terms (Table 5). Based on the Bayes' theorem, new test criteria in exercise testing have been developed. The theoretical background of this approach has been widely presented in papers by several authors [3, 6, 14, 16]. Clinical application of probability theory has become routine in today's cardiology. Most exercise performance laboratories use the terms and the related strategy proposed in Table 5. The model of Bayes' theorem has been extended by combining stress testing with other techniques such as fluoroscopy, thallium scintigraphy, and cardiokymography [3].

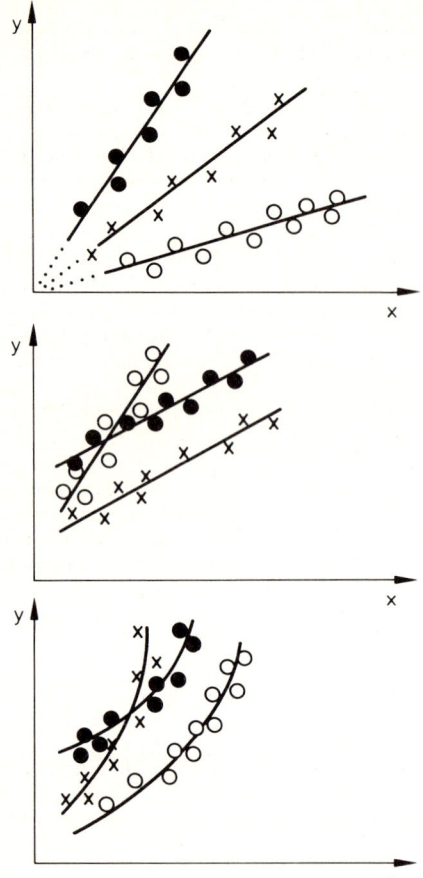

Fig. 1. Examples for correct and incorrect composing of a quotient. *Above:* Quotient correctly formed; *middle, below:* quotient not allowed, as the relationship is not linear or the intercept does not pass zero [12]

The procedures increase the discriminatory power of predicting by stress testing whether coronary artery disease is present or not. Predictability and quality of stress testing is greatly enhanced by these procedures.

Though the wide variations of results of exercise tests have been decreased by decision analysis and use of probability models, the mathematical analysis remains dubious due to methodological uncertainties [6]. This again raises the point of quality control.

A review of 33 studies (with 7501 patients) dealing with stress testing and coronary angiography demonstrated that of "seven methodological standards" (Table 6) "for research design, only one received general compliance" [15]. This means that one study met only five standards, eight met four, and fewer than half of the quoted studies failed to meet two or more standards [15]. These studies have mostly been reported in renowned journals. The critical review by Philbrick et al. [15] emphasizes the need for quality control in research design as well as in the instrumentation used in the research.

Table 5. Glossary of terms of test criteria in use in ergometry

Simplicity	Tests with a complement of brevity for the operator (sometimes requires complex apparatus)
Acceptability	Tests with absence of any cooperative maneuvers by the subject or motivation. Test which is rather simple and not unpleasant for the subject
Objectivity	Results obtained by the test should be independent of instruments used, of emotional participation of the subjects, and of the personality of the operator (lack of subjective influences)
Reproducibility	Variance of test results after repeated measurements. Characterized by confidential interval, by standard deviation, and by coefficient of variation. Error exists due to instrumental or observer bias or randomly due to biological variability. Bias is avoided or decreased by multiple, or at least duplicate, measurements. Repeatability should be small in relation to its absolute magnitude (coefficient of variability preferably not exceeding 8%)
Reliability	Physical, including manufacturing tolerances of a test, overall variability of a test, including *intra*- and *inter*-individual deviations. Includes ease of maintainance, availability of spares
Discrimination	Test should describe an attitude of function which may become impaired during the course of a particular disease. Means: difference between healthy and diseased subjects should be demonstrated distinctly. (Or: discriminatory power). There should be little overlap between the "normal" and "abnormal" values, meaning that the scatter of results is small
Validity	Physiological appropriateness. The appropriate test for detecting or for proving the impaired function or disease
Sensitivity	Probability of an abnormal test result in patients with disease
Specificity	Probability of a negative test result in patients without disease
False positive	Patient with abnormal test results in whom a disease is not present
False negative	Patient with normal test results in whom a disease is present
Predictive accuracy	Percent of positive results in patients with disease (results which are true positive); (= predictive value of a positive test)
Pretest likelihood	Probability of disease in a patient to be tested: $$\frac{\text{Number of patients with disease in the test population}}{\text{Total number of patients in the test population}}$$
Posttest likelihood	$$\frac{\text{Number of patients with disease showing a given test result}}{\text{Total number of patients showing the test result}}$$
Likelihood ratio	Ratio of the true-positive ratio to the false-positive ratio
Prevalence	Number of patients with disease in the number of patients in the study group (or in a population of 100000 subjects)
Risk ratio	Percent of subjects with a positive test who manifest a disease in relation to percent of subjects with negative test who manifest a disease
Summary	TP, true positive; TN, true negative; FN, false negative; FP, false positive

$$\text{Sensitivity (\%)} = \frac{\text{TP}}{\text{TP} + \text{FN}} \times 100$$

$$\text{Specificity (\%)} = \frac{\text{TN}}{\text{TN} + \text{FP}} \times 100$$

$$\text{False positives (\%)} = \frac{\text{FP}}{\text{TP} + \text{FP}} \times 100$$

$$\text{False negatives (\%)} = \frac{\text{FN}}{\text{TN} + \text{FN}} \times 100$$

$$\text{Predictive value (\%)} = \frac{\text{TP}}{\text{TP} + \text{FP}} \times 100$$

$$\text{Risk ratio} = \frac{\text{TP}}{\text{TP} + \text{FP}} \bigg/ \frac{\text{FN}}{\text{FN} + \text{TN}}$$

Table 6. Standards of the diagnostic value of tests for detecting coronary artery disease by means of exercise testing [15]

1. Adequate identification of the study group
2. Adequate analysis of anatomic lesions
3. Consideration of chest pain syndromes
4. Inclusion of patients who might have false-negative for false-positive results
5. Avoidance of bias in clinical and laboratory evaluation
6. Blinding of the interpreters of the angiograms to exercise test results
7. Blinding of the interpreter of the exercise test to the angiographic results

To summarize, quality control in ergometry is an essential part of the exercise testing laboratory. Recommendations on quality assurance should be elaborated and checklists have to be presented.

References

1. Blackburn H (1968) The exercise electrocardiogram: differences in interpretation. Am J Cardiol 21: 871–876
2. Cramer E (1984) Draft of a recommendation for type testing of ergometers for footcranking work. In this volume
3. Diamond GA, Forrester JS (1979) Analysis of probability as an aid in the clinical diagnosis of coronary artery disease. N Engl J Med 300: 1350–1358
4. Detre KM, Wright E, Murphy ML (1975) Observer agreement in evaluating coronary angiograms. Circulation 52: 979–983
5. Ellestadt M, Blomquist CG, Naughton JP (1979) Standards for adult exercise testing laboratories. Circulation 59: 421 A–430 A
6. Ederer F (1975) Patient bias, investigator bias and the double-masked procedure in clinical trial. Am J Med 58: 295–299
7. Eissing G (1982) Eicheinrichtung für Fahrradergometer. Biomed Tech 27: 84–86
7a. Franetzki M, Kresse H: Überblick über die Verfahren der Pneumotachographie mit neuen Meßansätzen. Jahrestagung der Dtsch Ges f Biophysik und Med Technik, 6.10. 1972, Erlangen (Abstract and manuscript)
8. Fröhlicher VF (1983) Exercise testing and training: clinical applications. J Am Coll Cardiol 1: 114–125
9. Goffloo K (1984) Quality criteria and power calibration on ergometers. In this volume
10. Hellerstein HK (1979) Specifications for exercise testing equipment. Circulation 59: 849 A–854 A
11. Hoffmann K, Kuhlmann E (1983) Meßtechnische Untersuchungen an Fahrrad-Ergometern – Zur Kalibrierung von Ergometern. In: Mellerowicz H, Franz I-W (eds) Standardisierung, Kalibrierung und Methodik in der Ergometrie. Perimed, Erlangen, pp 14–22
12. Lange H-J, Hertle FH (1968) Zum Problem der Normalwerte. In: Hertz CW (ed) Begutachtung von Lungenfunktionsprüfungen. Thieme, Stuttgart pp 44–64
13. Löllgen H (1983) Kardiopulmonale Funktionsdiagnostik. Documenta Geigy, Wehr
14. McNeil BJ, Adelstein SJ (1976) Determining the value of diagnostic and screening tests. J Nucleid Med 17: 439–448
15. Philbrick JT, Horwitz RI, Feinstein AR (1980) Methodologic problems of exercise testing for coronary artery disease: group, analysis and bias. Am J Cardiol 48: 807–812
16. Rifkin RD, Hoop WB (1977) Bayesian analysis of electrocardiographic exercise stress testing. N Engl J Med 297: 681–686
17. Smidt U, v Nieding G, Löllgen H (1976) Methodische Probleme der Atemgasmessung. Biomed Tech 21: 102–114

18. Tonks DB (1963) Quality control systems in clinical chemistry laboratories. Postgrad Med J 34: A55–A70
19. Ulmer H-V (1970) Eine Einrichtung zur schnellanzeigenden, analogen und digitalen Registrierung der Stoffwechselgrößen sowie weiterer ergometrischer Daten. Int Z Angew Physiol 28: 292–320
20. Wagner HN (1982) Bayes'theorem: an idea whose time has come? Am J Cardiol 49: 375–377
21. Wilmore JH, Constable SH, Stanforth PR, Buono MJ, Tsao YW, Roby FB, Lowdon BJ, Ratcliff RA (1982) Mechanical and physiological calibration of four cycle ergometers. Med Sci Sports Exerc 14: 322–325
22. Wollschläger H, Löllgen H, Zeiher A, Wieland B, Just H (1984) Significance of longitudinal variance of ergometric measurements. In this volume
23. Mellerowicz H, Franz I-W (eds) (1983) Standardisierung, Kalibrierung und Methodik in der Ergometrie. Perimed, Erlangen

Methodological Aspects and Quality Control in Ergometry - Exercise Testing During Invasive Studies: Reproducibility and Effect of Posture

U. Thadani

Introduction

Bicycle exercise testing is employed to assess cardiac function and the effects of therapy during both noninvasive and invasive studies. Bicycle ergometry can be performed in the supine or upright position. However, it is often not appreciated that a period of exercise may modify subsequent hemodynamic data at rest and during exercise and that the posture by itself modifies hemodynamics. In this communication, the methodology of exercise testing during invasive studies is discussed and the literature on reproducibility and the effects of posture on hemodynamics at rest and during exercise is reviewed.

Exercise Testing During Invasive Studies

In the cardiac catheterization laboratory, bicycle exercise testing is usually performed in the supine position, although sitting position is utilized in some laboratories. Apart from the usual precautions necessary for factors which are known to affect exercise performance (i.e., time of day, environment, temperature, diet, drugs, preliminary testing, anxiety, clothing [1]), many other important factors can modify the hemodynamics during exercise. These factors are: (a) use of inadequately calibrated ergometers, (b) use of manually versus electronically controlled brakes, (c) level of ergometer pedals above or below the table level, (d) comparison of exercise data with resting data with the legs in the horizontal position on the table top or with the resting data obtained with legs on the bicycle pedals in the exercise position, (e) use of fixed work load versus multiple work loads, and (f) variations in zero reference level for the pressure measurements.

It should also be recognized that occasionally motion artifact and catheter movement may make the interpretation of pressure recordings difficult or impossible. Although high-fidelity pressure recordings during exercise can be obtained with catheter tip transducers, most of the laboratories utilize external strain gauge transducers, which have their inherent problems.

Circulatory changes during exercise stabilize after 2–3 min of exercise and many workers prefer to exercise patients for 4–6 min at a given work load. However, this often leads to early fatigue and limits exercise to a single, rather than multiple, work load in many of the patients with heart disease. Therefore, many investigators increase work load progressively every 3 min during exercise and obtain data between the 2nd and 3rd minute of exercise at each work load.

In the supine position, the zero reference level is invariably fixed at 5 cm below the level of angle of Louis, while in the sitting position, a zero reference level at the fourth intercostal space at the sternal border is the usual site [2].

Posture and Hemodynamics: Normal Subjects

Hemodynamics in the Supine Position

The values for resting data obtained in 50 patients in the supine position are shown in Table 1 [3]. These values are in agreement with previous reports [4–6]. The upper limit of normal for left ventricular end-diastolic pressure in the supine position is usually considered to be 12 mmHg. However, a minority of subjects may have values as high as 17 mmHg [3].

The values for hemodynamic data during submaximal exercise in 50 normal, un-trained subjects are shown in Table 1. During supine bicycle exercise, heart rate and cardiac output increase, but the changes in stroke volume have been variable: either little increase in stroke volume [7] or an increase of 5%–20% during exercise has been reported [3, 5, 6, 8, 9]. Increase in cardiac output is dependent upon the work load and this should be taken into consideration when one compares one group of subjects with another [3].

Pulmonary arterial pressure invariably increases during exercise. Changes in left ventricular end-diastolic pressure are variable (Fig. 1); no change, decrease, and in-crease have all been reported. Upper limit of left ventricular end-diastolic pressure during exercise is considered to be 20 mmHg, although higher values may be occasionally encountered [3]. Some workers have compared changes in hemodynamics and pressures during exercise with the resting position in the horizontal position,

Table 1. Summary of hemodynamic data in supine position in normal subjects [2]

	Rest		Exercise		p Values
Heart rate (beats/min)	71	±11	125	± 7	<0.001
Brachial arterial pressure (mmHg)					
Systolic	128	±14	167	±18	<0.001
Diastolic	72	± 8	82	± 8	<0.05
Mean	93	± 9	113	±12	<0.01
Pulmonary arterial pressure (mmHg)					
Systolic	23	± 5	36	± 9	<0.001
Diastolic	9	± 3	17	± 5	<0.001
Mean	14	± 4	24	± 6	<0.001
Right ventricular end-diastolic pressure (mmHg)	4	± 2	6	± 3	<0.05
Left ventricular end-diastolic pressure (mmHg)	9	± 3	13	± 5	<0.05
Cardiac output, $1 \, min^{-1}$	6.3	± 1.5	12.0	± 3.0	<0.001
Stroke volume, cm^3	82	±22	95	±23	<0.05
Left ventricular stroke work, $J \, beats^{-1}$	0.93±	0.27	1.26±	0.34	<0.001

$n=50$; mean ± SD

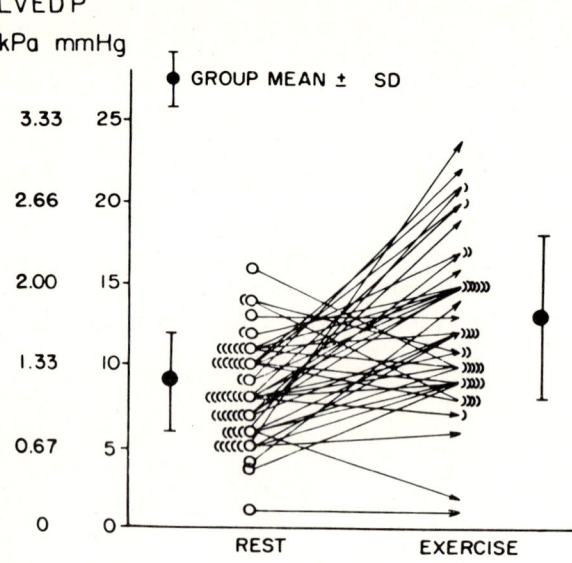

Fig. 1. Individual and group values of left ventricular end-diastolic pressure (LVEDP) during exercise in normal subjects [3]

while others have compared them with the values obtained at rest in the exercise position with feet either above or below the horizontal. This, in all probability, accounts for variable results from different laboratories.

Effect of Posture

The effect of posture on cardiac output and stroke volume has been extensively studied. When a person stands or sits, venous pooling in the legs occurs and this leads to reduction in venous return, a fall in cardiac output and stroke volume, and an increase in heart rate [2–6, 9–11]. Beginning at the lower baseline in the sitting position, stroke volume increases significantly from the resting values during exercise and approaches the values obtained during supine rest [4, 8–11]. During exercise in the sitting position, lower values for cardiac output and stroke volume have been reported in normal, well-trained subjects, but untrained subjects were found to have similar cardiac outputs in the supine and upright positions [10].

The absolute values of mean pulmonary capillary wedge pressure and left ventricular end-diastolic pressure are lower in the sitting than in the supine position both at rest and during exercise (Fig. 2) [10]. At rest, left ventricular stroke work is lower in the sitting position and increases steeply during exercise.

When one compares the changes from rest to exercise, the increases in heart rate, systolic blood pressure, mean systemic arterial pressure, mean pulmonary capillary wedge pressure, left ventricular end-diastolic pressure, cardiac index, stroke index, and left ventricular work index (Fig. 3) were reported to be similar in the supine and sitting positions [10].

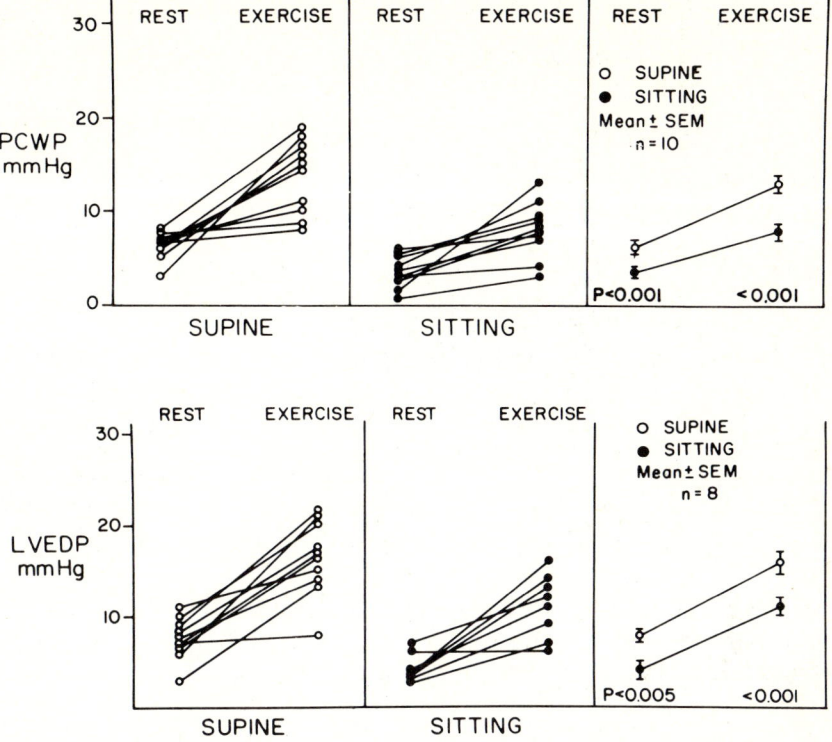

Fig. 2. Individual and group values of mean pulmonary capillary wedge pressure (PCWP) and left ventricular end-diastolic pressure (LVEDP) at rest and during exercise in normal subjects [10]

Posture and Hemodynamics: Coronary Artery Disease

Hemodynamics in the Supine Position

At rest, the values for various circulatory and hemodynamic data in patients with coronary artery disease overlap those reported in normal subjects. However, in some patients values for pulmonary capillary wedge pressure and left ventricular end-diastolic pressure may be higher than 17 mmHg. During exercise, heart rate, systemic and pulmonary arterial pressures, pulmonary capillary wedge pressure, and left ventricular end-diastolic pressure increase, and values of left ventricular end-diastolic pressure above 25 or 30 mmHg are not uncommon [2, 12, 13]. Cardiac output during exercise increases in the majority of patients, but stroke volume either does not change or decreases [2, 12, 13].

Effect of Posture

The resting values for cardiac output, stroke index, and stroke work index are lower in the sitting than in the supine position. The values for pulmonary capillary wedge

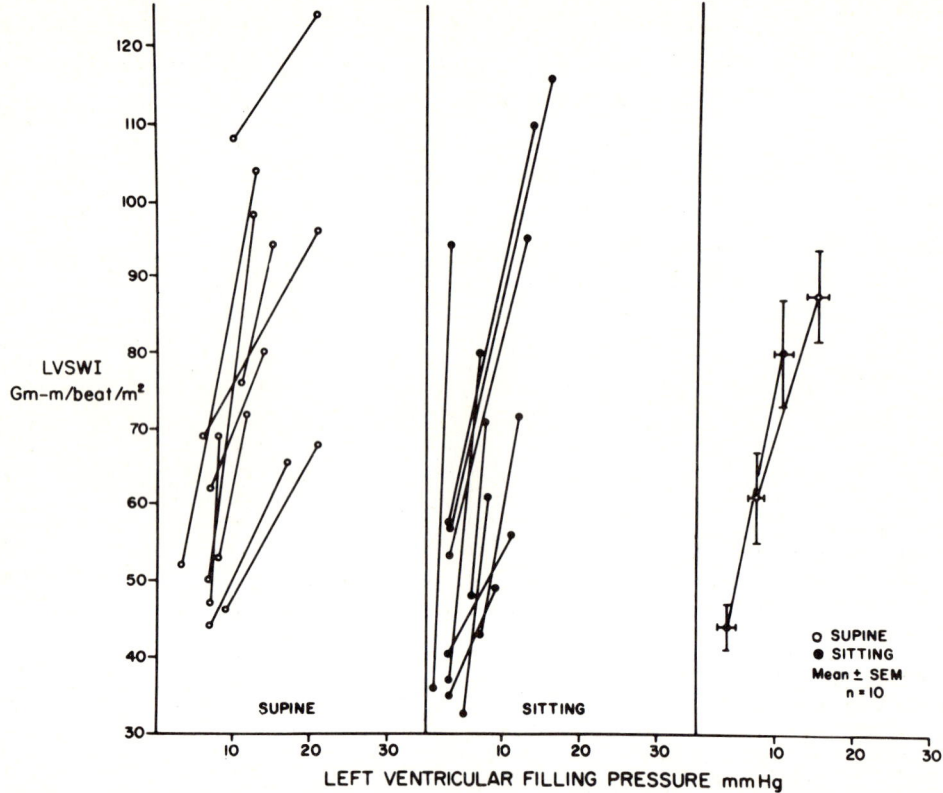

Fig. 3. Relationship between left ventricular filling pressure and stroke work (LVSWI) at rest and during exercise in the supine and sitting position in normal subjects. The function curves show a steep rise during both sitting and supine exercise [10]

and left ventricular end-diastolic pressure in the sitting position overlap those obtained in normal subjects [2]. During exercise in the sitting position, heart rate, systemic and pulmonary arterial pressures, left ventricular end-diastolic pressure, cardiac output, stroke volume, and left ventricular stroke work increase significantly compared with resting values [2]. In the sitting position, values for pulmonary arterial mean pressure, pulmonary capillary wedge pressure, and left ventricular filling pressure have been reported to be lower in comparison with supine position at identical work loads and at the onset of angina (Fig. 4) [2]. However, patients can often exercise to higher heart rates, cardiac index, stroke index, and rate pressure product before experiencing angina in the sitting than in the supine position [2, 14, 15].

Values for left ventricular end-diastolic pressure above 20 mmHg during supine exercise have been considered abnormal, but some 15%–20% of the patients with coronary artery disease may experience angina without elevation in left ventricular end-diastolic pressure above 20 mmHg [2]. In the upright position in some 40%–50%, left ventricular end-diastolic pressure may not increase above 15 mmHg [2]. Some patients who have abnormal left ventricular end-diastolic pressure in the supine position may have normal values during exercise-induced angina in the sit-

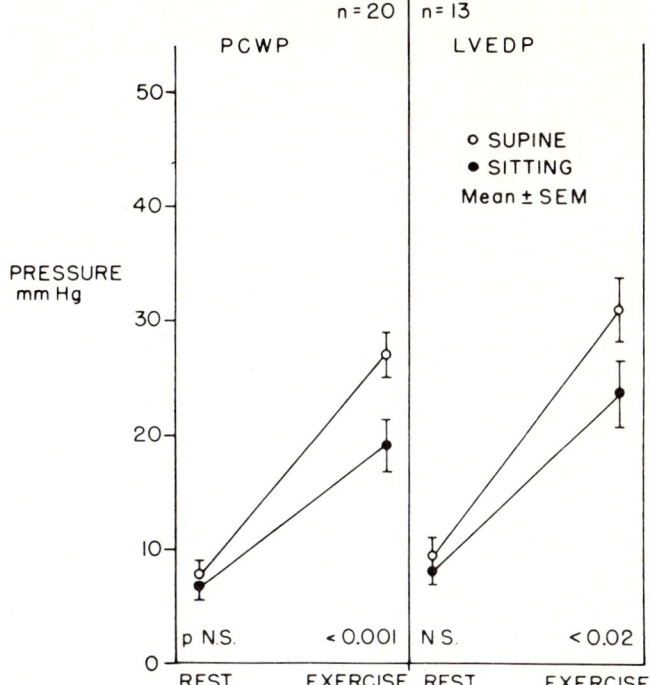

Fig. 4. Group mean values for pulmonary capillary wedge pressure (PCWP) and left ventricular end-diastolic pressure (LVEDP) at rest and during exercise-induced angina. Values during exercise are lower in the sitting than the supine position [2]

ting position (Fig. 4) [2]. It has also been shown that there is a good correlation between the left ventricular end-diastolic pressure and mean pulmonary capillary wedge pressure both at rest and during exercise in both the supine and sitting positions [2].

When one compares the changes from rest to exercise, the changes in mean pulmonary arterial pressure, mean pulmonary capillary wedge pressure, and left ventricular end-diastolic pressure have been reported to be lower in the sitting position (Fig. 4). However, higher levels of heart rate; systolic, diastolic, and mean systemic arterial pressures; cardiac output; and left ventricular stroke work index have been reported in the sitting position. Stroke volume increases in the sitting position but not in the supine position (Fig. 5) [2].

Reproducibility of Hemodynamics During Exercise

The reproducibility of the anginal threshold and circulatory changes have been well documented during noninvasive treadmill and bicycle exercise testing. Very few studies have been carried out regarding the reproducibility during invasive testing.

In normal subjects during upright exercise, significant differences in heart rate, blood pressure, and stroke volume during two successive periods of exercise separated by a period of 30 min rest were reported [16]. However, other workers have reported good reproducibility during two successive periods of exercise [17].

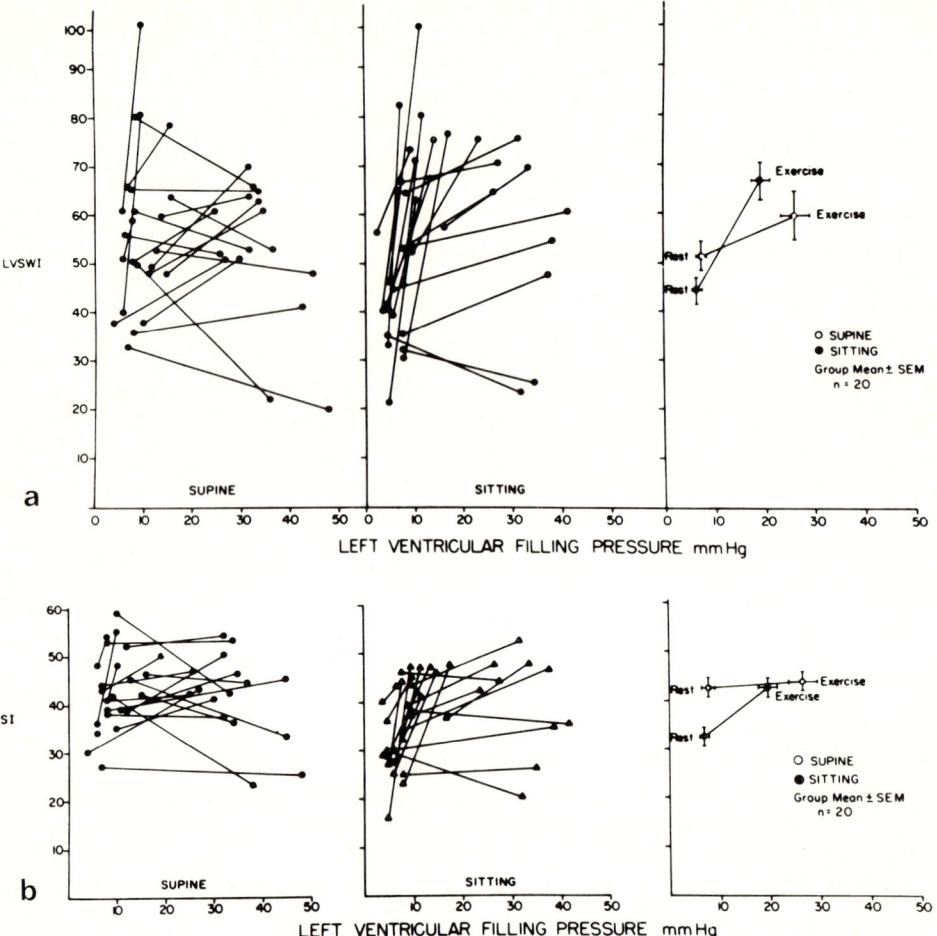

Fig. 5. a Relationship between left ventricular filling pressure and left ventricular stroke work index (LVSWI [g–m/beat per m²]) and ***b*** stroke index (SI [ml/beat per m²]) at rest and during exercise – induced angina. At rest the group values for LVSWI and SI are significantly lower in the sitting position. However, during exercise in the sitting position, LVSWI is significantly higher and left ventricular filling pressure significantly lower than during supine exercise, whereas SI is similar in the two positions [2]

In patients with coronary artery disease, during two successive exercise periods separated by 20–30 min, time-to-angina may not be reproducible and values for brachial arterial, systolic, and mean pressures, pulmonary arterial mean pressure, and left ventricular end-diastolic pressure (Fig. 6) have been reported to be lower during the second exercise study [18]. However, angina was reproduced at the same heart rate. Rate pressure product and cardiac output during exercise were reported to be similar [18]. Lower values for pulmonary arterial pressure, pulmonary capillary wedge pressure, and oxygen uptake have been reported when two exercise periods have been separated by 1 h [19, 20].

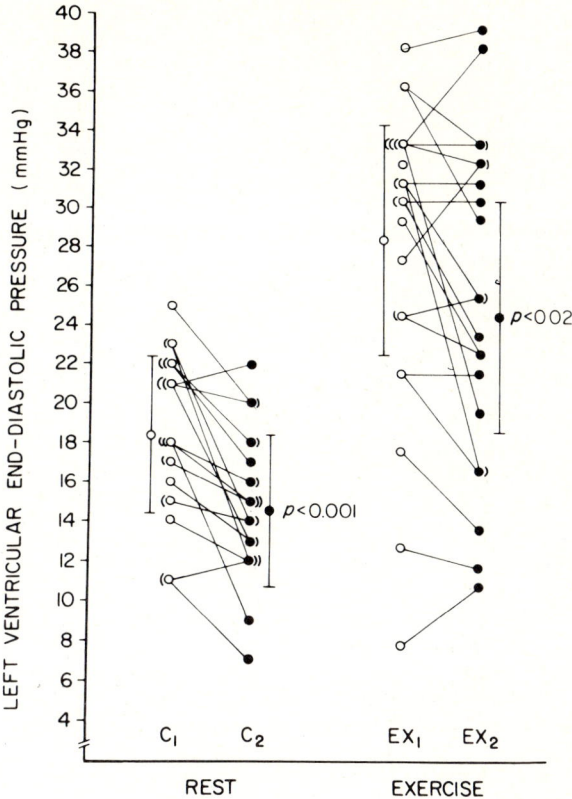

Fig. 6. Individual and group values for left ventricular end-diastolic pressure (LVEDP) during two rest (C$_1$, C$_2$) and exercise (Ex$_1$, Ex$_2$) studies. The group values for LVEDP are lower during the second study both at rest and during exercise [18]

In individual patients, values for various parameters during two successive periods of exercise might be highly variable [18].

In one study, we evaluated the effects of repeated exercise in the supine posture in 12 patients with coronary artery disease. The initial period of exercise modified the subsequent hemodynamics at rest and during successive periods of identical exercise testing [21]. The second period of exercise, however, modified subsequent hemodynamics only slightly (Fig. 7). Hemodynamic data during second and third exercise periods were similar except for systemic arterial pressure which was lower during the third exercise period [21].

In summary, various studies have shown that hemodynamics are modified by posture and this should be taken into consideration when one is evaluating the results of a hemodynamic study. Further, hemodynamics during two successive periods of exercise are not always reproducible, and this should be taken into account when effects of therapy or intervention are being evaluated during successive studies.

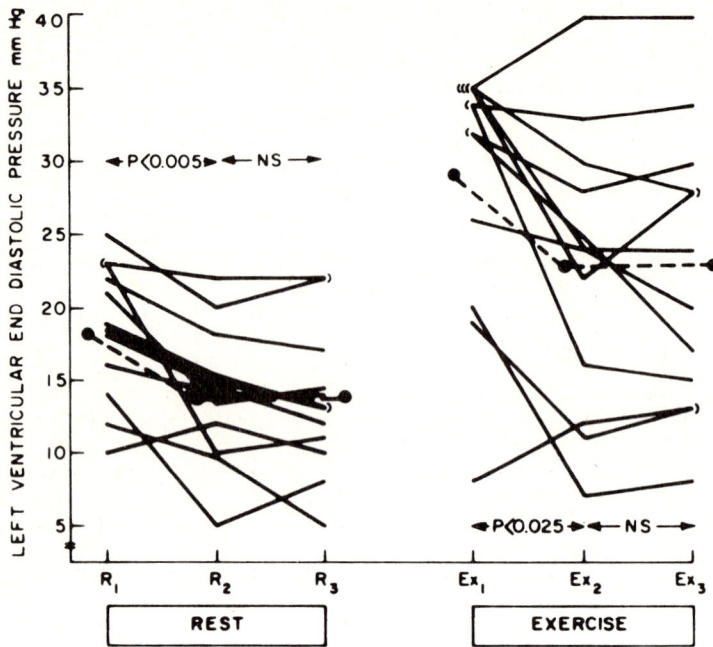

Fig. 7. Individual and mean (●) values for left ventricular end-diastolic pressure during three rest periods (R_1, R_2, and R_3) and three exercise periods (Ex_1, Ex_2, and Ex_3). Mean values are lower during R_2 than during R_1, but similar during R_2 and R_3. Mean values are lower during Ex_2 than during Ex_1, but similar during Ex_2 and Ex_3 [21]

Summary

Bicycle exercise is widely utilized to assess cardiac function during invasive studies. The factors which can modify hemodynamics, the effect of a period of exercise on subsequent hemodynamics at rest and during subsequent exercise, and the effects of posture on the hemodynamics were reviewed.

Studies show that posture modifies hemodynamics both in normal subjects and in patients with coronary artery disease. Further, a period of exercise modifies circulatory and hemodynamic data obtained during subsequent rest and exercise periods. Both the effect of posture and reproducibility of data during two exercise periods should be taken into consideration when one is evaluating the effects of exercise on left ventricular function and hemodynamics during serial studies.

References

1. Andersen KL et al. (1971) Fundamentals of exercise testing. World Health Organization, Geneva
2. Thadani U et al. (1977) Hemodynamics at rest and during supine and sitting bicycle exercise in patients with coronary artery disease. Am J Cardiol 39: 776–783
3. Parker JO, Thadani U (1979) Cardiac performance at rest and during exercise in normal subjects. Bull. Europ de Physiopathol Respiratoire 15: 935–949

4. Bevegard S, Freyschuss U, Strandell T (1966) Circulatory adaptation to arm and leg exercise in the supine and sitting position. J Appl Physiol 21: 37–46
5. Bevegard S, Holmgren A, Jonsson B (1960) The effect of body position on the circulation at rest and during exercise, with special reference to the influence on stroke volume. Acta Physiol Scand 49: 279–298
6. Granath A, Jonsson B, Strandell T (1960) Circulation in healthy old men, studied by right heart catheterization at rest and during exercise in supine and sitting positions. Acta Med Scand 176: 425–446
7. Rushmer RF (1959) Postural effects on the baselines of ventricular performance. Circulation 20: 897–905
8. Chapman CB, Fisher JN, Sproule BJ (1960) Behavior of stroke volume at rest and during exercise in human beings. J Clin Invest 39: 1208–1213
9. Wang Y, Marshall RJ, Shepherd JT (1960) The effect of changes of posture and of graded exercise on stroke volume in man. J Clin Invest 39: 1051–1061
10. Thadani U, Parker JO (1978) Hemodynamics at rest and during supine and sitting bicycle exercise in normal subjects. Am J Cardiol 41: 52–59
11. Stenberg J (1967) Hemodynamic response to work with different muscle groups, sitting and supine. J Appl Physiol 22: 61–70
12. Weiner L, Dwyer E, Cox W (1968) Left ventricular hemodynamics in exercise-induced angina pectoris. Circulation 38: 240–249
13. Parker JO, Digiorgi S, West RO A hemodynamic study of acute coronary insufficiency precipitated by exercise. Am J Cardiol 27: 470–483
14. Lecerof H (1971) Influence of body position on exercise tolerance, heart rate, blood pressure, and respiratory rate in coronary insufficiency. Br Heart J 33: 78–83
15. Thadani U, Parker JO (1978) Influence of glyceryl trinitrate during supine and upright exercise in patients with angina pectoris. Br Heart J 40: 1229–1236
16. Burkart F, Barold S, Sowton E (1967) Hemodynamic effects of repeated exercise. Am J Cardiol 20: 509–515
17. Epstein SE, Robinson BF, Kahler RL, Braunwald E (1965) Effects of beta-adrenergic blockade on the cardiac response to maximal and submaximal exercise in man. J Clin Invest 44: 1745–1753
18. Thadani U, Lewis JR, Manyari D, Boroomand K, Cohen J, West RO, Parker JO (1980) Are the clinical and hemodynamic events during exercise stress testing in invasive studies in patients with angina pectoris reproducible? Circulation 61: 744–750
19. Widimsky J, Berglund E, Malmborg R (1963) Effects of repeated exercise on the lesser circulation. J Appl Physiol 18: 983–986
20. Malmborg RO (1965) A clinical and hemodynamic analysis of factors limiting cardiac performance in patients with coronary artery disease. Acta Med Scand 177 [Suppl 426]: 1–94
21. Thadani U, West RO, Parker JO (1982) Effects of repeated exercise on hemodynamic profile in patients with stable angina pectoris. J Cardiac Rehab 2: 207–215

Advantages of the Computerized Breath-by-Breath Method for the Interpretation of Spiroergometric Data

J. Stegemann and D. Essfeld

There are still only a few laboratories using breath-by-breath methods for the determination of spiroergometric data. The reason for this may be that commercially produced equipment – if available – is expensive and complicated to handle. This is a surprising fact because computers and microprocessors are starting to govern the world in all other areas. Most of the breath-by-breath equipment, therefore, are homemade like the methods described by Beaver et al. [1], Smidt and Finkenzeller [9], and Stegemann [10] using digital computers and the method of Wigertz [11] and Linnarson and Lindberg [7] who performed the computation by an analog device.

All of these early programs wasted a lot of useful information, for instance, the duration of inspiration and expiration or the amount of nitrogen inspired in relation to the amount expired, which allows the regularity of the ventilation to be quantified. Since these early beginnings of the breath-by-breath method, refined versions have been developed. In the following, I will report on some improvements and applications of our device since 1976.

For those who are not familiar with the mathematical basis of the method, some remarks on the main principles should facilitate understanding. If expired air is divided into equal-volume fractions, then each fraction will contain a gas mixture different from the previous one. The first fraction corresponds to the inspired air because it is filled only with gas from the dead space which does not participate in the gas exchange. With each successive fraction, the gas concentrations approach the values of the alveolar air. If the expired volume (V_E) is divided into n equal sub-volumes (ΔV_i; $i = 1, 2 \ldots n$):

$$V_E = \Delta V_1 + \Delta V_2 + \ldots \Delta V_n \tag{1}$$

then the expired volume of CO_2 during this breath is

$$V_{CO_2} = \Delta V_1 \frac{C_1}{100} + \Delta V_2 \frac{C_2}{100} + \ldots \Delta V_n \frac{C_n}{100} \tag{2}$$

where the C_1 denotes the instantaneous concentrations of CO_2 (expressed as percentage). $C/100$ gives the fractional concentration F.

The time dependence of the process can be incorporated into Eq. (2) by multiplying with $\Delta t / \Delta t$.

$$V_{CO_2} = \frac{\Delta V_1}{\Delta t} F_1 \Delta t + \frac{\Delta V_2}{\Delta t} F_2 \Delta t + \ldots \frac{\Delta V_n}{\Delta t} F_n \Delta t \tag{3}$$

Pneumotachygraphs measure the instantaneous gas flow dV/dt. If the time interval Δt is short with regard to the duration of expiration, then a single reading $\left(\dfrac{dV}{dt}\right)_i$ may be taken as an estimate for $\Delta V/\Delta t$ during the interval Δt_i. In this case, Eq. (3) can be transformed into

$$\dot{V}_{CO_2} = \left(\frac{dV}{dt}\right)_1 F_1 \Delta t + \left(\frac{dV}{dt}\right)_2 F_2 \Delta t \ldots + \left(\frac{dV}{dt}\right)_n F_n \Delta t \tag{4}$$

or using the summation sign:

$$\dot{V}_{CO_2} = \sum_{i=1}^{n} \left(\frac{dV}{dt}\right)_i F_i \Delta t \tag{5}$$

This procedure, demonstrated for an expired volume of CO_2, can also be applied to the inspired volume of CO_2 as well as to the inspired or expired volumes of O_2 and N_2. The difference between the inspired O_2 volume per breath and the expired O_2 volume per breath is, of course, the O_2 uptake per breath. The lowest value of Δt that can reasonably be chosen is limited by the resolving power of the setup. Usually, it is the response time of the gas-measuring device which puts a limit to this choice.

In our laboratory we are using a mass spectrometer of the quadrupole type (Leybold, Köln) for this purpose. We countinuously record five analog channels, i.e., three channels from the mass spectrometer, where the voltage is proportional to O_2, N_2, and CO_2 concentration respectively; one channel from the differential manometer connected to a Fleisch sensor; and one channel from an ECG recorder. Every 10 ms all of these data are converted to digital values by a multiplexed analog-to-digital converter. Timing is controlled by a quartz clock which initiates an interrupt of the background computer program. During each respiratory cycle, 30 different data are stored for the inspiratory and expiratory phase. For each respiratory cycle, processed data such as PiO_2, $PiCO_2$, $PETO_2$, $PETCO_2$, as well $\dot{V}_E$, respiratory frequency, $\dot{V}O_2/\dot{V}CO_2$, RQ, load, heart rate, delay, and time are displayed on the monitor in order to facilitate the supervision of experiments.

We use a thin, Teflon tube attached to the flow sensor to transport the gas samples to the gas analyzer. This transport is a time-consuming process, causing a time delay of about 500 ms between the records of ventilatory flow and the records of gas concentrations. Unfortunately this delay varies with the respiratory rate and the moisture inside the Teflon tube. From Eq. (4) it follows that the synchronization between the instantaneous flow and the instantaneous gas concentration is essential for obtaining reliable results. Another crucial problem is the accurate determination of the starting points of inspiration and expiration. On the condition that the pneumotachygraph actually reads zero when no flow is present, every transition through the zero line indicates that either inspiration or expiration starts. The sign of the flow signal after passing the zero line allows one to decide whether it is expiration or inspiration that has started.

Another problem is to determine the time delay between the mass spectrometer and the pneumotachygraph reading. It can be seen in Fig. 1 that there is a sharp de-

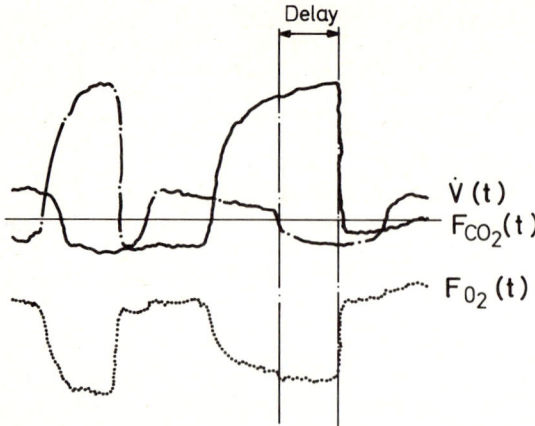

$\dot{V}(t)$
$F_{CO_2}(t)$

$F_{O_2}(t)$

Fig. 1. The determination of the time lag between pneumotachygraph and mass spectrometer readings. *Vertical lines* indicate the sharp change in gas signals and the inspiratory baseline transition of the flow signal. Both events represent starting points of inspiration and have to be synchronized before calculating the gas exchange values

crease in the CO_2 signal as well as an increase in the O_2 signal, both signals indicating the beginning of an inspiration. Due to the delayed gas transport, these events occur later in time than the respective inspiratory transition of the pneumotachygraph's baseline. The program determines the time of the CO_2 decrease and compares it with the point in time at which the flow signal indicates the start of inspiration. This time difference gives the delay time. Since the digitized readings of the flow signal are stored continuously, each reading of the mass spectrometer can be assigned to a pneumotachygraph signal occurring one time delay prior to the latest mass spectrometer reading.

Routine checks by the main sampling program have to make sure that no irregularities such as coughing or breath holding disturb the computing algorithm. Additionally, the pneumotachygraph's baseline has to be checked and corrected. Particularly during long-term experiments, a shifted baseline may become the major source of incorrect measurements.

The starting points of both inspiration and expiration are now defined and the background program can monitor the actual state within the respiratory cycle by means of a status register. After each interrupt by the clock routine, the program sums up the readings of the flow and the products of flow with the O_2, N_2, and CO_2 concentrations respectively. During the change from inspiration to expiration the inspiratory sums are stored and now the corresponding sums can be formed for the expiration period. In order to eliminate water vapor, the algorithms of Scheid et al. [8] are applied. Furthermore, the maximum and minimum pressure of each gas is determined in order to get inspiratory and the end-tidal pressures.

One advantage of integrating over the whole respiratory cycle is that the determination of gas exchange parameters becomes independent of the inspiratory gas concentrations. Therefore, experiments can be performed during hypoxia or hypercapnia without additional correction or adaptation of the program.

Thus, the basic program can be easily expanded by additional routines allowing the concentration of inspired gases to be controlled by means of stepmotor-driven valves as described in an earlier paper [10].

The disadvantage of integrating over the whole respiratory cycle is the resulting susceptibility to intrabreath asymmetries between inspired and expired volumes.

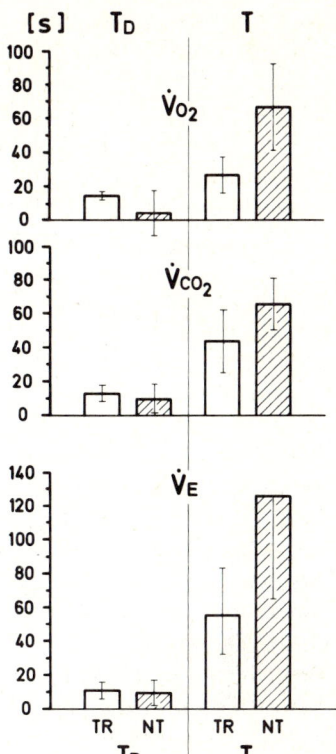

Fig. 2. Mean values ($\pm$ SD) of time constants (T) and time lags (T_D) for oxygen uptake ($\dot{V}O_2$), CO_2 output ($\dot{V}CO_2$), and ventilation ($\dot{V}_E$) kinetics after a step increase of work load. T and T_D were determined by nonlinear regression using the model: $F(X) = K[1 - EXP(-(X - T_D)/T)]$. T ($\dot{V}O_2$) values are significantly smaller in the trained (TR, $n = 8$) than in the reference group (NT, $n = 7$)

For this reason, the data trace would show a considerable amount of scatter. To smooth this effect, we use a "fudge" technique for the on-line display and for the standard plots. The quotient of N_2 uptake and N_2 output is calculated for each breath and used as a correction factor for the expired volumes of $\dot{V}_E$, $\dot{V}O_2$ and $\dot{V}CO_2$ before adding them to the corresponding inspiratory values. Additionally, the intra-breath N_2 balance can be utilized to quantify the asymmetry of a respiratory cycle and to detect systematic instrumental errors, such as baseline shiftings of the pneumotachygraph, which lead to a permanent asymmetry between inspired or expired volumes of N_2. Hoffmann and Bittner [6] from our laboratory have published a study on this application of the N_2 balance measurements.

In order to check the reliability of the system, we have constructed a testing apparatus, consisting of a box in which combustible materials can be burned. The box can be ventilated at different "respiratory" volumes and frequencies through an outlet in which the Fleisch sensor and the gas-sampling device can be inserted. From the amount of material burned and its RQ, the accuracy of the system can be determined.

Due to the limited space of this paper, I can report only briefly about some applications of the system.

From a theoretical point of view [2] the kinetics of oxygen uptake following a stepwise increase of work load should depend on the concentration of mitochondrial enzymes within the working muscles. These enzymes are known to be increased in endurance-trained athletes. A comparison of the time constants between a group of female sports students and high-performance female rowers demonstrated this difference. Figure 2 represents the results, which were recently published by Essfeld et al. [3]. This method can be used as a test of the specific muscular endurance.

The determination of the kinetics of ventilation, heart rate, O_2 uptake, and CO_2 output can be improved by using pseudorandom binary testing signals instead of classical signals such as step-and-ramp functions. The pseudorandom technique is less sensitive to random fluctuations of the respiratory parameters and less dependent on the subject's cooperation [5].

Another field of application should be finally mentioned: the investigation of respiratory stimuli at constant ventilation. By changing two different respiratory stimuli – or groups of stimuli – in opposite directions, the strenghth of the stimuli can be compared at constant ventilation. This method requires additional servo loops to be incorporated into the basic system. We have developed different servo loops to keep the end-tidal values of PO_2 and PCO_2 constant at arbitrary levels by varying the inspiratory gas mixture. Furthermore, a ventilation homeostat was developed with work load being the controlling element. If the actual and set values of ventilation differ, then the load of the ergometer will be changed automatically until ventilation reaches the set value again. If now the $PETCO_2$ is clamped at a level higher than normal, then the work load will be reduced to keep ventilation constant. Thus the interaction of PCO_2 stimuli and exercise-induced effects on ventilation can be studied without a disturbing change in ventilation. The results of these experiments have recently been puslished by Essfeld and Stegemann [4].

We are still using on old-fashioned minicomputer (PDP 12). Our next aim is to minimize the size of the system by means of a modern microprocessor.

Summary

The paper gives a short introduction to the basic principles and the materialization of a breath-by-breath method. Both the reliability and the effects of correcting algorithms can be studied by means of a testing device which allows different tidal volumes, respiratory frequencies, and respiratory quotients to be simulated. Respiratory gas exchange kinetics and the dynamics of ventilatory control are important fields of application of the breath-by-breath technique. Two examples of such applications are briefly demonstrated.

References

1. Beaver WL, Wassermann K, Whipp BJ (1973) On-line computer analysis and breath-by-breath graphical display of exercise function tests. J Appl Physiol 34: 128–132
2. Ceretelli P, Pendergast C, Paganelli WC, Rennie DW (1979) Effects of specific muscle training on $\dot{V}O_2$ on response and early blood lactate. J Appl Physiol 47: 761–769
3. Essfeld D, Hoffmann U, Stegemann J (1982) Influence of aerobic capacity on time delays and time constants of gas exchange kinetics measured on a breath-by-breath basis. Pflügers Arch [Suppl] 322: R 32

4. Essfeld D, Stegemann J (1983) CO_2-H^+-stimuli and neuromuscular drive to ventilation in dynamic exercise: comparison of stimuli at constant levels of ventilation. Int J Sports Med (to be published)
5. Essfeld D, Stegemann J (1983) Evaluation of $\dot{V}O_2$ kinetics on the basis of pseudo-random binary sequences of work load. Naunyn Schmiedebergs Arch. Pharmacol [Suppl] 322: R 79
6. Hoffmann U, Bittner H (1983) Correction of erroneous pneumotachograms in breath-by-breath analysis by means of the intrabreath N_2 balance. Naunyn Schmiedebergs Arch. Pharmacol [Suppl] 322: R 51
7. Linnarsson D, Lindberg B (1974) Breath-by-breath measurement of respiratory gas exchange using on-line analog computation. Scand J Clin Lab Invest 34: 219–224
8. Scheid P, Slama H, Piiper J (1971) Electronic compensation of the effects of water vapor in respiratory mass spectrometry. J Appl Physiol 30: 258–260
9. Smidt U, Finkenzeller P (1972) Ein Computerprogramm für die Ergometrie. Pneumologie 147: 245–250
10. Stegemann J (1976) Rechnergesteuerte Spiroergometrie nach der Methode der Einzelatemzuganalyse. Sportarzt Sportmed. 27: 1–7
11. Wigertz O (1971) Dynamics of respiratory and circulatory adaptation to muscular exercise in man. Acta Physiol Scand [Suppl] 363: 5–32

Quality Control in Exercise Testing, with Special Reference to Computer Processing of Exercise Electrocardiograms

M. L. Simoons

Introduction

The value of an exercise test, as well as any other diagnostic procedure, is dependent on the quality of the test itself as well as the knowledge of the physician who ordered the test. The indications for the test should be clear, while the physician should be able to correlate the test results with other information on the patient.

In the last 10 years it has become evident that the optimal exercise test consists of stepwise or continuous work load increments, either on a bicycle ergometer or on a treadmill. The work load should start at a low level and continue until symptoms become apparent which necessitate the patient or the physician to terminate the test. While one ECG lead may be sufficient in patients who do not suffer from coronary artery disease, multiple leads are mandatory for diagnosis and evaluation of patients with coronary disease. Adequate multiple lead systems include:

- pseudoorthogonal lead systems such as the combination of V_2, V_5, and lead II
- multiple chest leads
- corrected orthogonal lead systems (Frank leads) analyzed with the aid of a computer system

In addition to careful recording and analysis of the electrocardiogram, changes in heart rate and blood pressure during exercise should be noted, as well as any symptoms which may occur during the test. A proper report of the test includes all this information.

Proper quality of the electrocardiogram can be achieved in virtually all patients if the skin preparation is meticulous. The skin should be abraded with sandpaper or a dental drill. Special electrodes should be used which prevent motion artifacts. The ECG cables should be of special design which prevents artifacts due to cable motion. Finally, amplifiers should be used with high input impedance.

If a proper combination of skin preparation, electrodes, cables, and amplifiers is used, a real, stable baseline can be achieved throughout the test in most subjects.

Computer Processing of Exercise Electrocardiograms

In order to facilitate interpretation of the electrocardiogram during exercise, the noise level in the signal can be reduced with the aid of special computer programs. It should be stressed again that the main emphasis should be on prevention of excessive noise as described above. However, even under optimal conditions muscle noise will interfere with the electrocardiogram during exercise. The noise level can then be reduced by signal averaging.

Although earlier systems used single-lead computer analysis during exercise, it is now evident that multiple leads are required for optimal results.

We shall now describe the subprograms used for exercise ECG processing. Analog-digital conversion should be carried out at a minimum rate of 200 samples per second. In the system used at the Thoraxcenter, 250 samples per second are used. The QRS complexes can be detected with the aid of a combination of the derivatives of multiple ECG leads. Such programs use the characteristic of the signal during ventricular activation when large voltage changes occur in all leads more or less simultaneously. On the other hand it is unlikely that noise will be present simultaneously in all leads. The various methods for QRS detection and signal processing have been reviewed recently [1, 2].

In order to determine the dominant type of QRS complex, all detected beats should be classified. For the purpose of exercise ECG processing, it is adequate to separate normal and abnormal beats. Abnormal beats may be due to premature ventricular or supraventricular complexes, or they may be normal beats distorted by excessive noise or baseline drift. Such classification is not designed to perform a real rhythm analysis. If that would be required, further subclassification of the abnormal beats is mandatory.

Beat classification can be performed by one of three methods:

– comparison of selected features of all beats, such as the preceding R-R interval, R wave amplitude, and QRS width [3]
– correlation of the waveforms of all detected complexes
– correlation of the detected complexes with a reference beat obtained from the same patient at rest

Signal averaging can be performed at preselected intervals, for example, during 20 s of each minute, or continuously. In the latter method each incoming beat is compared with the existing averaged complex and added to that averaged complex if it has a similar shape. Again various methods for averaging have been developed [1, 2]. Satisfactory results can be obtained with all of them. Accordingly, the choice can be made by the designer of the system, dependant on the number of leads which should be analyzed and the computational speed of various algorithms. Before various beats are averaged, they should be properly aligned. Furthermore, special tests should be included to reject beats with excessive baseline drift, excessive high-frequency noise, or distortion through overflow of the analog-digital convertor [3, 4, 5].

In order to obtain measurements from the electrocardiogram, certain measurement points should be selected, such as the baseline in the P-R interval, the S-T segment, and possibly the R wave. In several commercially available systems, the baseline and the S-T segment are defined at fixed intervals before and after a single fiducial point in the QRS complex. Since this method will fail in many patients with abnormal electrocardiograms at rest, it is more appropriate to define the proper onset and end of the QRS complex. Precise definition of QRS onset and end is only possible if a combination of multiple leads is used [1, 2, 3, 5].

Optimal Measurements for Interpretation of the Electrocardiogram During Exercise

In earlier studies we compared the clinical value of a wide range of ECG measurements. These included the amplitudes at various intervals after the QRS complex, slope measurements in the S-T segment, the area of the S-T segment below the baseline, and a large number of measurements from the QRS complex. It turned out that optimal detection of coronary artery disease was achieved by a combination of S-T amplitude, S-T slope, and heart rate [6]. These results were reproducible when applied to a larger set of patients who had been studied in an other center in the Netherlands [7].

Recent studies from our laboratory indicate that small but significant differences are present between exercise electrocardiograms in normal males and females. Since most studies so far have included only male subjects, special criteria should be developed for quantitative interpretation of the exercise ECG in female patients with chest pain.

In Table 1, five studies have been summarized where computer-assisted interpretation of the electrocardiogram was compared with visual ECG reading in the same patients. The results are presented as the sensitivity of the test, which is the fraction of patients with an abnormal ECG response, and the specificity, which is the fraction of normal subjects with a normal ECG response. In one study, using the area of the ST segment below the baseline, the results of computer processing were similar to those of visual ECG reading [8]. In the others, computer processing was superior. These programs used measurements of ST amplitude and slope [9, 10], in combination with heart rate [6] and treadmill time [11].

In four of these studies the coronary arteriogram was used as a reference method. One study [9] used other clinical data as a reference, while two studies [6, 11] included normal volunteers without coronary arteriography. Most studies reported

Table 1. Comparison of visual and computer-assisted ECG interpretation

Refer-ence	Visual		Computer		Lead system and computer criteria
	Sensitivity [%]	Specificity [%]	Sensitivity [%]	Specificity [%]	
9	10/35 (29)	25/26 (96)	31/35 (89)	22/26 (85)	CC_5 ST amplitude and slope
10 L	15/57 (25)	30/30 (100)	40/57 (70)	27/30 (90)	CC_5 ST amplitude and slope
T	11/39 (28)	21/21 (100)	25/29 (64)	20/21 (95)	
6 L	26/52 (50)	81/86 (94)	44/52 (85)	78/86 (91)	X, ST amplitude and slope, heart rate
T	22/43 (51)	41/43 (95)	36/43 (84)	38/43 (88)	
11	44/59 (75)	9/11 (82)	50/59 (85)	10/11 (91)	V_5 ST amplitude and slope, heart rate, treadmill time
8	35/59 (59)	44/48 (92)	34/59 (58)	42/48 (88)	V_5 ST area

Results of 5 studies which compared computer-assisted interpretation of exercise electrocardiograms with visual reading of the same tracings. In two studies [6, 10] results obtained in a learning set *(L)* were verified in independant test series *(T)*. The numbers in parentheses are the sensitivity and specificity in percentages. The actual group sizes and numbers of abnormal (sensitivity) and normal (specificity) findings are presented. See further explanation in test.

the test results as either "normal" or "abnormal." In two studies [11, 12] measurements were presented on a continuous scale which represents the degree of abnormality in the ECG. Such measurements can be used to estimate the likelihood of coronary disease in a given patient.

Computer-Assisted Exercise Testing in General Practice

A computer system for exercise testing can be used to advantage by the practicing cardiologist. Such systems can aid in the performance of the test according to one or several standard protocols. In addition to processing of the electrocardiogram, the computer can regulate the work load of the bicycle ergometer or the speed and slope of the treadmill and prepare a report of the test. It should be remembered that the report of an exercise test should include a large amount of information, such as illustrated in Fig. 1.

Unfortunately, several commercially available systems for computer-assisted exercise testing do not meet the requirements, as stated in a recent editorial [1]. Therefore, the cardiologist who wishes to buy a system should check the design and performance of the systems under consideration. Key questions which should be asked include:

1. *How many leads are analyzed simultaneously?*
   A minimum of three leads is required for adequate signal processing.
2. *Is ECG analysis performed continuously or intermittently, for example, during 20 s of each minute?*
   Although both methods may be adequate, continuous (running) averaging is to be preferred, since it provides beat-by-beat monitoring of ECG changes during the test.
3. *How can I verify the averaging procedure?*
   Errors due to averaging, for example, from improper alignment of QRS complexes or inclusion of beats with an excessive noise level, can be recognized easily by visual inspection of both the original ECG and the averaged beats. For this comparison these two signals should be available, preferably using the same amplitude and time scales.
4. *Which measurements are included for ECG interpretation, and how can I check these?*
   The system should define the proper onset and end of the QRS complex. In the computer report the averaged beats should be provided with markers which indicate where these points have been defined. If these markers are not in the right position, the whole ECG analysis should be disregarded. In Fig. 1 the use of such markers is illustrated.
5. *Has the system been tested rigorously under clinical conditions?*
   In our opinion, the buyer should ask for a "consumer's report" on the systems under consideration. The systems should be tested in clinical practice. Such tests should include a description of the system performance in a large number of patients with various abnormalities, and a summary of the diagnostic performance of the system in comparison with an independant method, such as coronary arteriography or thallium scintigraphy. Unfortunately, to our knowledge, the latter type of testing has not been performed.

```
NAME :
DATE OF BIRTH :                              DATE OF TEST :

HISTORY : INFARCT (1980)
SYMPTOMS : TYPICAL ANGINA
INDICATION : EXERCISE CAPACITY, CORONARY INSUFFICIENCY ON EXERCISE
MEDICATION : β -BLOCKER
ECG AT REST : SINUS RHYTHM, ANTERIOR INFARCTION,
              SECONDARY REPOLARISATION CHANGES
LENGTH : 171 CM      WEIGHT : 70 KG
STEPWISE INCREMENTS 10 WATTS/MIN ON BICYCLE ERGOMETER
MAXIMAL LOAD : 110 WATTS (NORMAL=133)   EXERCISE TOLERANCE : REDUCED
```

	REST	ONSET SYMPTOMS	MAXIMUM	6 MIN RECOVERY
HEARTRATE	68	118	121	79
BLOODPRESSURE	145 – 80	180 – 100	190 – 95	130 – 80

```
PATIENT TERMINATED TEST : ANGINA
SYMPTOMS DURING TEST (STARTED 90 WATTS, UNTIL 5 MIN. RECOVERY) :
TIGHTNESS ON CHEST (MODERATE)

ACCORDING TO PATIENT TEST WAS : MAXIMAL
ARRHYTHMIAS : ISOLATED PVC
```

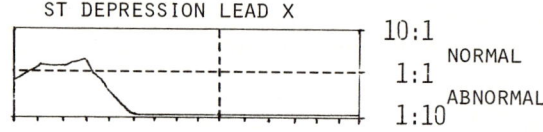

```
ST DEPRESSION LEAD X
                                              10:1
                                                     NORMAL
                                              1:1
                                                     ABNORMAL
                                              1:10
```

```
ST ELEVATION X
ST ELEVATION Y                                        ST DEPRESSION Y
ST ELEVATION Z · · · · · · · · · · ·                  ST DEPRESSION Z
ABNORMAL T WAVE · · · · ·                             ABNORMAL ST CHANGES
QRS PROLONGATION                                      I.V. CONDUCTION DEFECT
```

```
GREATEST PROBABILITY OF EXERCISE INDUCED ISCHEMIA (LEAD X)
                                    999.9 : 1  (PROBABLE)
ST CHANGES LEAD (Y, Z) COMPATIBLE WITH MYOCARDIAL ISCHEMIA
```

| REST | GREATEST ST ABNORMALITY | 110 WATTS | 6 MIN. RECOVERY |

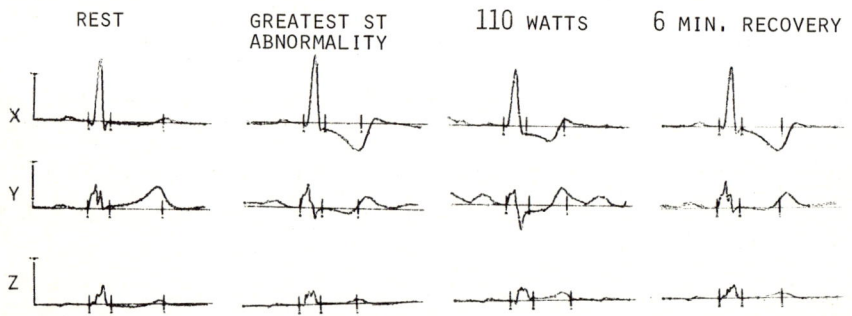

Fig. 1. Summary of an exercise test report from the computer developed at the Thoraxcenter. The picture contains an English translation of the original computer report written in Dutch. The report contains from *top* to *bottom:* a summary of the patient's history and medication and a description of the ECG at rest, heart rate, and blood pressure at various stages of the test. It lists a description of the symptoms during the test.

Graphical display of the likelihood ratio of an abnormal ECG response in lead X [12] as well as an indication of the presence of other ECG changes is indicated.

Conclusions

Computer hardware is becoming so cheap that computer processing of exercise electrocardiograms can now be offered to the practicing cardiologist with little extra costs. However, such computer processing will only help in detection and management of patients with heart disease if it is designed according to the state of the art. We hope that the information in this chapter will aid the cardiologist or internist in the selection of an optimal system for computer-assisted exercise electrocardiography in his practice. The quality of the exercise electrocardiogram can indeed be improved by computer processing. However, quality controls remains a major task for the technician and physician who are responsible for the test.

References

1. Simoons ML, Hugenholtz PG, Ascoop CA, Distelbrink CA, Lang PA de, Vinke RVM (1981) Quantitation of exercise electrocardiography. Circulation 63: 471–475
2. Bhargava V, Watanabe K, Froelicher VF (1981) Process in computer analysis of the exercise electrocardiogram. Am J Cardiol 47: 1143–1151
3. Wolf E, MacInnis PJ, Stock S, Helppi RK, Rautaharju PM (1972) Computer analysis of rest and exercise electrocardiograms. Comput Biomed Res 5: 329–346
4. Simoons ML, Boom HBK, Smallenburg E (1975) On-line processing of orthogonal exercise electrocardiograms. Comput Biomed Res 8: 105–117
5. Jonson B, Werner O, Jansson L, Johannson K, Olsson LG, Westling (1976) A system for computer-assisted ECG recording at rest and exercise. In: Computer assistance in the ECG laboratory – a new look. Berlingska Boktryckeriet, Lund, pp 7–21
6. Simoons ML (1977) Optimal measurements for detection of coronary artery disease by exercise electrocardiography. Comput Biomed Res 10: 483–499
7. Simoons ML, Ascoop CA, Block P, Distelbrink C, Lang PA de, Vinke RVM (1977) Computer processing of exercise electrocardiograms. In: van Bemmel J, Willems J (eds) Trends in computer-processed electrocardiograms. North Holland, Amsterdam
8. Sketch MH, Mohiuddin SM, Nair CK, Mooss AN, Runcho V (1980) Automated and nomographic analysis of exercise tests. JAMA 243: 1052
9. McHenry PL, Stone AE, Lancaster MC (1968) Computer quantitation of the ST-segment response during maximal treadmill exercise; clinical correlation. Circulation 38: 691
10. Ascoop CA, Distelbrink CA, de Lang PA (1977) Clinical value of quantitative analysis of ST slope during exercise. Br Heart J 39: 212–217
11. Hollenberg M, Budge WR, Wisneski JA, Gertz EW (1980) Treadmill score quantifies electrocardiographic response to exercise and improves test accuracy and reproducibility. Circulation 61: 276–285
12. Simoons ML, Hugenholtz PG (1977) Estimation of the probability of exercise-induced ischemia by quantitative ECG analysis. Circulation 56: 552–559

◁ **Fig. 1** *(continued)*

At the *bottom* representative, averaged ECG complexes are given at rest, at the moment with the greatest S-T changes (likelihood ratio), at peak exercise, and at the end of the QRS complex as well as the maximum spatial magnitude of the T wave.

In this case, S-T segment depression in lead *X* is present during the last 5 min of exercise and the 1st min in the recovery period.

The spatial magnitude of the changes of the S-T vector from rest to exercise is abnormal during the last 6 min of exercise and the first 4 min of the recovery period, while S-T depression occurs in lead *Y* and S-T elevation occurs in lead *Z*

Some General Functions and Their Differential Use

G. Borg

Introduction

In ergometry the usual test validations can be complemented by a broader kind of construct and content validation that utilizes information from three different kinds of effort variables, viz, physiological, perceptual, and performance variables. Most importantly, estimates of reliability and validity can be obtained by analyzing the internal consistency of the total system of response variables. For a top athlete the test criterium is very simple: a medal in the Olympic games, preferably a gold medal. For predictions of performances of ordinary people engaged in industrial work, sports, or other leisure activities, there are, however, no such simple performance criteria.

In a short-term efficiency perspective, people are selected or trained for maximum achievements. However, in practical affairs the question is often one of optimizing performance and performance costs with regard to the possible long-term effects of loads and strain. Costs and strains must therefore be measured not only in physiological but also in psychological terms. Thus we discover a need to integrate many variables in order to obtain a satisfactory criterium.

No research paradigm is sufficient if it only uses simple performance criteria or focuses on a reductionistic model, where "explanation" means referring to data on a lower descriptive level. A better and more complete validation has to integrate information from all important variables and their specific interactions. We have to reckon with a person × situation interaction that complicates the possibilities of explanations and predictions.

In a systems approach to the validation problem, we have to admit that man is a very complex system interacting with the physical and social world around him. We cannot understand man by reading his responses like we read a book. Or as Goethe so poetically expressed it: *"Ich bin kein ausgeklügelt Buch, ich bin ein Mensch mit seinem Widerspruch."* In the beginning of this century the German Gestalt psychologists strongly emphasized the importance of a holistic view. They introduced the Gestalt concept and argued that we must understand the whole as more than a simple addition of its parts. Performances and experiences are just such complicated "gestalts," which cannot be understood by means of an oversimplified, explanatory model. We must take into consideration such factors as differing techniques and levels of fitness, emotion, motivation, and personality, as well as an individual's idiosyncratic reaction to environmental factors. A simple initial improvement in the study of ergometric problems can be achieved if we try to complement our physiological variables with measures of perceived effort or other subjective symptoms and, furthermore, if we maintain a better control over motivational factors.

Parallel to developing a broader systems approach to these complex problems we must also simplify the situations and objects of our study, in order to build up a body of disparate knowledge that we hopefully may later integrate in a meaningful way. An important aim of such limited research studies is to try to identify, to a greater degree than is done at present, the general relationships that exist among the various types of effort variables. It is necessary to find invariant relationships between different subvariables and then integrate the information into relatively simple models.

A General Function: The Power Law

To me one of the most important general laws describing the variation of physiological and perceptual data is the power law. In psychophysics, S. S. Stevens [30] proposed the power function as the most general function, describing how perceptual intensity varies with the physical stimulus. Prior to this most functions had usually been described by Fechnerian log functions. Gustaf Fechner proposed in 1860 in *Elemente der Psychophysik* that sensation generally increases with the logarithm of the stimulation [19]. A hundred years later Stevens published the article [31] "To Honour Fechner and Repeal His Law." Stevens stressed Fechner's immense importance for psychophysics, but nevertheless declared that the power function gives a considerably better description in most sensory modalities.

A simple example of a power function is given by the sensation of speed when driving a car, which, roughly speaking, increases with the square of the speed. Another example is change in taste sensation following stimulation of the tongue with taste solutions. As shown by Borg et al. [13], the physiological nerve response following taste stimulation can be described by a power function which closely follows the perceptual response.

A general formulation of the power function to fit most perceptual and physiological variables was proposed by Borg [2, 3]:

$$R = a + c\,(S - b)^n$$

where R is the intensity of the response, S is the stimulus intensity, a and b are constants showing the starting point of the function or the absolute threshold, c is the measure constant, and n is the exponent.

Maximal Performances and Their Variation over Time in Brief Exercises

As early as 1937, Grosse-Lordeman and Müller proposed that the relation between maximal performance time *(T)* and power in bicycle exercise *(W)* may be described by a power function [20]:

$$T = c\,W^n$$

In a study by Tornvall [33], the exponents n = −4.79 and n = −4.96 were obtained for two different groups of young men, performing "maximal" exercise from a few to several minutes in duration.

In the cycling strength test (CST) developed by Borg [2, 3], maximal performance thresholds are determined by increasing the power linearly with time during cycling until the person cannot maintain a stipulated pedaling rate. With a power increase of $10 \, W \, s^{-1}$, a pedaling rate of $60 \, r \, min^{-1}$, and a "braking rate" of $40 \, r \, min^{-1}$, a performance threshold (in watts) is usually reached after 30–60 s. In a study on maximal performances where the power increase was varied from about $2 \, W \, s^{-1}$ to about $67 \, W \, s^{-1}$, a power function was found to describe the relation between power *(W)* and time *(T)* with an exponent, n = −0.25. The reciprocal value describing how *T* varies with *W* is then n = −4.0 [12].

In a short-term test at a constant load it is impossible to also keep time constant as an independent variable. If, however, previous information is used to choose a power level that is just right for a given individual, the variation in time over individuals may be kept fairly small. If a subject deviates in his performance time from the desired reference level (e. g., 10 s, 30 s, or 2 min), it is easy to correct the set power level by using the information given by the general *W-T* function. In a recent study on 20 subjects this method was found to work well and to give very reliable estimates of performance capacity [10].

Perceived Exertion as an Indicator of Work Strain

One area where ratings of perceived intensity serve as an excellent aid in diagnosis, training, and rehabilitation is exertion in physical exercise. The subjective feeling of exertion constitutes a comparatively reliable measure of individual physical exertion. As such, it can be used purely "diagnostically" but can also be used for governing the intensity of exercise or for evaluating the "value" of exercise.

A scale for the determination of perceived exertion, which is now used rather often, is the RPE scale (ratings of perceived exertion) [3 and 6]. The scale consists of numbers ranging from 6 to 20 (to cover a variation in heart rate from about 60 to 200) and in which all odd numbers are anchored with verbal expressions such as "rather light," "very hard," etc. The scale is constructed as a nonsymmetrical category scale so that the ratings increase approximately linearly with work load for cycling on the bicycle ergometer and, thereby, also with heart rate and oxygen consumption.

Figure 1 shows how a psychological variable can serve as an indicator of physical quantity. The independent variables are the factors to which the subject is exposed

VARIABLE	"STIMULUS" (I V)	"RESPONSE" (D V)
PHYSIOLOG.		
PSYCHOLOG.		

Fig. 1. The figure aims to show how both a physiological dependent variable *(DV)*, such as heart rate, and a psychological dependent variable, such as perceived exertion, can alternately serve as a "measure" or "indicator" of physiological or psychological load as independent variable *(IV)* (see text)

– for example, load on the bicycle ergometer, distance over which he/she competes, the temperature on the tennis court, etc. The dependent variables can be those used as a measure of performance, e.g., speed when running 5000 m, which is then dependent on the individual's running ability and the prevailing conditions. If we study how prior food intake affects heart rate while running, then we are studying how one physiological variable affects another. If we take the same individual but subject him to some fear-arousing situation while running (for example, an angry dog) and let the individual's perceived fear constitute the dependent variable, then we have a purely psychological study. We can also make use of physiological measures as dependent variables to psychological factors, e.g., take heart rate as a measure of fear. In a comparable way we can also use self-ratings as a measure of physiological load. Thus, for example, perceived exertion can be a measure of the degree of physical exertion.

In *Ergonomic Principles in the Design of Work Systems* [22] the following two paragraphs are of special interest. In 4.1.2 it is stated that: "The design of the work shall be such as to avoid unnecessary or excessive strain in muscles, joints, ligaments, and in the respiratory and circulatory systems." In 4.2, concerning the design of the work environment, it is stated that: "The work environment shall be designed and maintained so that physical, chemical and biological conditions have no noxious effect on people but serve to ensure their health, as well as their capacity and readiness to work. Account shall be taken of objectively measurable phenomena and of subjective assessments."

It is especially important to notice that "account shall be taken of objectively measurable phenomena and of subjective assessments." As a complement to physical or physiological determinations, it is of importance to carry out subjective assessments. Primarily it is a question of using psychophysical methods to determine perceptual intensities over and above the physiological ones.

It can be fruitful in this connection to regard *work strain* as a *hypothetical construct*. In a stricter sense, therefore, we are not able to actually measure strain but rather develop good or less good indicators of it. Merely determining the intensity of a single variable thus gives a too crude and sometimes completely misleading picture of the real intensity of the strain. We should rather try as often as possible to improve validity by determining strain intensity in as many variables as possible: physiological, behavioral, and perceptual. This also applies to the concept *dangerous strain*. In prevention of a heart attack or in cardiac rehabilitation, the perception of exertion is probably an equally good indicator of dangerous strain as is the heart rate. In the perception of strain an individual seems to be able to integrate different cues into a kind of configuration or gestalt of strain, or in some situations to give special weight to the more important or dangerous strain variables.

If we estimate the perception of exertion with psychophysical "ratio scaling" methods such as "magnitude estimation" [30, 32] or the new category scale with ratio properties [12], we will also find power functions in this modality. For work on the cycle ergometer a positively accelerating function has thus been obtained with an exponent of about 1.6 [3, 7]. For walking on a treadmill with zero grade for 4 min at several different speeds, the following power function was obtained:

$$R = 1 + 0.0125 \, (S - 1.5)^3$$

where R is the subjective intensity of perceived exertion when walking; 1 denotes the value of the basic perceptual noise level, which is equal to the R value arbitrarily set to 1 at $S = 1.5\,\mathrm{km\,h^{-1}}$; 0.0125 is the measure constant; S is the physical speed in $\mathrm{km\,h^{-1}}$; 1.5 shows the starting point of the curve; and 3 is the exponent.

The Increase of Lactate with Exercise Intensity

The intensity increase in many physiological variables during exercise also seems to follow the form of a power function. Most of these functions seem to be linear or positively accelerating.

In general the increase of blood lactate with power in bicycle exercise seems to follow one monotonously increasing power function. By analyzing data obtained by Holmgren and Ström [21], Borg [3] proposed a power function with an additive constant (a) related to the basic value at rest and with an exponent of about 2. For a group of normal, ordinary subjects the exponent was 2.2, and for a group of racing cyclists it was slightly higher with an extra "b" constant (of about 50 W) in the function showing the starting point when lactate starts to accumulate. There was no "braking point" observed and no need to divide the function into two (or three) different parts, but the function was monotonously increasing. This has been confirmed in several later studies. However, in some situations and for some individuals or groups of individuals, especially some well-trained athletes, there seems to be a need to divide the function into a first part with no or very little increase – or even a decrease, as in walking – and one subsequent part, where the increase is very large.

The muscle lactate also seems to follow a power function but with a faster acceleration in the muscle than in the blood and an exponent of about 3 [26].

A Psychophysiological Study

Many interesting psychophysiological studies have been performed in which the perception of exertion has been manipulated to vary with various physical factors and simultaneously appearing physiological responses have been collected [25]. Most of these previous studies have been reviewed in articles by Borg and Noble [16], Borg [9], Mihevic [24], and in papers presented at the 1981 American College of Sports Medicine (ACSM) Symposium in Florida on perceived exertion by Borg [11], Cafarelli [17], Robertson [28], Pandolf [27], and Noble [26]. One interesting experiment on muscle metabolites, force, and perceived exertion was performed by the chairman of this conference [23].

An experiment in which perceived exertion, heart rate (HR), and blood lactate (BLA) were determined for three different types of physical work, namely, walking, running, and cycling, has just been completed. The perceived exertion has been determined with two different methods, the RPE scale [6] and the newly constructed category scale with ratio properties [12]. A preliminary analysis of the results shows that the changes are best described by a power function. The perception of both exertion and heart rate measured by the RPE scale increases linearly with work load,

Table 1. Exponents of the power function ($R = a + cS^n$, see text) for heart rate *(HR)* and blood lactate *(BLA)* and ratings of perceived exertion according to two different scales, the *RPE* scale and the R_{C-R} scale (see text), obtained in an experiment on cycling, running, and walking

Variable	HR	BLA	RPE	R_{C-R}
Bicycling	1.0	3.9	1.0	1.9
Running	1.0	3.5	1.0	2.0
Walking	2.4	>3	2.0	3.5

whereas the ratio scale technique gives a positively accelerating function for exertion. Blood lactate appeared to follow a more pronounced, positively accelerating function. As can be seen from the exponents in Table 1, we found in this experiment relatively invariant relationships among the measured variables for such different types of physical work such as walking, running, and cycling. As Borg [3] has suggested earlier, a good prediction of perceived exertion can be made for these types of work from a combination of heart rate and blood lactate.

The Range Model

In the above description of psychophysiological functions we were interested in general changes and not in individual differences. There are, however, rather large differences among individuals or groups of individuals who are characteristic in certain respects. This applies to perceptual as well as to physiological functions.

In order to facilitate interindividual comparisons in perceptual intensity obtained with the so-called ratio scaling methods, Borg [2, 3, 6] has proposed the following model. It is assumed that subjective range and perceptual intensity at a maximum intensity level are about the same for all individuals in spite of the fact that the stimulus range, from zero or the b value (see above) might differ very much. In the general psychophysical equation, the measure constant c can be solved for each individual in the following way:

$$c = \frac{R_t - a}{(S_t - b)^n}$$

where R_t is the subjective intensity at the maximal (terminal t) level and $R_t - a$ is assumed to be equal for all subjects.

According to this range model, the intensity of a response is determined by its position in each individual's range. For a valid comparison of different individuals, it is then of paramount importance to be able to determine reliable values of the constants of the equation (see above). The a value is closely related to a basic "noise" value or a value at rest (a basic metabolic rate). It may also reflect the hypothetical rest value that best satisfies the mathematical equation or the absolute threshold. If we do not have a good empirical rest value, e.g., heart rate at rest, we may extrapolate one from the individual data points obtained in an exercise test with several loads (excluding the lowest power level that might be influenced by anxiety, etc.).

As a rule we shall naturally try to determine the various constants in the equation purely empirically. This applies not least to the "maximum value," which we set the same for all subjects. It is often difficult or actually impossible to determine the maximum level reliably. One possibility in this case is to use the covariation among different types of functions. Thus, for example, we can make use of the RPE scale's fixed end points (6 and 20) and, from a series of submaximal relations, e. g., between HR and RPE, extrapolate maximum HR values to the maximum RPE values.

The b constant is normally equal to 0 except for a few functions or in certain limited cases, e.g., certain lactate functions. The b value can also sometimes be replaced with the absolute threshold. This can then also be used as a unit in a relative energy scale in intermodal comparisons.

The meaning of the measure constant "c" should also be specified. As recently pointed out by Borg and Marks [15], this constant can have at least 12 distinct referents. These depend upon standardized (SI) units of measurement of the physical stimulus, choice of experimental method, group and individual differences in responses, modality- or quality-specific differences, differences in the transformation of distal to proximal stimuli, pathological differences, and other things.

Since in this connection we are using data on a ratio scale level, it is often practical to express the results in percental units. It is, however, important here not to use percent of the maximum level but rather percent of the range. Since the constants in the equation differ for different variables, e. g., the a constant for heart rate, oxygen consumption, and blood lactate, it is only percent of the range that gives comparable values.

Some Special Estimates of Physical Working Capacity

Since working capacity normally refers to the maximum capacity to work, attempts are often made to obtain direct measures of this maximum. Not only are such measures difficult, in fact, to obtain, but it is not altogether certain that they give a valid picture of the capacity for working under everyday conditions. One person can, for example, have a better capacity than another for working at 50% of the maximum.

If one tries to estimate working capacity from submaximal measures of exertion, then measures based on perceived exertion can be important complements to those based on heart rate. Thus, just as W_{170} (or PWC_{170} after Sjöstrand [29] and Wahlund [34]) constitutes an estimate of working capacity, so does W_{R17} (i.e., load in watts corresponding to rating 17 on the RPE scale, Borg [3]). Working capacity estimated from perceived exertion in this way has equally good predictive power as that based on heart rate mentioned above or after Åstrand [1]. This applies under the assumption that the subjects have the normal ability to carry out the ratings. In heterogeneous age groups, W_{R17} (or W_{R15}) gives a better estimate than W_{170}.

In order to correct heart rate for age changes, a reference level defined in percent of the range, e.g., 80%, can be used rather than a fixed level, such as 170 beats min^{-1}. We know that 170 for a 20-year-old corresponds roughly to 150 beats min^{-1} for a 50-year-old from the formula:

$$HR_{0.80} = 170 - \tfrac{2}{3}(A - 20)$$

where A is the relevant person's age in years [14]. Instead of W_{170} we can then use the age-corrected $W_{0.80}$. If we wish to use a lower reference level, for example, 66% of the range and $W_{0.66}$, we can use the equation:

$$HR_{0.66} = 150 - \tfrac{1}{2}(A - 20)$$

In differential studies of working capacity one sometimes tries to obtain at least two empirical test values, one on either side of the reference level. But since we can make use of the general function found earlier, we need, in principle, only one value. We then can extrapolate to the reference level by means of the general equation (or group-specific equations). The load should obviously be chosen so that the test value comes as close as possible to the arbitrarily chosen reference level, e.g., $4\,\mathrm{m}M\,\mathrm{l}^{-1}$ in blood lactate.

For quick and approximate estimates of working capacity, for example, in conjunction with military enrollment, information should be used from several strain indicators, such as heart rate, blood lactate, and perceived exertion. If we wish to keep the testing time short and fairly constant over subjects, we can abandon the rigid test proceedings with equality of stimuli and instead make the test as flexible as possible in order to obtain equality of subjective response. This can be achieved by controlling the testing conditions through feedback from test responses [4, 8] and utilization of the information we have stored in the form of general equations.

Such a flexible and individually adapted work test (IAT) has been compared in a large, empirical study with other tests and found to have good reliability and validity [18].

Conclusion

To improve the quality and the validity of methods and measurements in ergometry we should not restrict our studies to too few variables, but try to obtain information from all three "domains" of work and effort: the physiological, the behavioral, and the perceptual. We should try to find invariant relations among the different variables and try to express these mathematically. In differential studies and interindividual comparisons the data from each individual may then be complemented by using the information from the general functions.

References

1. Åstrand I (1960) Aerobic work capacity in men and women with special reference to age. Acta Physiol Scand 49 [Suppl 169]: 1–91
2. Borg G (1961) Interindividual scaling and perception of muscular force. K Fysiogr Sällsk Lund Förh 31 (12): 117–125
3. Borg G (1962) Physical performance and perceived exertion. Studia psychol paed, Series altera, Investigationes XI. Gleerup, Lund
4. Borg G (1968) Ett flexibelt arbetsprov med styrning av arbetsbetingelser. Nord Med 79: 85–95
5. Borg G (1970) Perceived exertion as an indicator of somatic stress. Scand J Rehabil Med (2–3) 2: 92–98

6. Borg G (1970) Relative response and stimulus scales. Reports from the Institute of Applied Psychology, vol 1. University of Stockholm, pp 1–8
7. Borg G (1972) A ratio scaling method for interindividual comparisons. Reports from the Institute of Applied Psychology, vol 27. University of Stockholm, pp 1–12
8. Borg G (1974) Psychological aspects of physical activities. In: Larson LA (ed) Fitness, health, and work capacity. International standards for the assessment. MacMillan, New York, pp 141–163
9. Borg G (1978) Subjective effort in relation to physical performance and working capacity. In: Pick HL et al. (eds) Psychology: from research to practice. Plenum, New York, pp 333–361
10. Borg GAV (1982) Ratings of perceived exertion and heart rates during short-term cycle exercise and their use in a new cycling strength test. Int J Sports Med 3: 153–158
11. Borg GAV (1982) Psychophysical bases of perceived exertion. Med Sci Sports Exerc 14 (5): 377–381
12. Borg GAV (1982) A category scale with ratio properties for intermodal and interindividual comparisons. In: Geissler HG, Petzold P (eds) Psychophysical judgment and the process of perception. Deutscher Verlag der Wissenschaften, Berlin
13. Borg G, Diamant H, Ström L, Zotterman Y (1967) The relation between neural and perceptual intensity: a comparative study on the neural and psychophysical response to taste stimuli. J Physiol 192: 13–20
14. Borg G, Linderholm H (1967) Perceived exertion and pulse rate during graded exercise in various age groups. Acta Med Scand 472: 194–204
15. Borg GAV, Marks LE (1983) Twelve meanings of the measure constant in psychophysical power functions. Bull Psychon Soc 21 (1): 73–75
16. Borg G, Noble BJ (1974) Perceived exertion. Exercise Sport Sci Rev 2: 131–153
17. Cafarelli E (1982) Peripheral contributions to the perception of effort. Med Sci Sports Exerc 14 (5): 382–389
18. Edgren B, Marklund G, Nordesjö LO, Borg G (1976) The validity of four bicycle ergometer tests. Med Sci Sports Exerc 8 (3): 179–185
19. Fechner GT (1860) Elemente der Psychophysik, vol I. (Available in English as Elements of psychophysics. Holt, Rinehart and Winston, New York, 1966)
20. Grosse Lordemann H, Müller EA (1937) Der Einfluß der Leistung unter der Arbeitsgeschwindigkeit auf das Arbeitsmaximum und den Wirkungsgrad beim Radfahren. Arbeitsphysiologie 9: 454–475
21. Holmgren A, Ström G (1959) Blood lactate concentration in relation to absolute and relative work load in normal men, and in mitral stenosis, atrial septal defect, and vasoregulatory asthenia. Acta Med Scand 163: 185–193
22. ISO Standard (1981) Ergonomic principles in the design of work systems. ISO 6385
23. Löllgen H, Graham T, Sjogaard G (1980) Muscle metabolites, force and perceived exertion bicycling at varying pedal rates. Med Sci Sports Exerc 12 (5): 345–351
24. Mihevic PM (1981) Sensory cues for perceived exertion: a review. Med Sci Sports Exerc 13: 150–163
25. Mountcastle VB, Poggio GF, Werner G (1963) The relation of thalamic cell response to peripheral stimuli varied over an intensive continuum. J Neurophysiol 26: 807–834
26. Noble BJ (1982) Clinical applications of perceived exertion. Med Sci Sports Exerc 14 (5): 406–411
27. Pandolf KB (1982) Differentiated ratings of perceived exertion during physical exercise. Med Sci Sports Exerc 14 (5): 397–405
28. Robertson RJ (1982) Central signals of perceived exertion during dynamic exercise. Med Sci Sports Exerc 14 (5): 390–396
29. Sjöstrand T (1947) Changes in the respiratory organs of workmen at an ore smelting works. Acta Med Scand [Suppl] 196: 687–699
30. Stevens SS (1957) On the psychophysical law. Psychol Rev 64: 153–181
31. Stevens SS (1961) To honour Fechner and repeal his law. Science 133: 80–86
32. Stevens SS (1971) Issues in psychophysical measurement. Psychol Rev 78: 426–450
33. Tornvall G (1963) Assessment of physical capabilities. Acta Physiol Scand [Suppl] 201: 58
34. Wahlund H (1948) Determination of the physical working capacity. Acta Med Scand [Suppl] 215: 1–68

Measurement and Interpretation of Lactate

T. E. Graham

Introduction

With the exception of heart rate and $\dot{V}_{O_2}$, blood lactate is probably the most common physiological measurement made in ergometry and sports medicine. It is commonly associated with theories of fatigue, pain, and perceived exertion and assumed to be diffusing across membranes passively and/or freely and to directly reflect muscle metabolism. Direct, conclusive evidence for many of these interpretations is lacking, and there is a growing body of knowledge that is demonstrating that our basic assumptions concerning lactate metabolism are simplistic. This report will attempt to review fundamental aspects concerning sampling and analyzing of lactate samples, interpretations of lactate levels, and some factors that can influence lactate levels independent of muscle activity.

Quality of the Measurement

Sources of Variation

Sampling

Before one even discusses environmental or endogenous factors that can influence lactate metabolism, the gathering, storing, and analyzing of the sample must be considered. Often the researcher is attempting to estimate the lactate concentration either in the venous blood draining the active muscle or in arterial blood, but elects to sample the blood at a different site for safety or for convenience. Margaria et al. [1] demonstrated that following 10 min of strenuous work the blood lactate levels in arterial (femoral) and venous (femoral or arm) blood were extremely similar. However, other workers [2, 3] have reported that femoral samples can be twofold greater than cephalic or antecubital venous samples during strenuous exercise and the early portion of recovery. The sample site is probably critical in other than steady state conditions and must be considered in interstudy comparisons. The more distant the sample site is from the source of blood lactate, the more important it is to maintain a high flow rate to the site via heat or exogenous vasodilator creams.

For the sake of convenience many workers used fingertip microsamples for analysis. The enzymatic assays are sensitive enough to quantitatively analyze small samples; however, one should appreciate that if the antecubital venous sample can underestimate the leg blood lactate level (at least in other than steady state conditions), then obviously the same can be true of the fingertip technique. Care should be taken to encourage rapid circulation to the finger prior to sampling, to sample during steady state conditions, and to collect the blood without contaminating it with perspiration (which can both dilute the blood volume and serve as a source of lactate) [4] or tissue fluid due to trauma or pressure on the finger [5].

Frequently, "arterialized" samples are obtained either from the earlobe [6] or a dorsal hand vein [7]. Under ideal conditions, either method will probably provide good approximations of arterial lactate concentrations. The earlobe technique is reliable as long as the flow is high (i.e., brisk enough to fill a 125 μl capillary tube within 30 s). However, the dorsal hand vein technique is often favored, since it usually approximates arterial values for lactate (and blood gases) without the potential complications of arterial catheters, is perhaps simpler to use, and can provide large volume samples. This technique has been used in our laboratory for several years and generally is a very satisfactory method. However, it should be noted that if the hand is not warmed adequately, the samples are distinctly venous (based on blood gases). Furthermore, some subjects have pronounced superficial vasoconstriction during or immediately following exhaustive exercise. At that time it may be impossible to obtain a sample or the sample is venous. For example, in a recent study, we took blood samples at rest, immediately after exhaustion, and at 2 and 4 min postexercise; the mean PO_2 values were 57.1, 28.8, 44.7, and 58.2 mmHg respectively. Thus, the immediately postexercise sample was a venous sample, taken during a time of pronounced vasoconstriction. If one wants an arterialized sample, one must demonstrate that the sample has a PO_2 high enough to maintain saturation.

The technique for maintaining patency of catheters is important. A heparin lock or a slow heparin infusion may introduce enough of the anticoagulant to influence fat metabolism [8, 9]. We have found a slow infusion of isotonic saline very satisfactory in preventing coagulation. Not only can the presence of an anticoagulant influence the lactate and metabolic response, but Dunn and Critz [8, 9] also demonstrated in anesthetized animals that the type of anesthetic can alter lactate-fat metabolic interactions.

These various problems associated with sampling should be considered when comparing data of other studies or when comparing data collected at various times throughout an exercise bout. It can lead to somewhat misleading data if there is inadequate equilibrium within the cardiovascular system due to factors such as blood distribution and rapid changes in lactate flux.

While blood samples are the focus of many lactate studies, muscle biopsies are also routinely assayed for lactate. There are two areas of concern: freezing time influences on the lactate level and regional differences within the muscle. Several studies have delayed the freezing of muscle samples [11, 13, 14] and found that the lactate changes are minimal for at least 10 s and perhaps even for as much as 1 min (biopsies are routinely taken and frozen in 3–5 s). The variability among samples of the muscle can be as much as 15%–20% [13, 14]. Interbiopsy variability in fiber type can be high [10, 15] and it may be a source of the variation in lactate. Several studies using single fiber dissection-analysis have reported that fast twitch (type II) fibers within a single biopsy can have greater lactate levels [12, 16, 17]. Regional differences may also be due to perfusion variation and/or differences in motor unit activity.

Storage and Analysis

Blood samples can be stored as serum, plasma, or whole blood (deproteinated in perchloric acid). The choice is partially dependent on whether additional measures are to be made on the samples. However, the deproteinated form has the advantage that its acidic nature will resist bacterial growth and contamination. The storage form can also be important in that differences in lactate concentration can exist between plasma and red blood cells (RBC) [19] (Welch, personal communication 1983). Normally these are small, but under some conditions Welch suggests that the difference can be quite large (e.g., approximately 5 mM) with higher levels in the plasma.

The assay methodology has been reviewed in detail [18]. Generally the method employed is the enzymatic (lactate dehydrogenase) reduction of NAD as described by Kragenings [18] and Bergmeyer [20]. The technique is simple, inexpensive, and reliable. With aqueous stock solutions we get 98% agreement with the true value and a highly linear relationship ($r = 0.99$). When these stock solutions are added to muscle samples, extracted with perchloric acid, and analyzed, the recovery is 75% and the response is linear ($r = 0.93$). Since lactate is present in perspiration [4], care should be taken in keeping glassware clean.

The assay uses a lactate dehydrogenase (L-lactate: NAD oxidoreductase, EC 1.1.1.27), which is specific for the L(+)-lactate stereoisomer. Thus, one actually measures only the L form, which is generally accepted as the biologically active form. However, recently several investigators have suggested that the D form may have metabolic significance [21–23]. This raises two potential problems: we may be ignoring a lactate pool (i.e., the D form), or if the lactate dehydrogenase is contaminated with some of the enzyme specific for D-lactate, we may be measuring both D- and L-lactate without realizing it. However, we have recently analyzed human blood that had high levels of L-lactate for D-lactate and found no detectable amounts.

Intraindividual Variation

The level of blood lactate attained during strenuous activity can vary considerably among subjects. This has been attributed to differences in fitness [24, 25], fiber type [26–28], and even menstrual cycle position in females [29]. However, seldom has the variability within subjects been examined. We [30] tested subjects on five to seven occasions and measured blood lactate at rest, at exhaustion, and 30 min into recovery. At all three data points the individual coefficients of variation were between 7% and 38% with the mean values being approximately 20%. This is approximately four times the intraindividual variation reported for $\dot{V}O_2$ max [30, 31]. The author is not aware of similar data for muscle lactate.

Interpretation

The Metabolic Role of Lactate

Evolutionary Considerations

Increases in blood or muscle lactate are frequently interpreted as direct evidence of anaerobic metabolism, inadequate oxygen availability (hypoxia), and reflections of lactate production and glycolytic rates [24, 32]. To better understand the complexity of lactate data, it is necessary to examine both the evolutionary aspects of lactate metabolism and its role and implications in cell metabolism.

The early, energy-consuming cells were anaerobes lacking mitochondria and the tricarboxylic acid (TCA) cycle, but with a primitive fermentation system [33, 34]. Both glycolytic and alcoholic fermentation were probably available. These systems are identical down to the level of pyruvate by which point the breakdown of hexoses has yielded a pair of three-carbon pyruvate molecules, two NADH molecules, and two ATP molecules. One might question why should the cell continue on to produce either lactate, in the case of glycolysis, or acetaldehyde and then ethanol in alcoholic fermentation? The critical function of either of these final states is the concomitant regeneration of NAD^+. NAD^+, a substance in low concentration, is a critical component of the glyceraldehyde-3-phosphate breakdown to 1,3-diphosphoglycerate (Fig. 1). Thus, without the recycling of the cofactor NAD^+, this reaction, and hence fermentation, would halt. In fact, it has been suggested that this step can limit maximal glycolytic flux [35, 36]. Hochachka and Somero [33] suggest that alcoholic fermentation was probably the first system, with glycolysis coming later. While it is conjecture on the author's part, it may be that many anaerobes "opted" for glycolysis, since the lactate procedure for regenerating the critical NAD^+ would be less detrimental to the internal environment than producing acetaldehyde and ethanol in quantity. Furthermore, once aerobic metabolism was acquired, it would be easier to use lactate as a fuel (by conversion to pyruvate) than to use ethanol to produce pyruvate.

Subsequently, as genetic mutations occurred, gradual modification took place. One such modification was duplication of the gene for lactate dehydrogenase [37, 38]. These were then capable of independent mutations producing gradual modifications of the originally redundant mechanism. Barker and Dayhoff [37] estimate that the original gene duplication occurred 650 million years ago, i.e., prior to the emergence of the vertebrates. Thus, two slightly different forms (H and M) of the same protein could be synthesized. The lactate dehydrogenase enzyme is a tetramer of varying numbers of these two proteins, making possible five isozymes. It is the specific characteristics of these isozymes which provide different tissues and fiber types with varying abilities to produce or take up lactate [39].

The Contemporary Scheme

In the contemporary metabolic scheme (Fig. 1), cells have acquired alternative processes for regenerating the vital cytoplasmic NAD^+. These shuttle systems not only

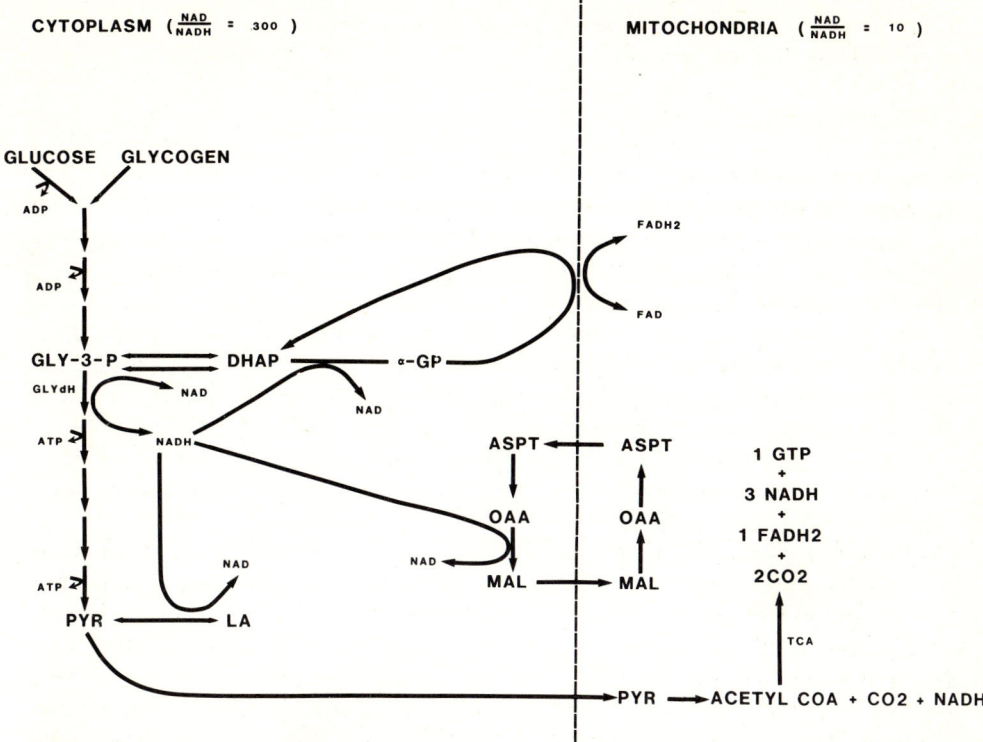

CYTOPLASM ($\frac{NAD}{NADH}$ = 300)   **MITOCHONDRIA** ($\frac{NAD}{NADH}$ = 10)

Fig. 1. A metabolic scheme demonstrating the recycling of cytoplasmic NAD^+. The *vertical line* represents the mitochondrial membranes. Availability of NAD^+ is critical to the glyceraldehyde-3-phosphate reaction and, hence, to glycolysis. NADH can be oxidized during lactate production or by the "shuttle reactions" of malate and α-GP. Note that the quantity of pyruvate that enters the TCA cycle is that portion of the glycolytic rate that is *not* represented in lactate production. It is also the minimal quantity of NADH that must be recycled by the shuttle reactions

produce cytoplasmic NAD^+, but they also transfer the electrons of the cytoplasmic NADH into the mitochondria for oxidative phosphorylation. Thus, these processes compete with lactate dehydrogenase for the cytoplasmic NADH.

The rate of production of this NADH is identical to that of pyruvate (i. e., the true glycolytic rate). Thus, during heavy exercise, NADH is produced rapidly. McGilvery [60] estimates the maximal glycolytic flux at $1.0\,\mu M g^{-1} s^{-1}$. The total NAD^+ concentration (approximately $0.5\,\mu M g^{-1}$) [13, 40] would only last $0.5\,s$ without these reactions. It is noteworthy that every unit of pyruvate to be decarboxylated by the TCA cycle represents an NADH unit that must be processed by the shuttle systems rather than by lactate production. Therefore, lactate production can be far less than pyruvate production or the glycolytic rate. In addition, every lactate unit that enters the cell and is oxidized to pyruvate represents an NADH unit that also must be handled by shuttle systems. During exercise the shuttle systems must be quantitatively more important than lactate production.

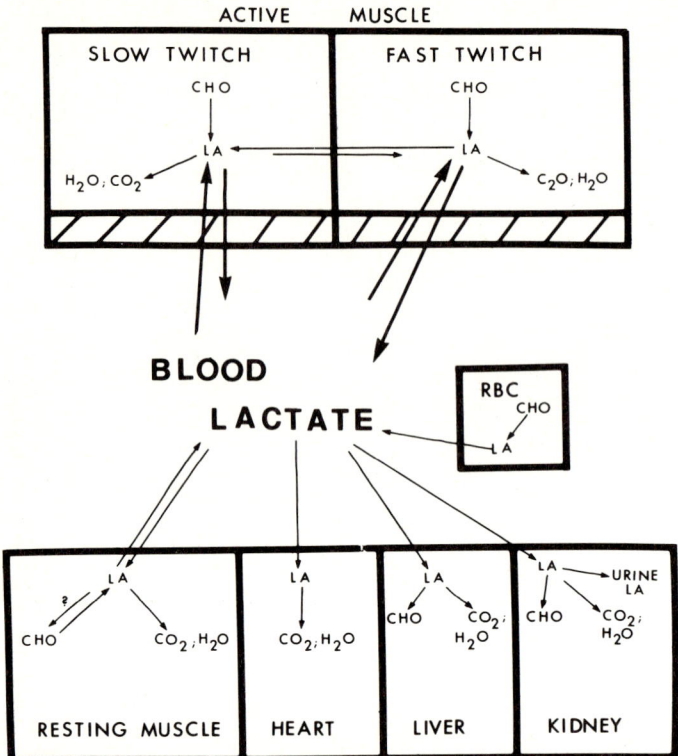

Fig. 2. A scheme representing the dynamic nature of the major tissues that influence blood lactate *(La)*. Sources include red blood cells *(RBC)* and resting muscle as well as active muscle. Within active muscle, diffusion can occur between fiber types, and the *hatched border* represents the semi-permeable nature of the muscle membrane. Heart, liver, and kidney metabolize lactate, with gluco-genesis occurring in liver and kidney. Resting muscle can either release or take up lactate, and the possibility of glucogenesis is uncertain at the present time

Lactate Concentration, Release, and Production

Nevertheless, lactate production is of metabolic significance. Unfortunately, it is very difficult to quantify, and measured changes in blood lactate concentration at a single sample site are, at best, a crude, qualitative indication of the production rate. One must appreciate that there are a variety of sources and sinks for lactate (Fig. 2), and the blood lactate pool is merely a reflection of the dynamic equilibrium of these processes.

Even if one measures the rate of lactate release from a tissue via the direct Fick principle, the production rate can be considerably underestimated. Sacks and Sacks [41] demonstrated that a large concentration difference exists between active muscle and the venous plasma. Work by Karlsson [14], Hirche et al. [42], and others has verified this early observation. It is apparent that lactate is not the freely diffusing substance postulated by Hill et al. [43]. Jorfeldt et al. [44] reported that at muscle lactate greater than $4 \, mM/kg$ (approximately 70% $\dot{V}O_{2\,max}$), the rate of release had no rela-

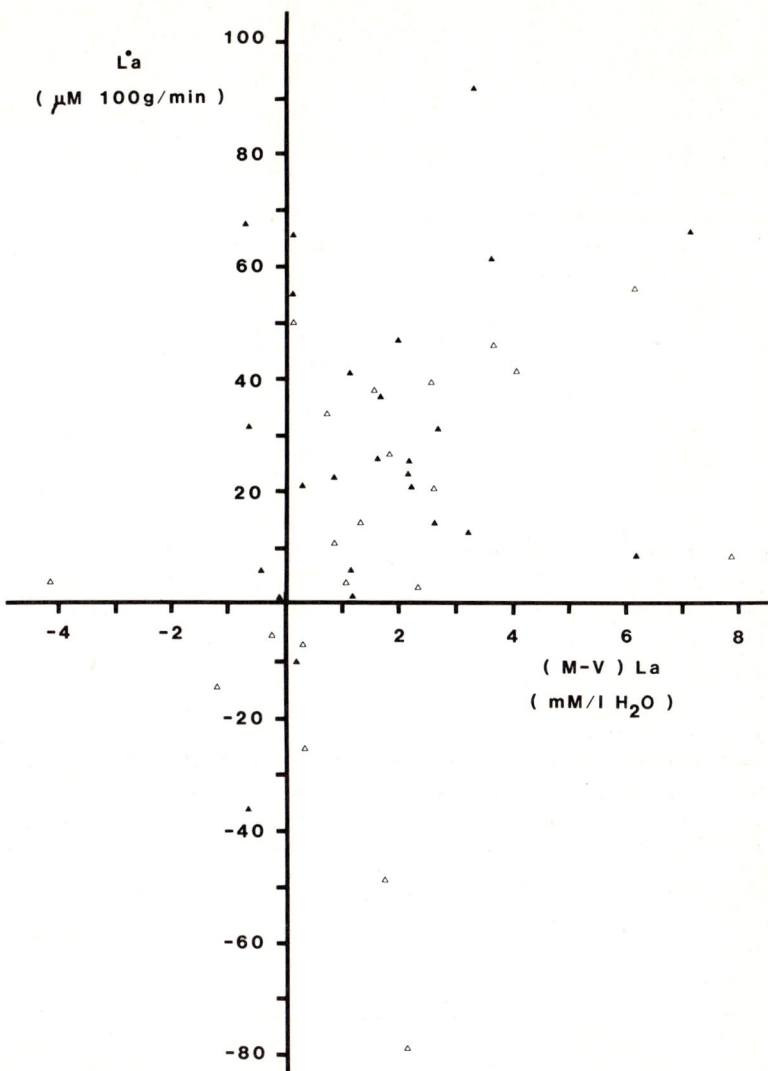

Fig. 3. The rate of lactate release *(L̇a)* from active, canine muscle and the muscle-venous blood lactate concentration difference [*(M-V)La*] that existed at that time. If lactate is released from muscle by passive diffusion then (a) a positive relationship should exist and (b) there should never be a data point in either the upper left or lower right quadrants. However, ten of 44 experiments are in these quadrants. Although it is immaterial to this paper, the *closed triangles* are controls and the *open triangles* are animals in respiratory acidosis. (Unpublished work by Graham, Barclay, and Wilson)

tionship to muscle levels. In an isolated muscle preparation [13], we failed to find any relationship between lactate release and muscle lactate levels. In a recent study (Fig. 3), we found that not only did the data support this earlier finding, but also 20%–30% of the muscle preparations had a net release or uptake of lactate against the direction of the muscle-venous lactate concentration difference. Lactate exchange is obviously complex; it is uncertain whether it crosses the membrane predominantly as La$^-$ or as the undissociated acid [45, 46]. Exchange may involve facilitated or active transport [23, 47, 48, 79]. If one wants to assess lactate production, both lactate release and intramuscular lactate storage must be considered.

The lack of agreement of a muscle-blood lactate relationship between subjects was demonstrated clearly by Löllgen et al. [26]. We found that subjects with similar blood lactate levels could have a threefold difference in muscle lactate levels. A dominant factor in deciding the muscle-blood relationship was fiber type. Similarly, Tesch et al. [27, 28] found that either capillaries per fiber or % ST plus capillary density correlated with onset of blood lactate (OBLA). This is in contrast with the work of Jacobs and Kaiser [49] who reported that a strong muscle-blood lactate relationship existed (although only at OBLA) and there was no relationship with fiber type. This contrast in findings may be due to protocol differences and/or because Jacobs and Kaiser [49] had a smaller range in fiber type than in the other studies. With the strong suggestion that subjects with different fiber types can have different muscle-blood relationships, intersubject comparisons of blood lactate become difficult to interpret. The strongest experimental design is to use the subject as his own control.

Even when one uses this design, it must be kept in mind that the muscle-blood relationship within a single subject can vary over time or according to treatment. Several studies have reported that the muscle-blood difference widens in heavy exercise and it returns to or goes below resting levels during recovery [14, 49]. An even more dramatic illustration of the intrasubject variability in this relationship is presented in Fig. 4. This subject performed extremely exhaustive bicycle exercise for 30 s [50] and recovered for 10 min. Obviously, the relationship between muscle and blood is changing very rapidly and to draw conclusions regarding the muscle lactate based on blood measures in this experiment could result in a very erroneous interpretation.

Lactate and Hypoxia

Short-Term Work

Lactate production is often classified as anaerobic glycolysis (it should be noted that all glycolysis is anaerobic and hence this phrase is redundant) and is thought to dominate as an energy source in high-intensity activity lasting less than 2 min or in fatigue situations when oxygen transport is limiting. While these interpretations are probably correct, due in part to the complicating factors discussed previously, these aspects require more investigation. The area of anaerobic metabolism is very poorly understood, although most exercise physiology texts summarize the time course and energy yield of high-energy phosphates and lactate production in a very confident, quantitative fashion. Nevertheless, recent studies are beginning to illustrate the limitations of these interpretations. Boobis et al. [51] and Jacobs et al. [52, 53] have used muscle biopsies to examine anaerobic metabolism during the Wingate test. They found large but less than maximal depletion of phosphagens and accumulation of muscle lactate at the end of the 30 s exhaustive test. Furthermore, their data after 6–10 s of work demonstrated clearly that muscle lactate accumulation is occurring very early in activity contributing to the energy supply prior to phospha-

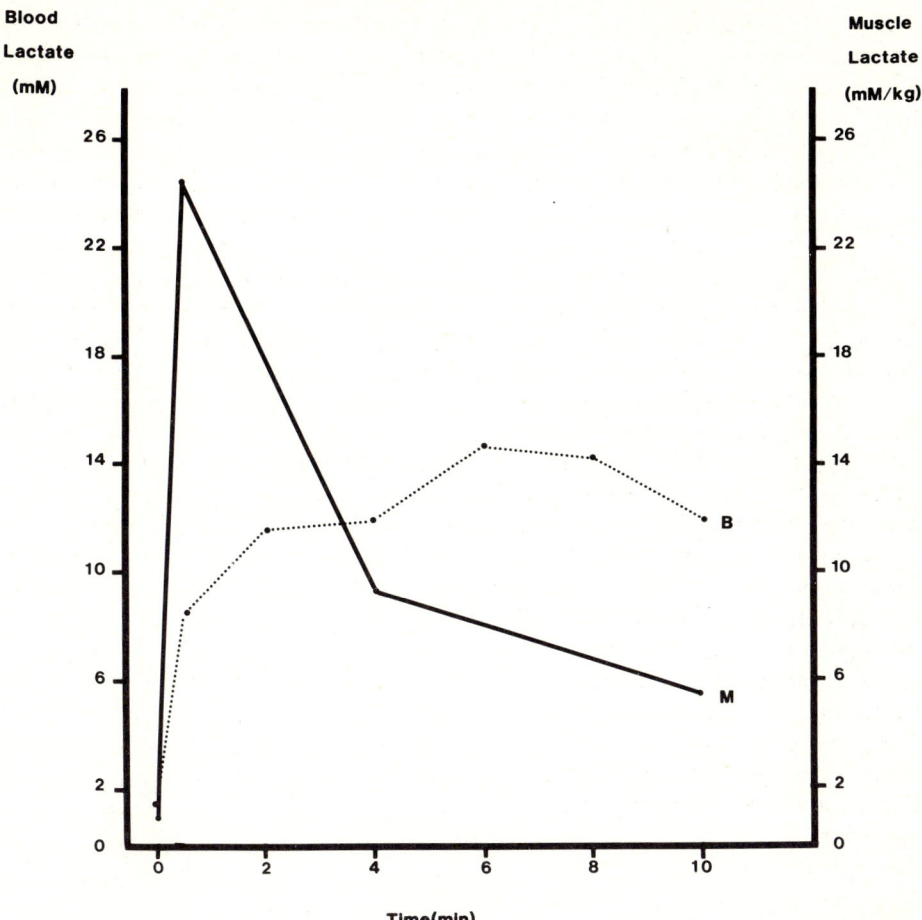

Fig. 4. The muscle *(M)* and blood *(B)* lactate responses in a single subject. The subject performed 30 s of exhaustive exercise and recovered for 10 min. Note the dramatic shifts in the muscle-blood relationship. (Unpublished work of Jones, McCarney, and Graham)

gens reaching a critically low value, in contrast with the theory of Margaria et al. [1] which was based on blood lactate results. The data from our own work (Fig. 4) also support the suggestion that lactate production occurs early in exercise to a very significant degree. However, the exact role of lactate production remains obscure, since neither the 10 s nor the 30 s lactate results correlated with power output [53].

We have recently compared indices of aerobic ($\dot{V}o_{2\,max}$) and anaerobic work (peak or maximum blood lactate, fast and slow components of O_2 debt, and performance on the Wingate test) in men and women ($n = 14$) (unpublished data). Peak lactate correlated poorly with total O_2 debt ($r = 0.59$) and the slow component ($r = 0.63$). Both the peak lactate and the fast component of the debt (presumably associated with phosphogen repayment [54, 55]) had nonsignificant correlations with either peak or total Wingate power. The energy sources for muscle activity must be complex in their integration, and it does not appear to be a straightforward transition of phosphogen use yielding to lactate as a prime energy source yielding to oxidative means.

Anaerobic Threshold

In recent years, a very popular measurement has been that of anaerobic threshold. Presumably this is a time point or an exercise intensity at which there is an increase in blood lactate and the ventilatory equivalent increases with a concomitant increase in end-tidal PO_2 and fall in end-tidal PCO_2. As this physiological phenomenon received more and more attention, terms such as "aerobic" and "anaerobic" threshold, lactate threshold, ventilation threshold, and OBLA ($>4 mM$ blood lactate) have all been introduced. The reader can refer to the recent review by Jones and Ehrsam [32] for a thorough treatment of the topic. It should be noted that the lactate of this measurement is no more or less anaerobic or a threshold than is a resting blood lactate level, and there is no evidence that either the whole body or the muscle is anaerobic or hypoxic at this point (see following section). Furthermore, it has not been demonstrated that the lactate and pulmonary observations are necessarily causally related.

Hughes et al. [56] demonstrated that altering either pedal frequency or glycogen level can divorce the apparent lactate-ventilation relationship. Similarly, inspiring 4% CO_2 hastens the ventilatory response and delays the blood lactate response [58]. Perhaps the most conclusive evidence is that of Hagberg et al. [59], who demonstrated that during progressive exercise patients with McArdle's disease showed no increase in blood lactate, while having a typical "ventilatory" threshold. McGilvery [60] suggests that during moderate and heavy exercise when anaerobic threshold and OBLA occur, lactate production is increased because ADP begins to stimulate glycolytic enzymes more than TCA enzymes. Thus lactate would be released in greater quantities and independent of oxygen availability. There is no disputing that both the blood lactate and ventilatory responses are interesting phenomenon, but they do not appear to be causally related.

Lactate and Oxygen Delivery

Associated with the anaerobic threshold concept is the assumption that lactate production is symptomatic of inadequate oxygen delivery. While there is no doubt that tissue hypoxia will evoke lactate production, lactate production does not necessarily signify hypoxia [24, 57]. Researchers have failed to show relationships between lactate production and venous muscle PO_2 [61, 62] or muscle NAD determinations [13, 40, 63]. Furthermore, hyperoxia has been found to lower exercising muscle lactate release (Fig. 5) without an increase in muscle $\dot{V}O_2$ (Fig. 6) [64, 65]. As outlined in the following section, hydrogen ion may be a key factor in understanding this mechanism. Muscle lactate production often occurs at rest and during mild as well as heavy exercise even while some aspects of the muscle take up and metabolize lactate [66–70]. Thus, lactate metabolism can occur independent of oxygen supply, it can be altered without changing $\dot{V}O_2$, and some components of muscle can oxidize lactate while other aspects are simultaneously producing it.

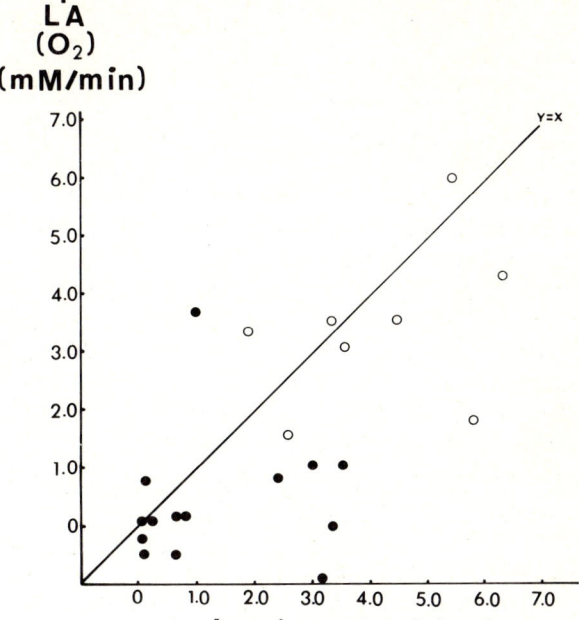

Fig. 5. Lactate release *(L̇a)* from the leg of exercising man; O_2 breathing plottet against air breathing. Each *point* represents a determination on a subject under the two conditions and the *line* is the line of identity. The *filled circles* represent 150 W of exercise and the *open circles* represent approximately 100% $\dot{V}_{O_2\,max}$. Note that L̇a is lower with hyperoxia, i.e., the majority of the points lie below the line of identity. Four subjects actually began to take up lactate (L̇a is negative) during hyperoxia [57]

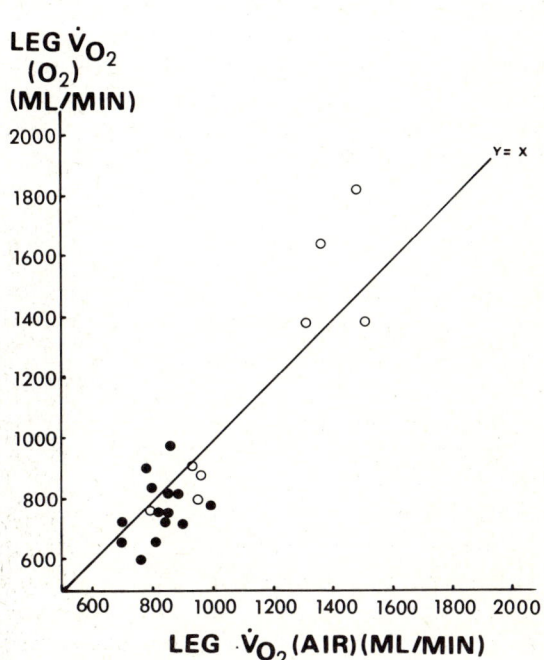

Fig. 6. Leg $\dot{V}_{O_2}$ from the leg of exercising man. The format is the same as in Fig. 5. If the lower L̇a of Fig. 5 was due to the removal of a hypoxic stress, the $\dot{V}_{O_2}$ should be increased in the hyperoxic condition, i.e., the points should lie above the line of identity $(y=x)$ [57]

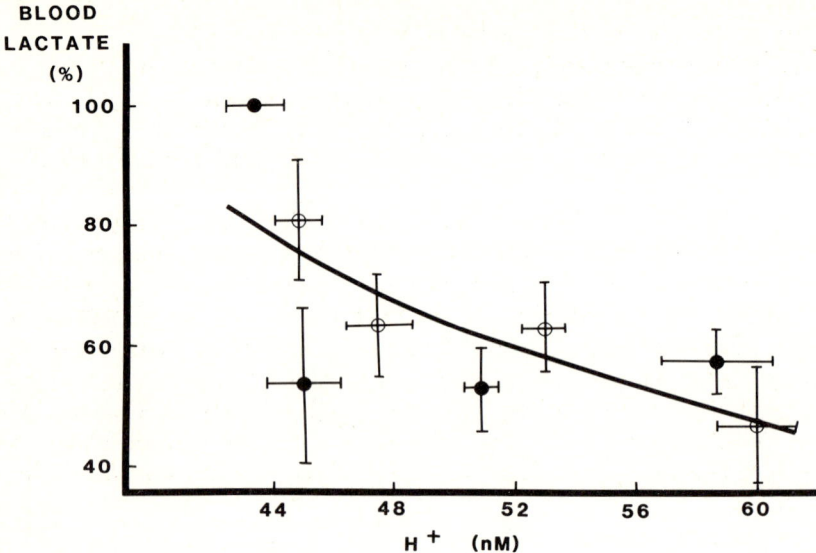

Fig. 7. Blood lactate level during steady state exercise. Subjects performed 30 min of work (60% $\dot{V}_{O_2 max}$) in eight conditions: four normoxic tests *(closed circles)* and four hyperoxic (60% O_2) tests *(open circles)*. The four tests used 0%, 2%, 4%, or 6% CO_2 with either 21% or 60% O_2. While hyperoxia lowered blood lactate, it was proportional to the arterial H^+ concentration. Furthermore, if H^+ was increased by hypercapnia independent of hyperoxia, the same depression in blood lactate was found [77]

Lactate and Hydrogen Ion

Acidosis is commonly associated with exercise and it has frequently been postulated that hydrogen (H^+) is a limiting factor in glycolytic and lactate production control [32, 46, 71]. A number of glycolytic enzymes are inhibited in vitro by H^+; however, the role(s) of H^+ in in vivo regulation are not clearly understood. A recent review by Hochachka and Mommsen [72] points out how complex the area of H^+ production is and how poorly it is understood. They note that H^+ and lactate are not necessarily produced either in a constant or in an equimolar relationship. External adjustment of the acid-base status has been shown to alter blood lactate levels. During exercise, alkalosis has been associated with elevated blood lactate levels [73, 74], while acidosis lowers the concentration [58, 74–76], independent of $\dot{V}_{O_2}$. In a recent study [77], we found that the steady state exercise blood lactate level was dependent on the arterial PCO_2 or the H^+ concentration and independent of whether the subject breathed 60% or 21% O_2 (Fig. 7). This supports the hypothesis of Adams and Welch [78] that the commonly reported lowering of blood lactate by hyperoxia was mediated by respiratory acidosis, not by the elevated PO_2. While the etiology of the response is not clear, it may be impaired membrane permeability for lactate and/or suppressed lactate production.

Summary

The measurement of lactate is straightforward and reliable. The sampling is normally dependable, although discrepancies between sample sites can occur if samples are drawn during other than steady state conditions or when blood flow to the site is low. Comparison of data between subjects should consider intra- and interindividual variation.

Interpretation of the significance of blood or even a muscle lactate concentration is extremely difficult and neither can be assumed to reflect rates of lactate release or production. Lactate metabolism appears to be very significant in short-term, high-intensity work lasting a few seconds, but the exact role remains to be established. Blood lactate and the pulmonary aspects of anaerobic threshold are not necessarily causally related and the significance of anaerobic threshold as a metabolic measure remains obscure. There are numerous examples of changes in blood or muscle lactate independent of oxygen availability, and factors such as H^+ concentration appear to play an important, although undefined, role in regulating lactate production and/or release.

Acknowledgements. The author gratefully acknowledges the contributions of the co-workers of the work reported. These are Drs. G. Andrew, F. Bonde-Petersen, A. Bonen, C. Chapler, B. Goslin, N. Jones, K. Klausen, H. Löllgen, N. McCartney, B. Saltin, M. Sample, N. Secher, D. Sinclair, G. Sjøgaard, J. Van Dijk, H. Welch, and B. Wilson. The work of the author has been supported by N. S. E. R. C. He also acknowledges the excellent work of Jan Van Dijk and Jenny Mlodozenec in preparing this manuscript.

References

1. Margaria R, Oliva D, Di Prampero PE, Cerretelli P (1969) Energy utilization in intermittent exercise of supramaximal intensity. J Appl Physiol 26: 752–756
2. Gisolfi C, Robinson S (1970) Venous blood distribution in the legs during intermittent treadmill work. J Appl Physiol 29: 368–373
3. Newton JL, Robinson S (1965) The distribution of blood lactate and pyruvate during work and recovery. Fed Proc 24: 590
4. Gordon RS, Thompson RH, Muenzer J, Thrasher D (1971) Sweat lactate is derived from blood glucose. J Appl Physiol 31: 713–716
5. Ricci B (1968) Measurement of oxygen debt and of blood lactate and pyruvate. In Knuttgen, H.G (ed) Physiological aspects of sports and physical fitness. Am Coll Sport Med and the Athletic Institute Philadelphia pp 12–15
6. McEvoy JDS, Jones NL (1975) Arterialized capillary blood gases in exercise studies. Med Sci Sports 7: 312–315
7. Forster HV, Dempsey JA, Thomson J, Vidruk E, DoPico GH (1972) Estimation of arterial PO_2, PCO_2, pH, and lactate from arterialized venous blood. J Appl Physiol 32: 134–137
8. Dunn RB, Critz JB (1975) Uptake of lactate by dog skeletal muscle in vivo and the effect of free fatty acids. Am J Physiol 229: 255–259
9. Dunn RB, Critz JB (1975) Effect of circulating FFA on lactate production by skeletal muscle during stimulation. J Appl Physiol 38: 801–805
10. Elder, GCB, Bradbury K, Roberts R (1982) Variability of fiber type distributions within human muscles. J Appl Physiol 50: 1473–1480

11. Essén B (1978) Studies on the regulation of metabolism in human skeletal muscle using intermittent exercise as an experimental model. Acta Physiol Scand [Suppl] 454: 1–67
12. Essén B, Haggmark T (1975) Lactate concentration in type I and II muscle fibers during muscle contractions in man. Acta Physiol Scand 95: 344–346
13. Graham TE, Sinclair DG, Chapler CK (1976) Metabolic intermediates and lactate diffusion in active dog skeletal muscle. Am J Physiol 231: 766–771
14. Karlsson J (1971) Lactate and phosphagen concentrations in working muscle of man. Acta Physiol Scand [Suppl] 358: 1–72
15. Sandstedt PER (1981) Representativeness of a muscle biopsy specimen for the whole muscle. Acta Neurol Scand 64: 427–437
16. Ball M (1982) Metabolic and mechanical correlates of muscle fatigue. M. Sc. thesis, University of Waterloo, Ontario
17. Tesch P (1980) Muscle fatigue in man with special reference to lactate accumulation during short-term intense exercise. Acta Physiol Scand [Suppl] 480: 1–91
18. Kragenings I (1979) Methodology of lactate assay. In: Bossart H, Perret C (eds) Lactate in acute conditions. International Symposium, Basel. Karger, Basel, pp 20–28
19. Kreisberg RA, Pennington LF, Boshell BR (1970) Lactate turnover and gluconeogenesis in normal and obese humans. Effects of starvation. Diabetes 19: 53–63
20. Bergmeyer HV (1974) Methods of enzymatic analysis, vol 3, 2nd ed. Academic, New York
21. Schumer W (1979) Ce-1 metabolism and lactate. In: Bossart H, Perret C (eds) Lactate in acute conditions. International Symposium, Basel. Karger, Basel, pp 1–9
22. Giesecke E, Wallenberg P, Fabritius A (1980) D(-) lactate acid – a physiological isomer in the rat. Experimentia 36: 571
23. Leichtweib HR, Schröder H (1981) L-lactate and D-lactate carriers on the fetal and the maternal side of the trophoblast in the isolated guinea pig placenta. Pflugers Arch 390: 80–85
24. Gollnick PD, Saltin B (1982) Significance of skeletal muscle oxidative enzyme enhancement with endurance trainging. Clin Physiol 2: 1–12
25. Holloszy JO, Booth FW (1976) Biochemical adaptations to endurance exercise in muscle. Annu Rev Physiol 38: 273–291
26. Löllgen H, Graham TE, Sjogaard G (1980) Muscle metabolites, force and perceived exertion bicycling at varying pedal rates. Med Sci Sports 12: 345–351
27. Tesch PA, Sharp DS, Daniels WL (1981) Influence of fiber type composition and capillary density on onset of blood lactate accumulation. Int J Sports Med 2: 252–255
28. Tesch PA, Daniels WL, Sharp DS (1982) Lactate accumulation in muscle and blood during submaximal exercise. Acta Physiol Scand 114: 441–446
29. Jurkowski JEH, Jones NL, Toews CJ, Sutton JR (1981) Effects of menstrual cycle on blood lactate, O_2 delivery, and performance during exercise. J Appl Physiol 51: 1493–1499
30. Graham TE, Andrew GM (1973) The variability of repeated measurements of oxygen debt in man following a maximal treadmill exercise. Med Sci Sports 5: 73–78
31. Katch VL, Sady SS, Freedson P (1982) Biological variability in maximum aerobic power. Med Sci Sports Exerc 14: 21–25
32. Jones NL, Ehrsam RE (1982) The anaerobic threshold. In: Terjung RL (ed) Exercise and sports sciences reviews. Franklin, Philadelphia (Am coll sport med series, vol 10)
33. Hochachka PW, Somero GN (1973) Strategies of biochemical adaptation. Saunders, Philadelphia
34. Schopf JW (1978) The evolution of the earliest cells. Sci Am 239: 110–138
35. Kobayashi K, Neely JR (1979) Control of maximum rates of glycolysis in rat cardiac muscle. Circ Res 44: 166–175
36. Newsholme EA, Crabtree B (1979) Theoretical principles in the approaches to control of metabolic pathways and their application to glycolysis in muscle. J Mol Cell Cardiol 11: 839–855
37. Barker WC, Dayhoff MO (1980) Evolutionary and functional relationships of homologous physiological mechanisms. Bioscience 30: 593–600
38. Sidell BD, Beland KF (1980) Lactate dehydrogenases of Atlantic hogfish: physiological and evolutionary implications of a primative heart isozyme. Science 207: 769–770
39. Sjödin B (1976) Lactate dehydrogenase in human skeletal muscle. Acta Physiol Scand [Suppl] 436: 1–82
40. Graham TE, Sjøgaard G, Löllgen H, Saltin B (1978) NAD in muscle of man at rest and during exercise. Pflügers Arch 376: 35–39

41. Sacks J, Sacks WC (1937) Blood and muscle lactic acid in the steady-state. Am J Physiol 118: 697–702
42. Hirche H, Hombach V, Langohr HD, Wacker V, Busse J (1975) Lactic acid permeation rate in working gastrocnemii of dogs during metabolic alkalosis and acidosis. Pflügers Arch 356: 209–222
43. Hill AV, Long CNH, Lupton H (1924) Muscular exercise, lactic acid, and the supply and utilization of oxygen – Parts I–III. Proc R Soc Lond [Biol] 96: 438–475
44. Jorfeldt L, Juhlin-Dannfelt A, Karlsson J (1978) Lactate release in relation to tissue lactate in human skeletal muscle during exercise. J Appl Physiol 44: 350–352
45. Mainwood GW, Worsley-Brown P (1975) The effects of extracellular pH and buffer concentration on the efflux of lactate from frog sartorius muscle. J Physiol (Lond) 250: 1–22
46. Roos A, Boron WF (1981) Intracellular pH. Physiol Rev 61: 297–434
47. Barac-Nieto M, Murer H, Kinne R (1982) Asymmetry in the transport of lactate by basolateral and brush border membranes of rat kidney cortex. Pflügers Arch 392: 366–371
48. Dubinsky WP, Racker E (1978) The mechanism of lactate transport in human erythrocytes. J Membr Biol 44: 25–36
49. Jacobs I, Kaiser P (1982) Lactate in blood, mixed skeletal muscle and FT or ST fibers during cycle exercise in man. Acta Physiol Scand 114: 461–466
50. McCartney N, Heigenhauser GJF, Sargeant AT, Jones NL (1983) A constant velocity cycle ergometer for the study of dynamic muscle function. J Appl Physiol 55: 212–217
51. Boobis L, Williams C, Wootton SA (1983) Human muscle metabolism during brief maximal exercise. J Physiol (Lond) 338: 21
52. Jacobs I, Bar-Or O, Karlsson J, Dotan R, Tesch P, Kaiser P, Inbar O (1982) Changes in muscle metabolites in females with 30 s exhaustive exercise. Med Sci Sports Exerc 14: 457–460
53. Jacobs I, Tesch PA, Bar-Or O, Karlsson J, Dotan R (1983) Lactate in human skeletal muscle after 10 s and 30 s of supramaximal exercise. J Appl Physiol 55: 365–368
54. Hultman E, Bergstrom J, McLennan-Anderson N (1967) Breakdown and resynthesis of phosphocreatine and adenosine triphosphate in connection with muscular work in man. Scand J Clin Lab Invest 19: 56–66
55. Margaria R, Edwards HT, Dill DB (1933) The possible mechanisms of contracting and paying the oxygen debt and the role of lactic acid in muscular contraction. Am J Physiol 106: 689–715
56. Hughes EF, Turner SC, Brooks GA (1982) Effects of glycogen depletion and pedaling speed on "anaerobic threshold." J Appl Physiol 52: 1598–1607
57. Graham TE (1978) Oxygen delivery and blood and muscle lactate changes during muscle activity. Can J Appl Sport Sci 3: 153–159
58. Graham TE, Wilson BA, Sample M, Van Dijk J, Bonen A (1980) The effects of hypercapnia on metabolic responses to progressive exhaustive work. Med Sci Sports Exerc 12: 278–284
59. Hagberg JM, Coyle EF, Carroll JE, Miller JM, Martin WH, Brooke MH (1982) Exercise hyperventilation in patients with McArdle's disease. J Appl Physiol 52: 991–994
60. McGilvery RW (1975) The use of fuels for muscular work. In: Howald H, Poortmans JR (ed). Metabolic adaptation to prolonged physical exercise. Proceedings of the second international symposium on biochemistry of exercise. Magglingen 1973. Birkhauser, Basel
61. Dempsey JA, Thomson JM, Forster HB, Cerney FC, Chosy LW (1975) HbO$_2$ dissociation in man during prolonged work in chronic hypoxia. J Appl Physiol 38: 1022–1029
62. Pirnay F, Dujardin J, Deroanne R, Petit JM (1971) Muscular exercise during intoxication by carbon monoxide. J Appl Physiol 31: 573–575
63. Jöbsis FF, Stainsby WN (1968) Oxidation of NADH during contractions of circulated mammalian skeletal muscle. Respir Physiol 4: 292–300
64. Welch HG, Bonde-Petersen F, Graham TE, Klausen K, Secher N (1977) Effects of hyperoxia on leg blood flow and metabolism during exercise. J Appl Physiol 42: 385–390
65. Wilson BA, Stainsby WN (1978) Effects of O$_2$ breathing on RQ, blood flow, and developed tension in in situ dog muscle. Med Sci Sports 10: 167–170
66. Corsi A, Zatti M, Midris M, Granata AL (1970) In situ oxidation of lactate by skeletal muscle during intermittent exercise. FEBS Lett 11: 65–68
67. Depocas F, Minaire Y, Chatonnet J (1969) Rates of formation of lactic acid in dogs at rest and during moderate exercise. Can J Physiol Pharmacol 47: 603–610
68. Hubbard JL (1973) The effect of exercise on lactate metabolism. J Physiol (Lond) 231: 1–18

69. Issekutz B, Shaw WAS, Issekutz AC (1976) Lactate metabolism in resting and exercising dogs. J Appl Physiol 40: 312–319
70. Jorfeldt L (1970) Metabolism of L(+)-lactate in human skeletal muscle during exercise. Acta Physiol Scand [Suppl] 338: 1–102
71. Sahlin K (1978) Intracellular pH and energy metabolism in skeletal muscle of man. Acta Physiol Scand [Suppl] 455: 1–64
72. Hochachka PW, Mommsen TP (1983) Protons and anaerobiosis. Science 219: 1391–1397
73. Edwards RHT, Clode M (1970) The effect of hyperventilation on the lactacidemia of muscular exercise. Clin Sci 38: 269–276
74. Sutton JR, Jones NL, Toews CJ (1981) Effect of pH on muscle glycolysis during exercise. Clin Sci Mol Med 61: 331–338
75. Graham TE, Wilson BA, Sample M, Van Dijk J, Goslin B (1982) The effects of hypercapnia on the metabolic response to steady-state exercise. Med Sci Sports Exerc 14: 286–291
76. Rizzo A, Gimenez M, Horsky P, Saunier C (1976) Influence d'une atmosphère de CO_2 à 4% sur le comportement métabolique à l'exercise d'hommes jeunes. Bull Eur Physiopathol Respir 12: 209–219
77. Graham TE, Wilson BA (1983) Effects of hypercapnia and hyperoxia on metabolism during exercise. Med Sci Sports Exerc (to be published)
78. Adams RP, Welch HG (1980) Oxygen uptake, acid-base status, and performance with varied inspired O_2 fractions. J Appl Physiol 49: 863–868
79. Cohen RD (1979) The production and removal of lactate. In: Bossart H, Penet C (eds) Lactate in acute conditions. International symposium, Basel. Karger, Basel, pp 10–19

OBLA Exercise Stress Testing in Health and Disease

J. Karlsson, A. Holmgren, D. Linnarson, and H. Åström

Introduction and State of the Art

During recent years there has been a marked increase in the number of published papers within the area of exercise stress testing based upon any expression for increased muscle metabolic acidosis (Fig. 1). Thus, the library at the Karolinska Institute has in its computerized files a total of 37 papers published during 1982 in recognized, international medical and related journals as compared with six 5 years before (1977) and non another 5 years earlier (1972). Although the majority of the papers published in 1982 were methodological or basic physiological in nature, as many as 11 (or 30%) were related to different clinical aspects of exercise stress testing.

Physiological Background

The appearance of lactate in muscle and blood with muscle contraction in experimental animals as well as in man is an old "Scandinavian exercise physiological" approach (Fig. 2) [1, 2]. However, its application and interpretation were more related to maximal exercise and evaluation of maximal oxygen uptake data ($\dot{V}O_{2 \, max}$): elevated blood lactate concentration (> 5 mmol l^{-1}) was the criterion for an established $\dot{V}O_{2 \, max}$ [3].

A more advanced physiological interpretation was introduced during the 1950s and 1960s and two papers can be cited: Holmgren and Ström [6] and Hollmann [4]. The latter named the curvilinear increase in blood lactates with increased exercise intensity O_2-Dauerleistungsgrenze, Punkt des optimalen Wirkungsgrades der Atmung oder PoW, aerob-anaerobe Schwelle, etc. [4, 5].

The appearance of elevated lactates has also been related to hyperventilation during acute exercise [7]. This phenomenon has been the basis for methodological considerations and developments that use respiratory variables to detect early onset of blood lactate accumulation [8]. As a result of these studies, a noninvasive technique has been introduced for the study of onset of lactate formation and accumulation with physical exercise, in addition to the classical approach, continuous monitoring of blood lactate.

What is then the reason for lactate formation during muscle contraction? To cover the energy need during physical exercise the energy flux through the very limited muscle pool of high-energy phosphates (ATP and CP) the muscle posseses two metabolic pathways: (a) ATP resynthesis by means of cellular respiration, with the

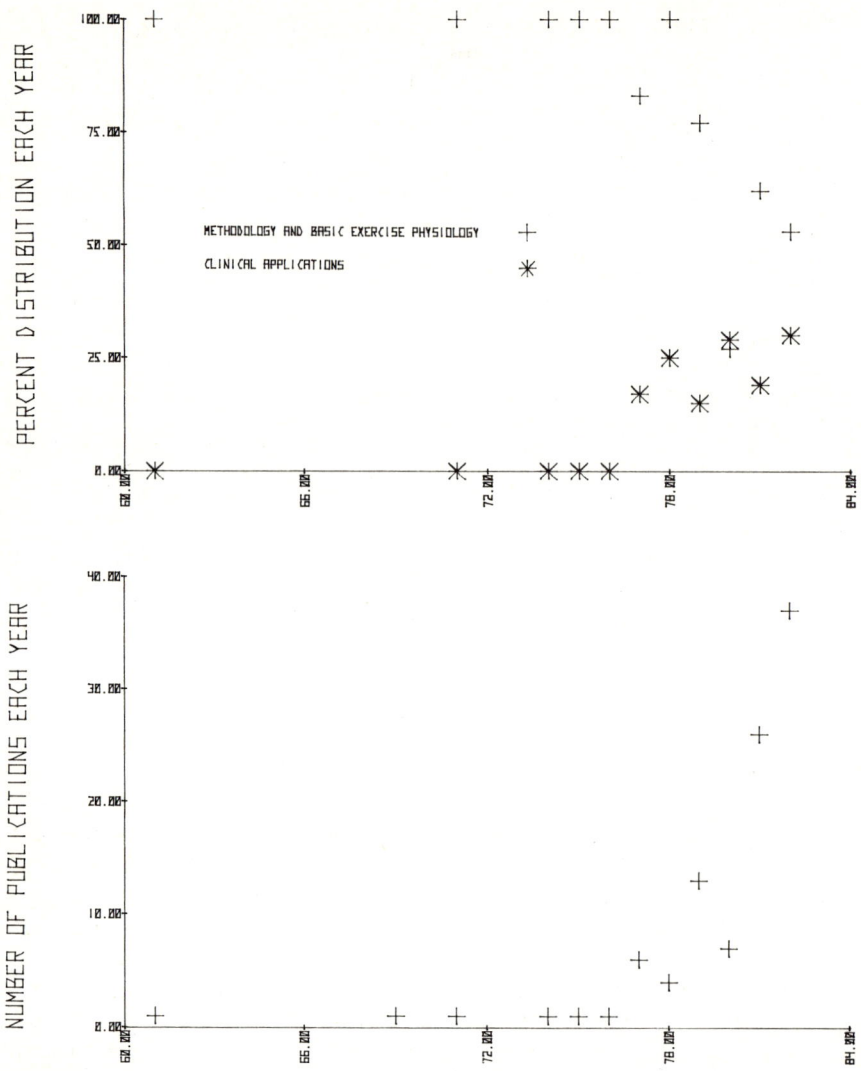

Fig. 1. The number of publications related to the concepts of OBLA, lactate threshold, anaerobic breaking point, etc. during 1960–1982, according to the library at the Karolinska Institute, Stockholm, Sweden

prerequisite of a molecular O_2 presence and (b) ATP resynthesis in the glycolytic pathway with muscle glycogen as the precursor and with lactic acid as the end product (anaerobic glycogenolysis) [9]. In addition, we have to consider the quality of the muscle expressed as muscle fiber composition and the fact that it is constituted of the two major muscle fiber types: (a) slow twitch or type I fibers and (b) fast twitch or type II fibers. Each has distinct differences in respect to their metabolic potentials (Fig. 3). Taken together, these differences will lead to the conclusion that the fast twitch is "faster" and stronger but more susceptible to fatiguing phenomena.

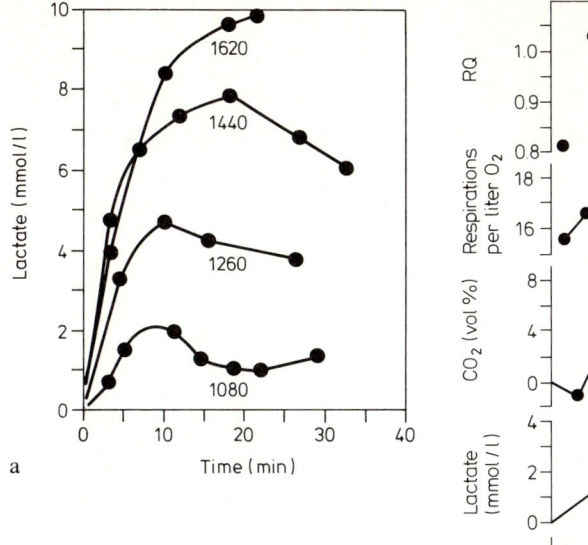

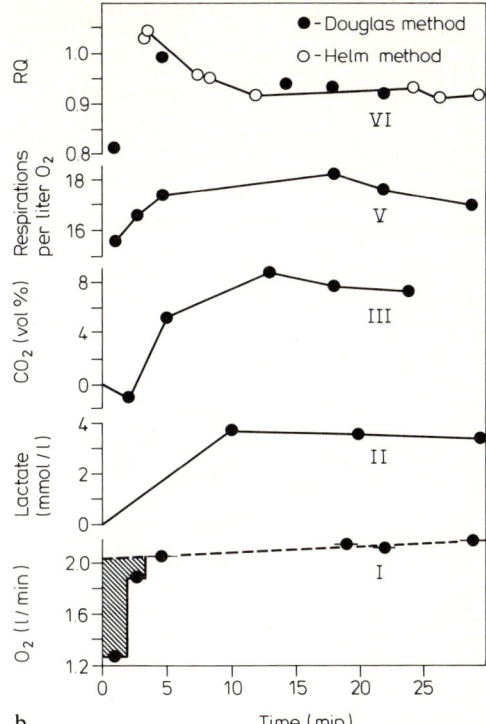

Fig. 2. a, b. Blood lactate concentration;
a at different work loads (kgm/min) and
b during prolonged exercise at the "highest sustainable" work load, according to
Christensen and Hansen [1, 2]

The slow twitch fiber, on the other hand, has all the characteristics necessary to promote endurance qualities [10–12].

It is well described in different animal preparations and in man that increased muscle activity affects regulation of central circulatory features such as heart rate and systemic blood pressure. These phenomena are suggested to be governed by peripheral chemosensors or receptors. They are referred to as ergoreceptors (Fig. 4) [13]. Recently it has also been shown that the muscle rich in fast twitch fibers as a response to stimulation causes a higher heart rate and blood pressure and that dilation is more dependent on local metabolic changes. The muscle rich in slow twitch muscle fibers, on the other hand, causes a less elevated heart rate and blood pressure response, and dilation is dependent on nervous phenomena [13–16].

In man it has been shown that the calculated peripheral resistance is lower in a subject rich in slow twitch muscle fibers both at rest and during exercise, due to both a lower systemic blood pressure and higher leg blood flow [11 and 17] (Fig. 4). The finding of a higher $\dot{V}O_{2max}$ and percent distribution of slow twitch muscle fibers in endurance athletes might very well reflect not only a parallelism between these two features but also a functional coupling (Fig. 5a). The same pattern is also present in subjects involved in conditioning activities of about the same quality and quantity on a weekly basis (Fig. 5b) [10].

To summarize, muscle metabolism in man is dependent on oxygen transport, muscle quality (expressed as muscle fiber composition and type), and intensity of

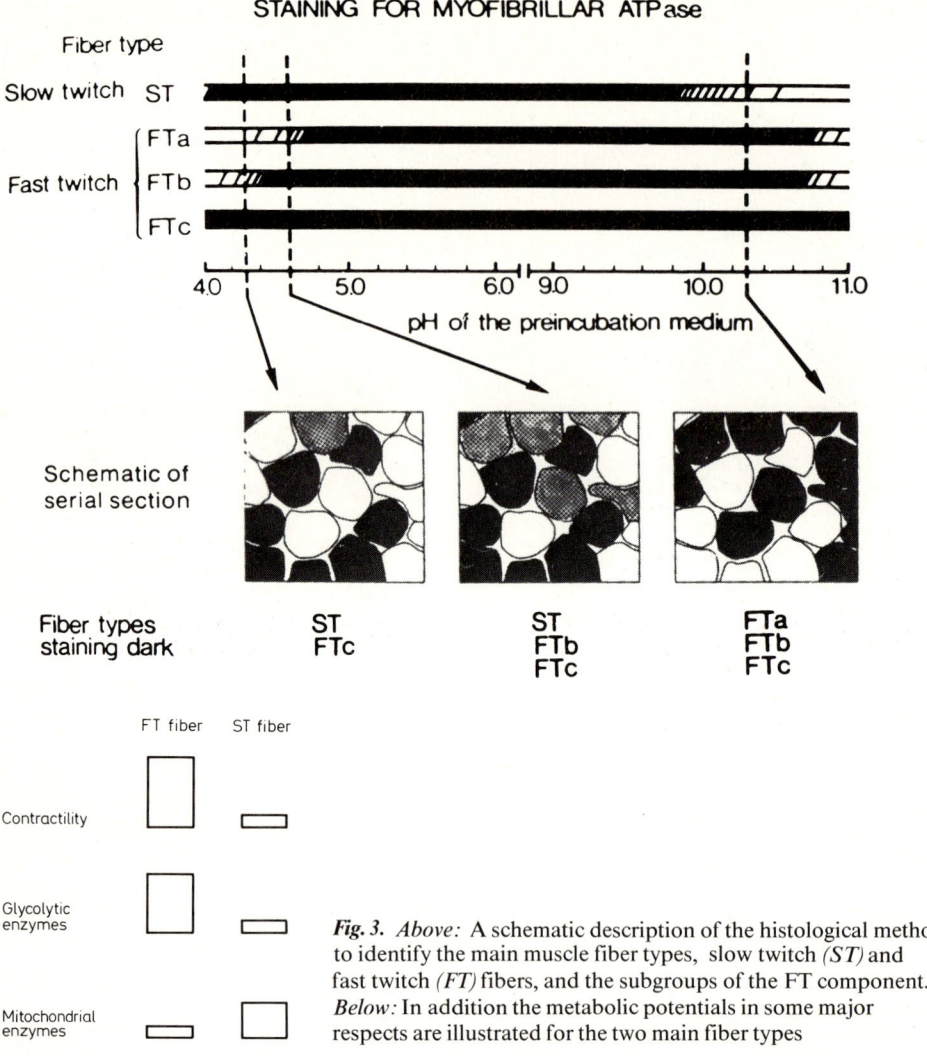

Fig. 3. *Above:* A schematic description of the histological methods to identify the main muscle fiber types, slow twitch *(ST)* and fast twitch *(FT)* fibers, and the subgroups of the FT component. *Below:* In addition the metabolic potentials in some major respects are illustrated for the two main fiber types

muscle activity. Moreover, it seems reasonable to include regulating features, which are muscle fiber related and seem to give priority to oxidative metabolism.

Lactate Formation and Accumulation in Muscle and Blood

That muscle and blood lactate accumulation is well related to the oxygen deficit is well documented [9]. Extremely low femoral venous oxygen saturation values in these experimental conditions have also led to the suggestion that the major reason for lactate formation is a true lack of molecular O_2 in the active muscles [9, 19]. It has been shown that in addition we have to consider lactate formation even under

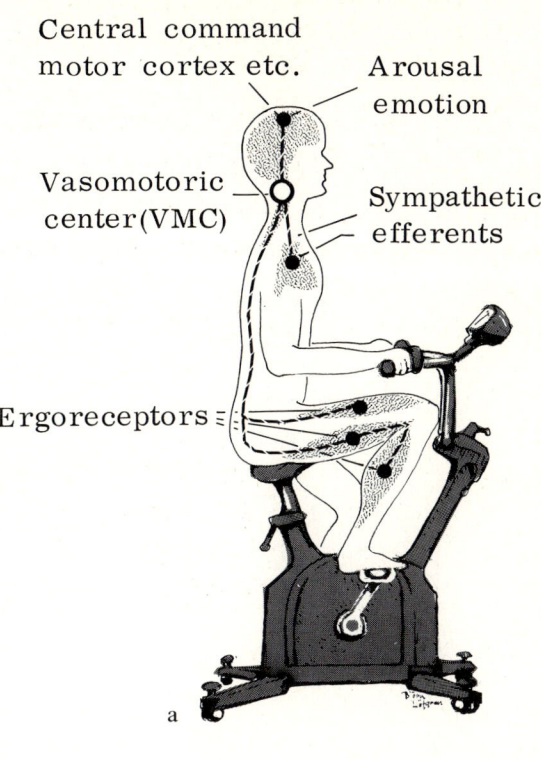

Fig. 4. a A schematic description of the peripheral chemoreceptors (ergoreceptors) and their afferents to the vasomotoric center (VMC). From there, efferents extend to the heart and can affect heart rate and heart muscle contractility, according to Shepherd et al. [13]. *b* Mean femoral artery pressures at rest and leg blood flow during exercise at one and the same submaximal work load presented in healthy young males *(upper panels)* and in patients with essential hypertension but without pharmacological treatment *(lower panels)*

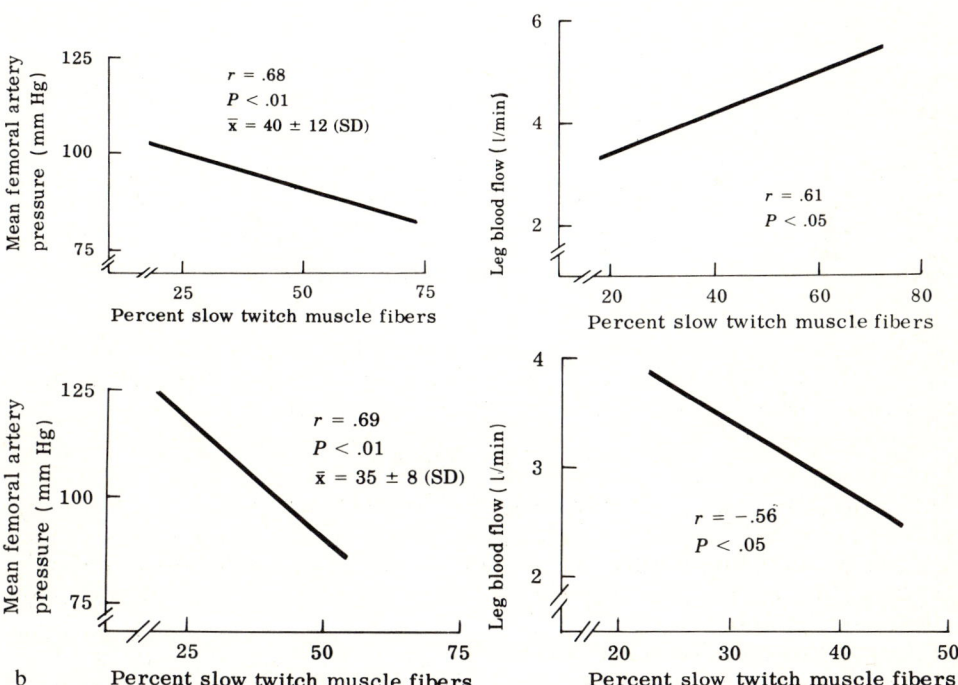

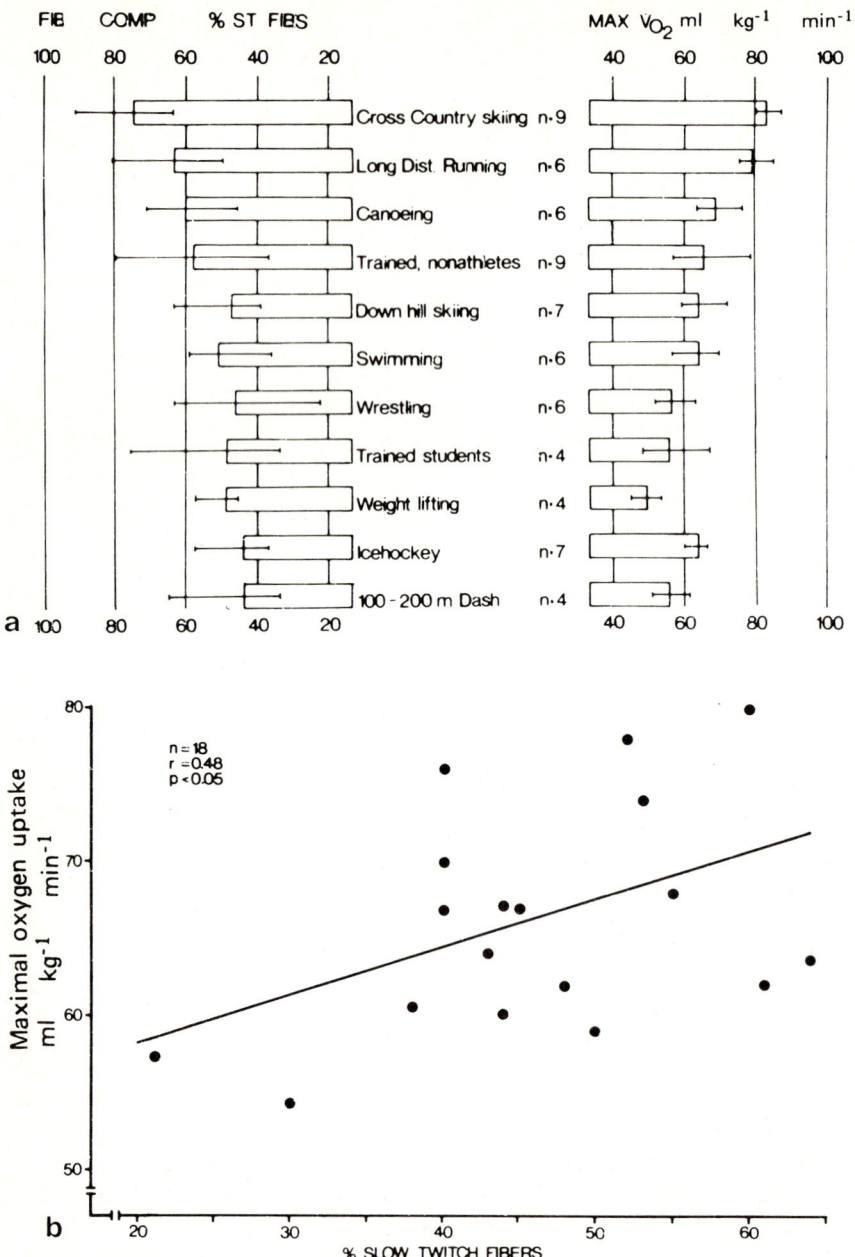

Fig. 5. a Mean values and SE for percent slow twitch *(ST)* fibers and maximal oxygen uptake in elite sport athletes and **b** in subjects participating in regular physical conditioning programs

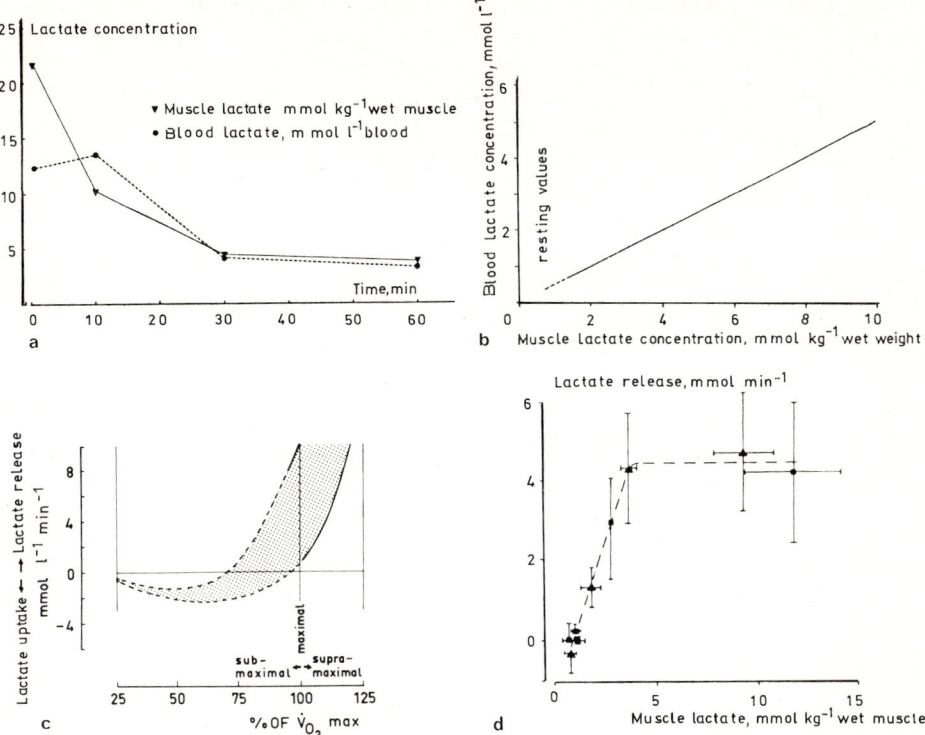

Fig. 6. a The relationship between muscle and blood lactate following a short time, exhaustive work load. **b** The relationship between blood and muscle lactate after submaximal work loads lasting for 6–10 min. **c** The concentration difference of lactate over the leg after approximately 15 min at different submaximal work loads. **d** The computed total lactate release per unit of time from the exercising leg for different muscle lactate concentrations

steady state conditions and that the most probable explanation for this is an "overflow" of the glycogenolytic pathway and a subsequent lactate formation [20]. Thus we have to take into account lactate formation basically in the two following situations:

1. Transient conditions with lactate formation due to an absolute or relative lack of molecular oxygen for metabolic purposes.
2. Steady state conditions with lactate formation presumably in the fast twitch muscle component due to an elevated activation of the glycogenolytic pathway and a subsequent "luxurious" glycogen breakdown, pyruvate, production and a subsequent lactate formation.

With dynamic exercise lasting for up to 10 min, a gradient for lactate is built up between muscle and blood (Fig. 6a) and blood lactate reflects muscle lactate rather well (Fig. 6b). With submaximal exercise at steady state, it is well documented that during mild exercise there is an uptake of lactate by the conctracting muscles, providing elevated blood lactate (Fig. 6c). With heavier exercise intensities, this uptake is shifted toward a release. The lactate release is linear to a muscle lactate concentration less than 4–5 mmol kg^{-1} muscle (Fig. 6 d). After that the increase in lactate release levels off [9, 20]. The reason why a muscle lactate increase above 4–5 mmol l^{-1} is not accompanied by a similar increase in lactate release is unknown.

To summarize, the lower oxygen transport capacity to the fast twitch muscle fiber, in addition to its higher glycogenolytic profile, will make this muscle component a very probable site for lactate formation during muscle activity, resulting in an earlier onset of lactate formation and higher concentrations as well, the higher the percentage of fast twitch fibers. In general, blood lactate concentration follows muscle lactate concentration, providing muscle blood flow is not restricted. A gradient exists though between muscle and blood, which might invalidate an attempt to quantify lactate formed from blood lactate concentrations. Lactate release follows a nonlinear function versus local muscle lactate, and at muscle lactates above 4–5 mmol kg^{-1}, it levels off.

Definition of Onset of Blood Lactate Accumulation or OBLA

Lactate release seems to be optimal and blood lactate seems best to reflect total lactate formation at a work load eliciting a blood lactate concentration of 4 mmol l^{-1}. This blood lactate concentration has then been chosen to reflect an onset of increased lactate formation during muscle activity (Fig. 7 a). For most healthy subjects and many patient categories this concentration is submaximal and is not in any respect hazardous to the subject's/patient's health or well-being. In addition it is the same concentration value originally chosen by Hollmann [4] as well as his later collaborators, Mader and Liesen, and their concentration choice based upon empirical considerations [21, 22].

In our laboratory we feel at present uncertain about the effect of different basal or "resting" blood lactate concentration values on the determination and interpretation of OBLA. In healthy subjects the 4 mmol l^{-1} criterion has one definite advantage, as it reduces the impact of normal variation in basal or "resting" blood lactates (Fig. 7 b).

To summarize, based upon a graded exercise protocol, basically two different procedures – one direct, based upon blood lactate determination, and one indirect, based upon studies of different respiratory variables – have been developed to study the shift in muscle metabolism with increased exercise intensity. Terms have been introduced, such as lactate breaking point or turn point, lactate or anaerobic threshold, respiratory compensation threshold, etc. As neither lactate formation nor "anaerobic metabolism" is unique for moderately heavy, submaximal exercise, but can already be present in a resting, healthy muscle, we have suggested the use of the terms *o*nset of *b*lood *l*actate *a*ccumulation over resting values, or OBLA, to describe the metabolic event occurring with increasing exercise intensity leading to a sharp increase in muscle and blood lactate concentration. This increase in lactate formation can be demonstrated in healthy subjects and in most patient categories as well.

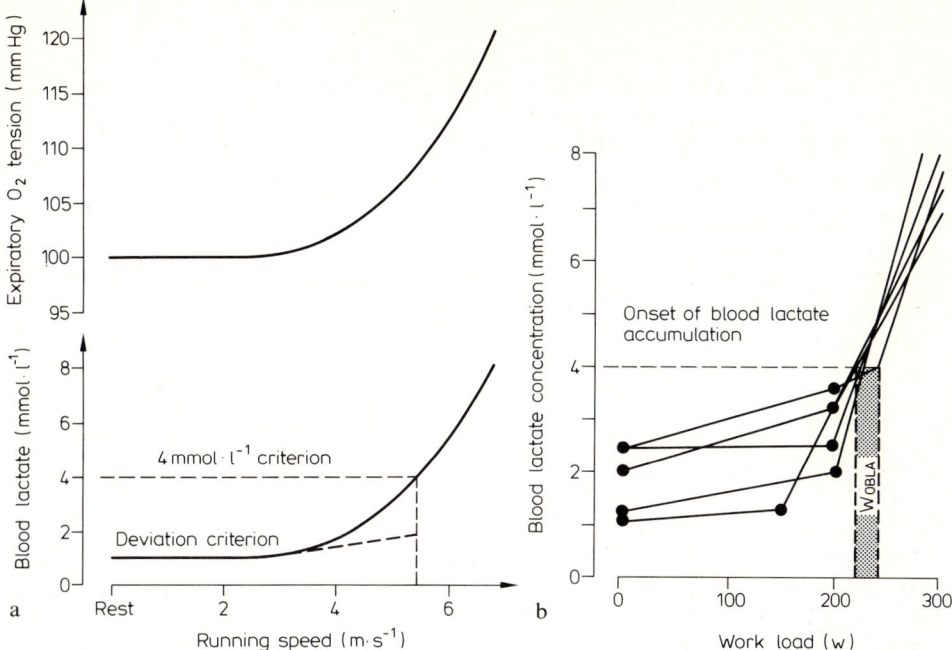

Fig. 7. a The schematic relationship between blood lactate concentration and a hyperventilation marker – expiratory O_2 tension – during graded exercise. The *lower panel* depicts the most frequent approach to the "turn or breaking point": 4 mmol l^{-1} blood. At present our laboratory is studying the possibility of using a deviation index to individualize the OBLA point. **b** Repeated experiments in one subject are illustrated to determine the OBLA point corresponding to the work load eliciting a blood lactate concentration of 4 mmol l^{-1}. In spite of rather large variations in "basal" blood lactates, a rather small variation was present in the estimated work load corresponding to OBLA (W_{OBLA}) providing the 4 mmol l^{-1} criterion

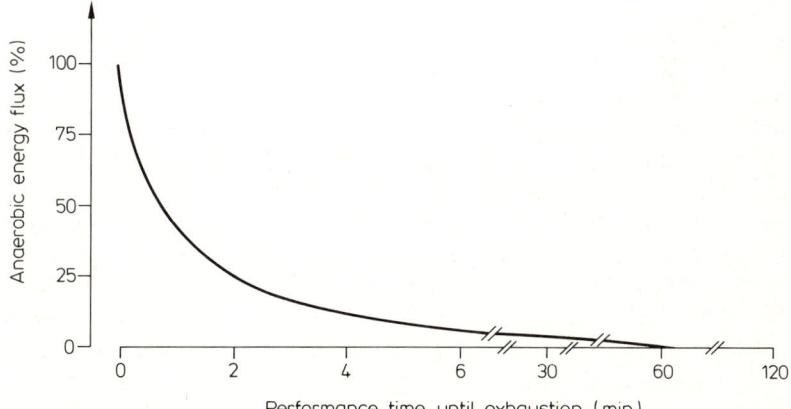

Fig. 8. The relative contribution of energy flux from "anaerobic sources" for exhaustive exercise tasks of different duration. If exhaustion time is approximately 1 min or longer, more than 50% of the total energy output is from "aerobic sources," i. e., combustive metabolism demanding the presence of molecular oxygen in the active muscle tissue

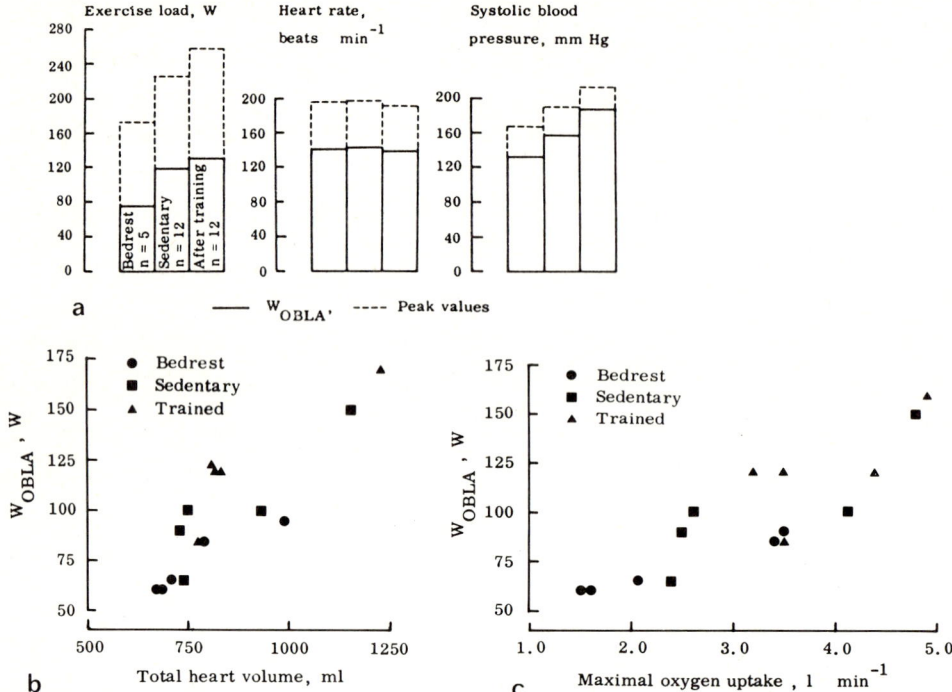

Fig. 9a–c. The so-called "bed rest study" by Saltin et al. [23] offers a possibility to compute the changes in the work load eliciting a blood lactate concentration of 4 mmol l^{-1} (W_{OBLA}) in the sedentary stage, after 20 days in bed, and after 55 days of training. *a* Mean W_{OBLA}, the corresponding heart rate, and systolic blood pressure means. *b* Relationship between individual W_{OBLA} and heart volume. *c* Individual relationship between W_{OBLA} and maximal oxygen uptake

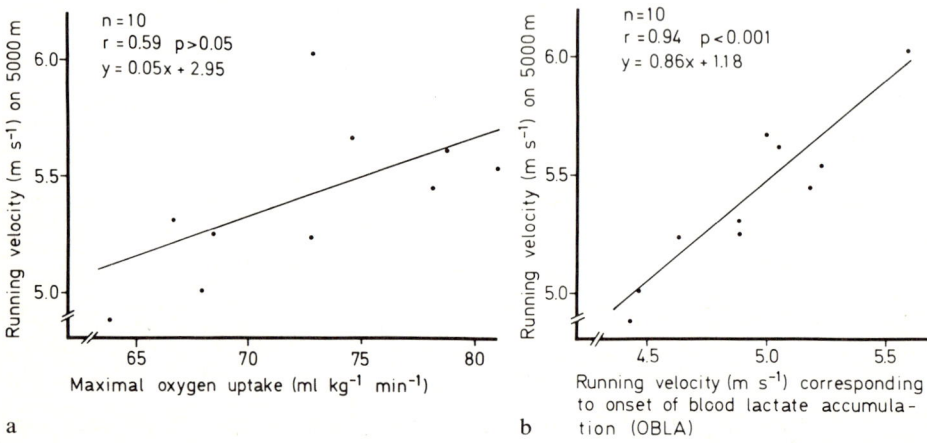

Fig. 10. *a* The individual relationships between the average running speed during a 5000 m race and the individual data for maximal oxygen uptake and *b* the running speed eliciting a blood lactate of 4 mmol l^{-1} (V_{OBLA}) on the treadmill

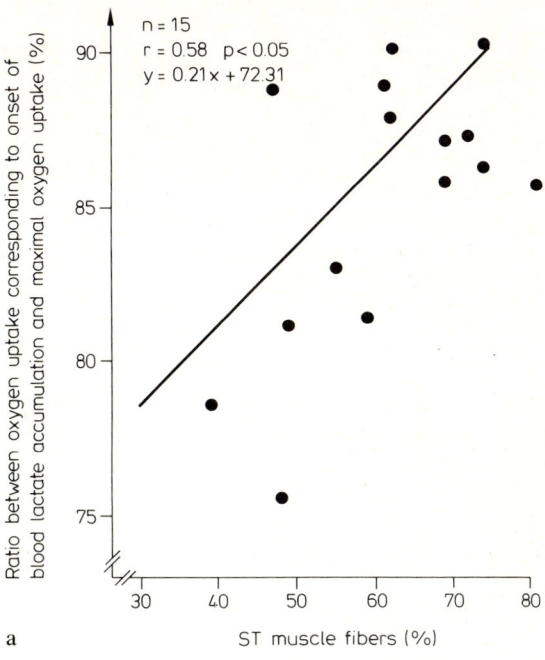

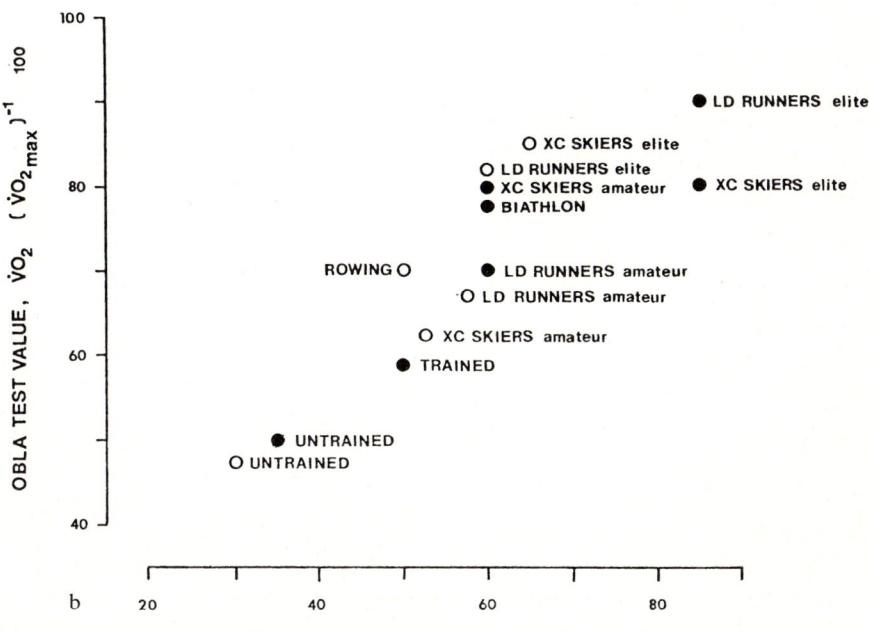

Fig. 11a, b. The oxygen uptake corresponding to a blood lactate concentration of 4 mmol 1^{-1} ($\dot{V}_{O_2OBLA}$) in percent of maximal oxygen uptake ($\dot{V}_{O_2max}$). *a* Versus the individual data for percent slow twitch *(ST)* or type I ("oxidative") muscle fibers. *b* Versus maximal oxygen uptake in some athlete groups. *XC*, cross-country skiers; *LD*, long-distance runners; *open circles*, females; *filled circles*, males

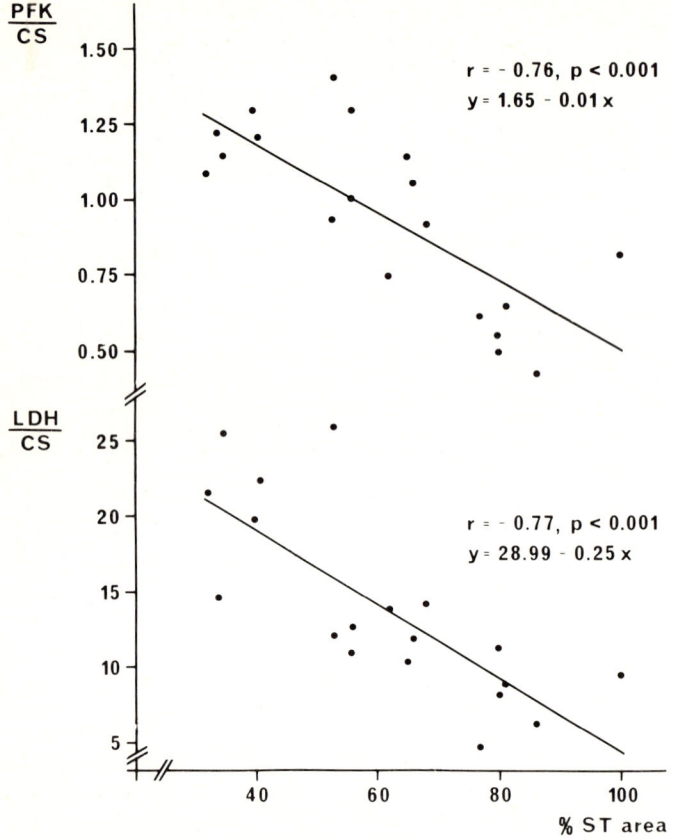

Fig. 12. a The relative glycogenolytic or "anaerobic" profile of muscles with different percentage of slow twitch *(ST)* muscle fibers. The enzymes phosphofructokinase *(PFK)* and lactate dehydrogenase *(LDH)* are markers of the glycogenolytic potential, whereas the enzyme succino-dehydrogenase *(SDH)* is a marker of the combustive or "aerobic" potential

Performance Evaluation in Recreational and Athletic Sports

Providing that an exercise task lasts for 45–60 s or more, the energy flux to 50% or more will be covered by "aerobic processes." This means, by definition, that an evaluation of endurance capacity will predict performance. It is also obvious from Fig. 8 that the aerobic component increases asymptomatically with increased performance time, i.e., that an endurance athlete can be better predicted than runners participating in 400 and 800 m track and field running.

As already pointed out, the endurance athlete is characterized by a large circulatory capacity leading to a high $\dot{V}o_{2max}$, a high percentage of slow twitch muscle fibers in the active muscles, which will ensure a high O_2 extraction capacity, and a high energy turnover with little or no lactate formation. It is well described that bed rest followed by intense physical training will change central circulatory features, such as heart volume, and subsequent changes in $\dot{V}o_{2max}$ with corresponding

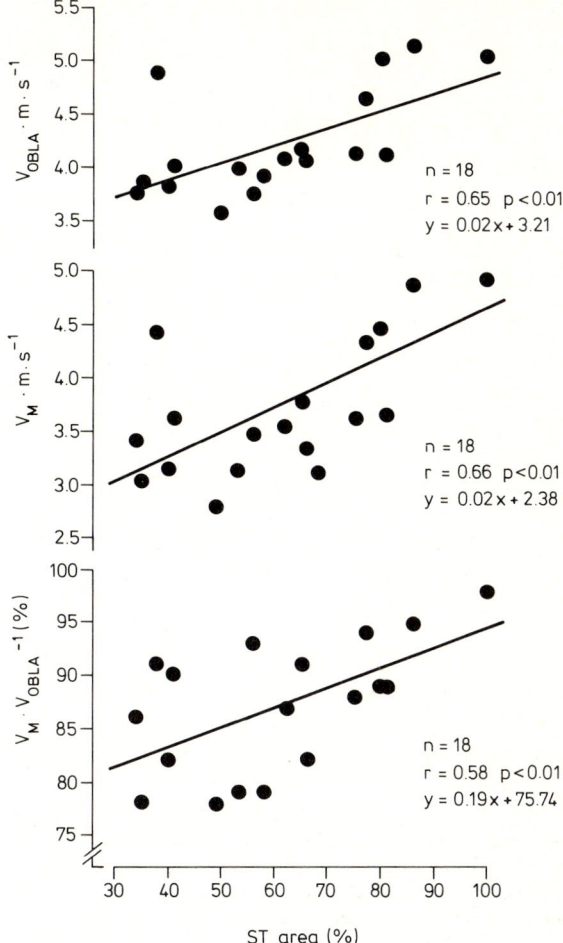

Fig. 12. b Relationship between physical performance expressed as V_{OBLA} (see Fig. 10), average speed during a marathon race (V_M), and V_M in percent of V_{OBLA} (compare with Fig. 11) and the individual muscle fiber composition in percent slow twitch *(ST)* muscle fibers

changes in lactate metabolism. Based upon simultaneous blood lactate values, changes in W_{OBLA} can be computed (Fig. 9 a–c) [11, 23]. If, in addition, performance is compared with $\dot{V}_{O_{2}max}$ and running speed corresponding to OBLA (V_{OBLA}), it is possible to conclude that V_{OBLA} is a better predictor [24] (Fig. 10 a, b). The better prediction is partly explained by the fact that fitter runners or runners with more slow twitch fibers in their leg muscles have their V_{OBLA} at a higher fraction of ther $\dot{V}_{O_{2}max}$ (Fig. 11 a). The same pattern is present when different athlete groups are studied with different proficiency levels in terms of running (Fig. 11 b). This shift in muscle metabolism to a higher proportion of aerobic energy output is explained by the higher combustive potential in a muscle rich in slow twitch fibers, resulting in a higher V_{OBLA} and performance (Fig. 12 a–c) [26, 27, 29]. In this context it has to be

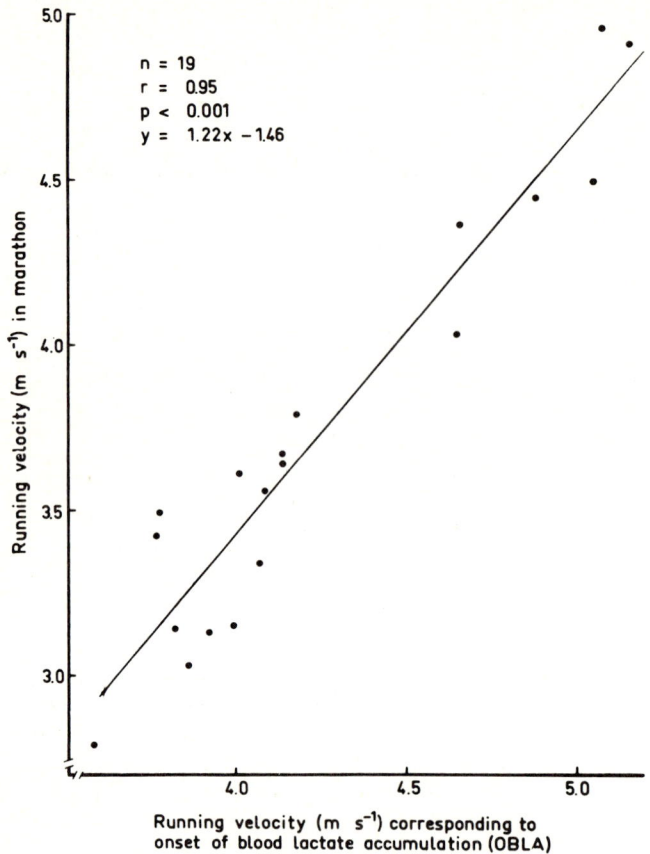

Fig. 12. c For the same individuals, the relationship between running performance of the marathon race and the predetermined V_{OBLA}. The correlation coefficient indicates that close to 90% of the individual variation during the race could be explained on the basis of V_{OBLA}

pointed out that female skeletal muscle has a different metabolic machinary than male skeletal muscle, which will be combined with other known sex differences in respect to performance as, e. g., circulatory capacity [28, 30]. Thus, for the same muscle fiber composition, females have a lower exercise capacity than males, expressed as W_{OBLA} (Fig. 13).

To summarize, the OBLA exercise stress test protocol offers a possibility to integrate for central circulatory (e. g., $\dot{V}_{O_2max}$) and peripheral metabolic potentials (e. g., muscle fiber composition). In respect to endurance performance, the predictive value is equal to or better than $\dot{V}_{O_2max}$ and corresponding concepts. The OBLA protocol offers, in addition, the advantage that it in all aspects is submaximal, i. e., that a top athlete can perhaps undergo the test on the day of an important event. It has to be pointed out, however, that males and females cannot be evaluated according to the same scale. Caution has to be applied even when evaluating pubertal and pre-pubertal patients as their lactate metabolism also differs as compared with that of adult males [31].

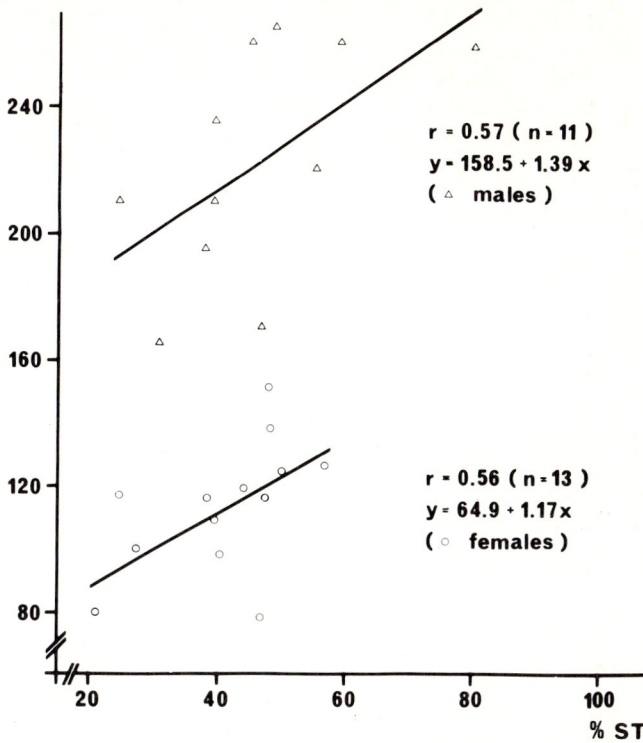

Fig. 13. The relationship between W_{OBLA} (see Fig. 7) as obtained on a cycle ergometer in males and females, respectively, and their muscle fiber composition expressed as percentage of slow twitch *(ST)* muscle fibers

OBLA Testing in Clinical Practice

Exercise stress tests have been used in clinical practice for decades [32], either as 1–2 consecutive, submaximal work loads or as a graded test [33]. In addition to exercise capacity as such, expressed as either the work load corresponding to a certain submaximal heart rate (170 or 130 bpm) or symptom-limited capacity (W_{max}), measures such as blood pressures, local and general rated exertion and rated anginal pain have been obtained [32, 34–36]. During recent years the exercise stress test concept, which the OBLA protocol represents, has gained rapidly increasing interest among clinicians (see Fig. 1).

During recent years, exercise stress tests have been more used in evaluation of blood pressure and the effect of exercise per se on pressure regulation [35, 37, 38]. It has been found that labile hypertension or exercise-induced exaggerated blood pressure response cannot be predicted by resting blood pressure as such. In a recent study, Criqui et al. [40] have further demonstrated the necessity of exercise stress testing to fully describe blood pressure regulation.

In our laboratory we have applied the OBLA exercise protocol to evaluate blood pressure response to exercise in healthy controls, hypertensive patients (WHO I–II), a risk group for essential hypertension ("contact sport athletes"), prediabetics,

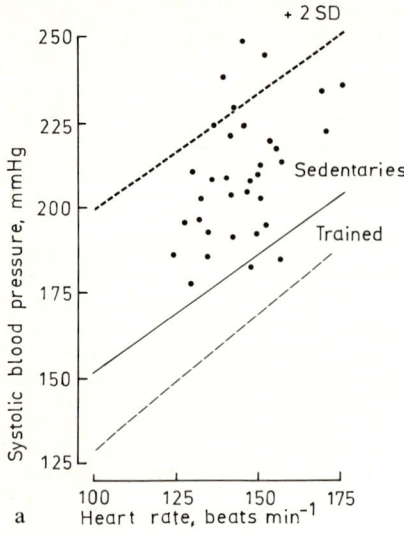

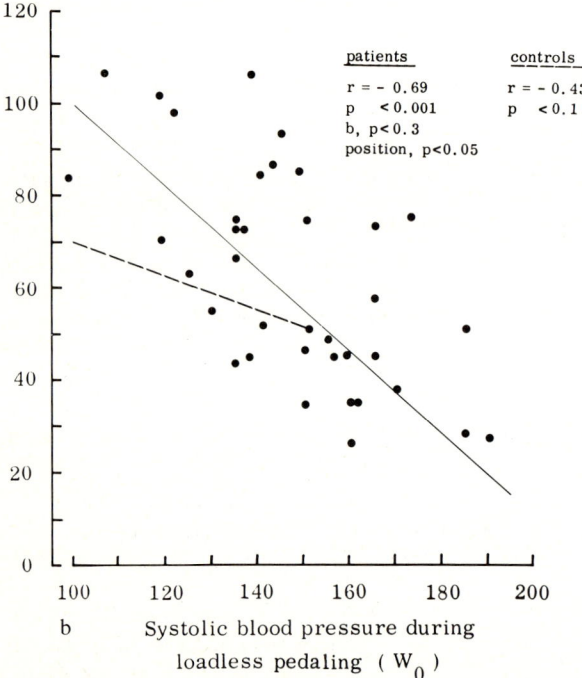

Fig. 14. a Individual relationship between exercise systolic blood pressure and heart rate at W_{OBLA} (see Fig. 7) for borderline or latent hypertensives. In addition, the regression lines of sedentaries according to Saltin et al. [23] plus twice SD and a regression line from trained subjects are noted [18]. **b** For the same patients, the systolic blood pressure increase from loadless pedaling (W_O) to (W_{OBLA}) versus the actual pressure at W_O. Systolic blood pressure at W_OL. Systolic blood pressure at W_O represents a "basal" level, including the possible impact from joints and tendons on blood pressure regulation and emotional stress related to the exercise test itself

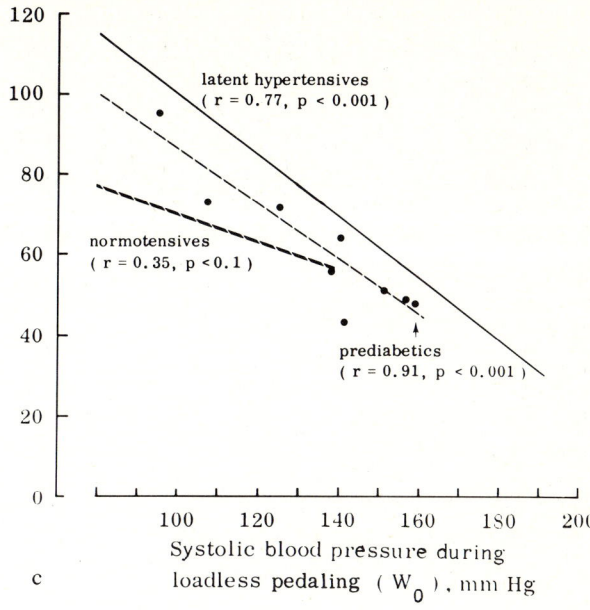

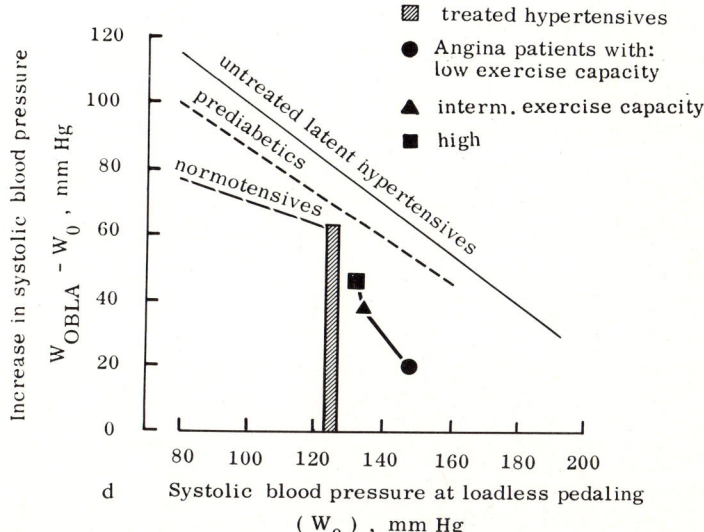

Fig. 14. c Patients with prediabetes shown for comparison. *d* Corresponding values for three patient groups with effort angina and with a combination of β-blockade and Ca-entry blockade as the pharmacological treatment (see Fig. 17) [34]

and patients with ischaemic heart disease (angina pectoris). We have based our evaluations on the pressure responses during loadless pedaling at the cycle ergometer [12, 18, 41, 42].

Patients with untreated essential hypertension show a relatively higher systolic blood pressure (SBP) during exercise than sedentary but healthy controls (Fig. 14a) for a certain exercise heart rate [42]. When the SBP increase is related to SBP at W_O, it seems that the blood pressure regulation is related to this basal level (Fig. 14b). This is in accordance with the Lipid Research Clinics Epidemiology Committee's findings [40]. It is also possible to conclude that prediabetics (Fig. 14c) and patients with effort angina (Fig. 14d) followed the same pattern: the higher the SBP at W_O, the smaller the increase to the work load eliciting a blood lactate concentration of

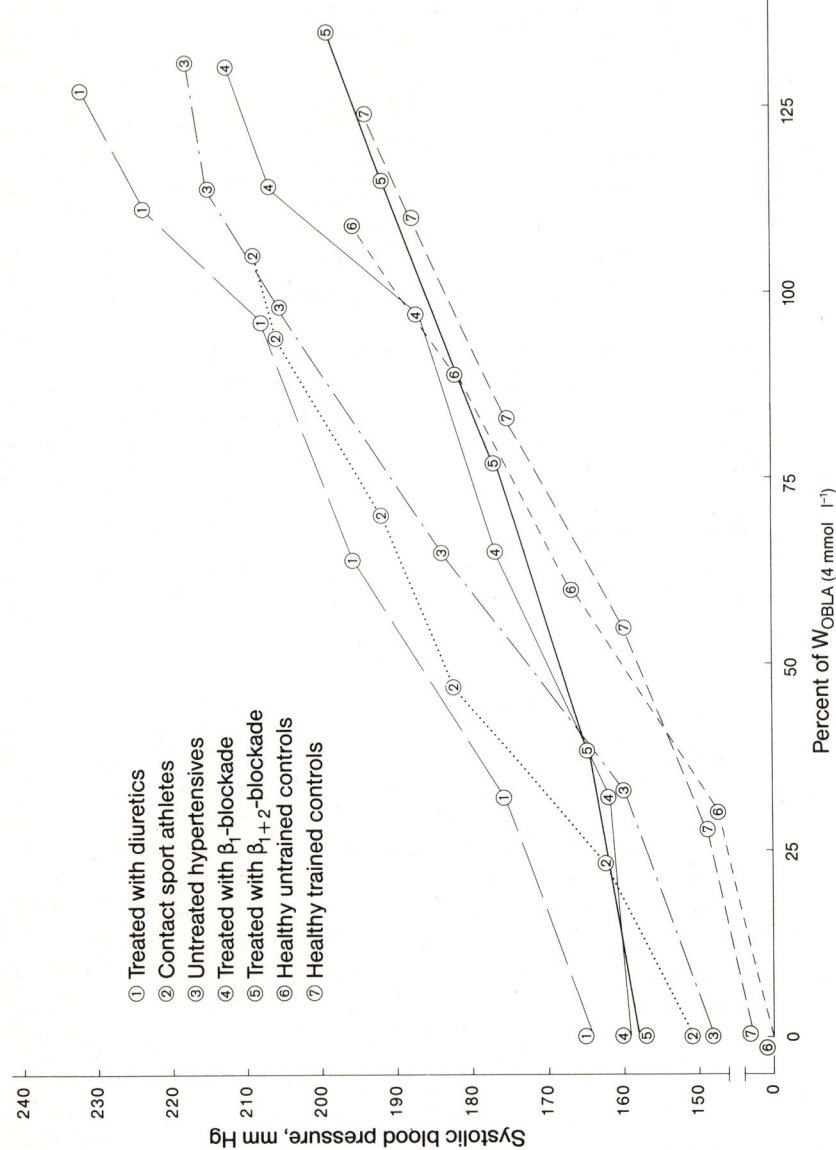

Fig. 15. a Systolic blood pressure during a graded exercise stress test in patients with different kinds of pharmacological treatments for their essential hypertension in relation to sedentaries, endurance trained, and athletes representing contact sports. The latter group is documented to be a risk group for essential hypertension [38]

4 mmol 1^{-1} or W_{OBLA}. It is also possible to see that the relationship in the angina patient group is shifted toward lower increases (Fig. 14 d). This shift might represent the effect of the pharmacological treatment, which was a combination of β-blockade and Ca-entry blockade [34]. That different medication programs might have effects on SBP during exercise is possible to conclude from Fig. 15 a [43], even if caution has to be applied, as the data are from cross-sectional materials.

A detailed analysis of the response to β_1-selective blockade as compared with the nonselective $(\beta_1 + \beta_2)$ blockade in Fig. 15 a shows that at higher work loads the SBP suppression was not as marked for the β_1-selective drug as for the nonselective. This has been further studied in a crossover model [18]. At W_{OBLA} a tendency was present to higher SBP, whereas the W_{OBLA} expressed in watts was significantly higher ($p < 0.01$) when the patients were treated with the β_1-selective drug atenolol as compared with the nonselective propranolol (Fig. 15 b).

It has earlier been shown that patients with essential hypertension have a distorted relationship between blood pressure at rest and muscle fiber composition (Fig. 4) [11, 17, 18]. Dlin et al. [38 or 39] have demonstrated that athletes with a normal resting blood pressure but exaggerated response during exercise constitute a risk group for essential hypertension. As compared with healthy, trained controls, this athlete group, frequently observed in contact sports, shows a different blood pressure response both at W_O and as SBP increases to W_{OBLA} (Fig. 16 a). This athlete group behaves differently during exercise even in other respects. Thus they rate fatigue higher for the same blood lactate concentration as compared with control athletes (Fig. 16 b) and have a higher SBP response to exercise as well (Fig. 16 c).

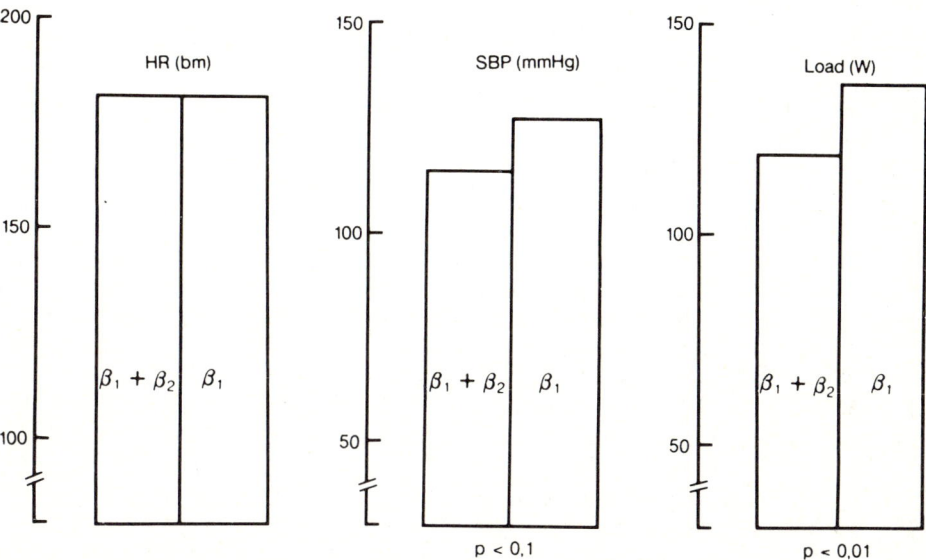

Fig. 15. b The effect of two different drug regimens on patients with essential hypertension is illustrated at W_{OBLA} (see Fig. 7). After a β_1- or heart-muscle-specific blockade, systolic blood pressure *(SBP)* tends to be higher for the same heart rate *(HR)*. The work load corresponding to W_{OBLA} is, however, significantly elevated as compared with the $\beta_1 + \beta_2$- or unspecific blockade

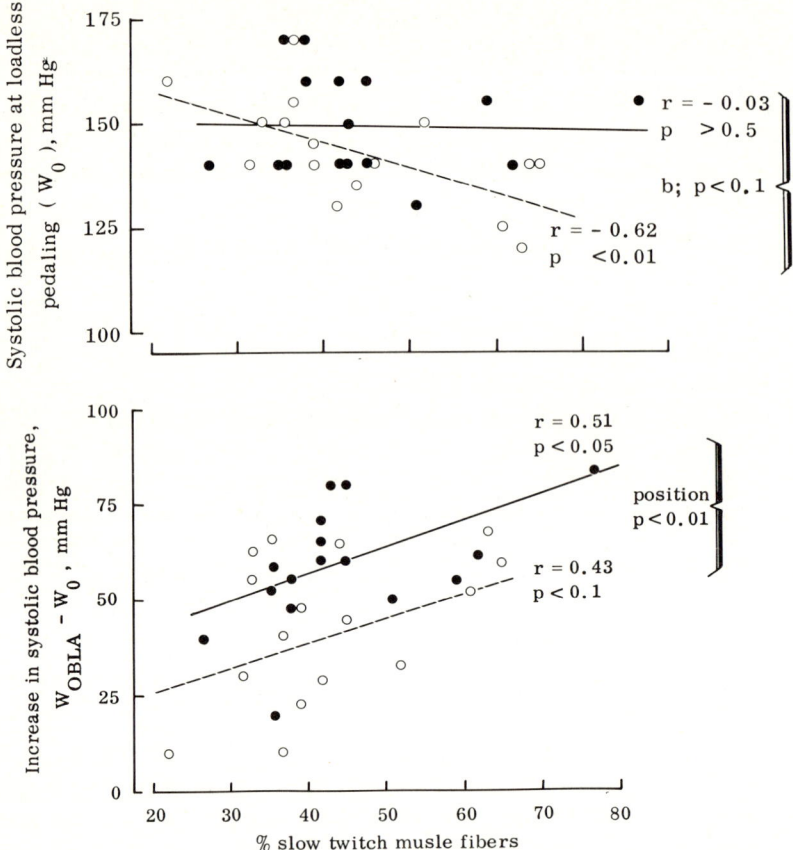

Fig. 16. a The systolic blood pressure at W_O (see Fig. 14) and the pressure increase between W_O and W_{OBLA} versus muscle fiber composition expressed as percentage of slow twitch muscle fibers (compare with Fig. 4b) in normal, healthy, trained subjects *(open symbols)* and those representing contact sports (a risk group for hypertension, see Fig. 15; *closed symbols*)

In another review in this volume, G. Borg covers the area of perceived exertion during exercise. The same type of scales can be used for pain evaluation in relation to effort angina [44]. If repeated blood lactate determinations are applied in the same type of graded exercise test and in patients with effort angina, it is possible to describe basically the same function for increase in blood lactate as for increase in anginal pain (Fig. 17a), in spite of large variation in symptom-limited exercise performance (W_{max}). Each curve represents the mean of more than nine patients [34]. As the result of the similarity between growth of blood lactate and anginal pain, almost the same relationship was present in the three patient groups between rated pain and blood lactate (Fig. 17b).

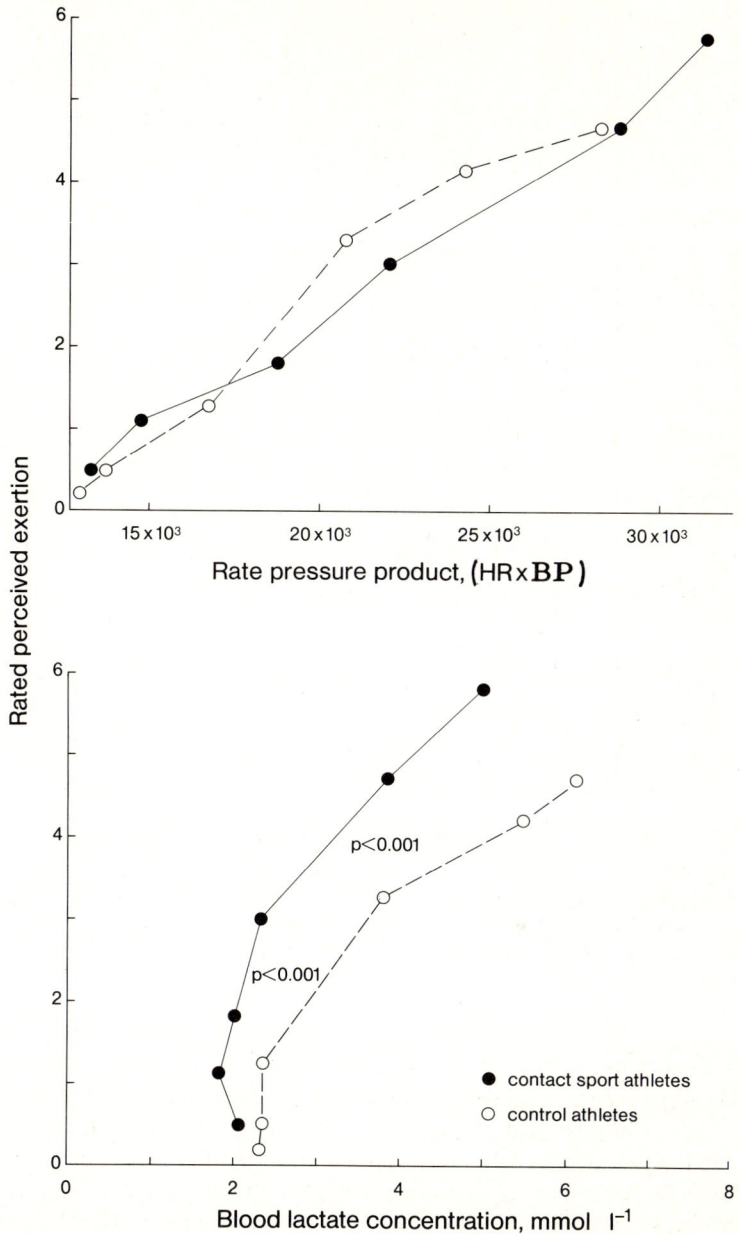

Fig. 16. *b* Greater susceptibility to experience of exertion for a certain blood lactate concentration in contact sport athletes as compared with endurance-trained athletes. This difference is abolished if rated perceived exertion is related to the rate pressure product, i.e., heart rate times systolic blood pressure *(HR × BP)*

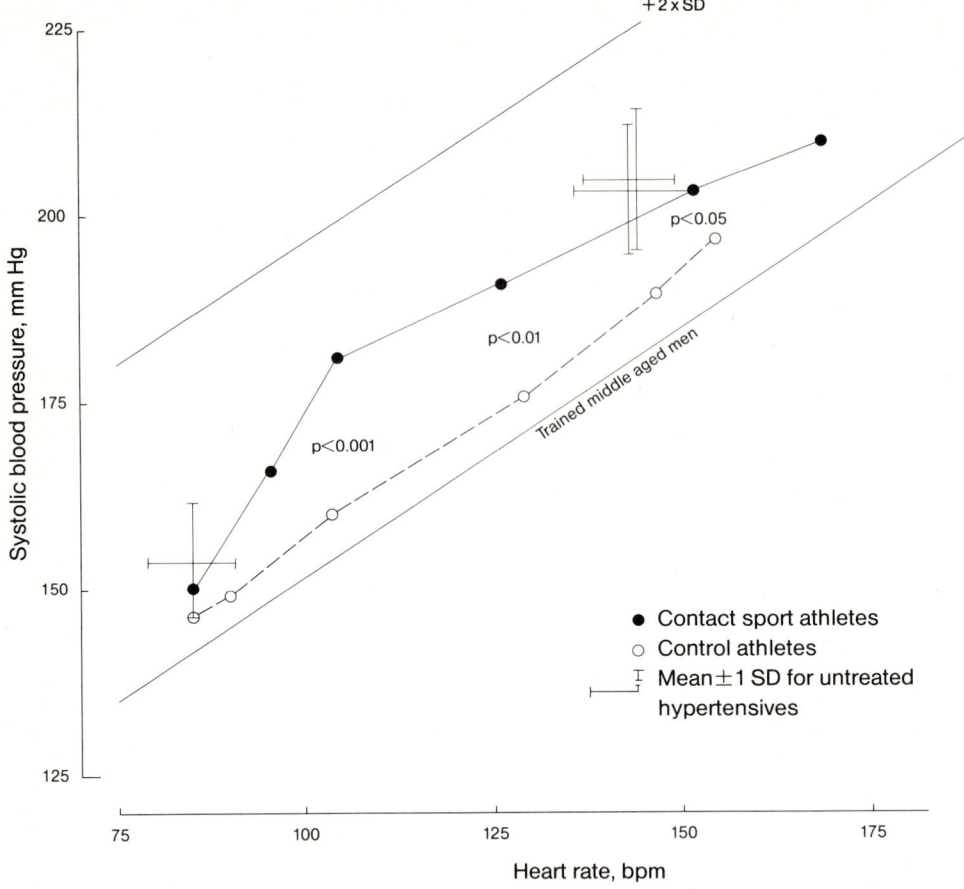

Fig. 16. c The "hypertensive" response in contact sport athletes (according to Fig. 14a) is illustrated

Conclusions

1. Central circulatory capacity and oxygen transport and muscle fiber composition are two different features which have to be taken into account when evaluating physical performance characteristics related to endurance exercise.
2. Graded exercise, with small increments to reduce transient "oxygen deficit"-related lactate formation, allows one to test for lactate formation related to "overstimulation" of the glycogenolytic pathway under "steady state conditions," i.e., sufficient transport of molecular oxygen. This form of lactate formation seems to be the main limiting factor for high exercise intensity over a long period of time – endurance capacity.
3. There are also reasons to assume that the appearance of "anaerobic metabolism" and related biochemical changes in the muscle are involved in the activation of peripheral sensors – ergoreceptors – which activate central circulation.

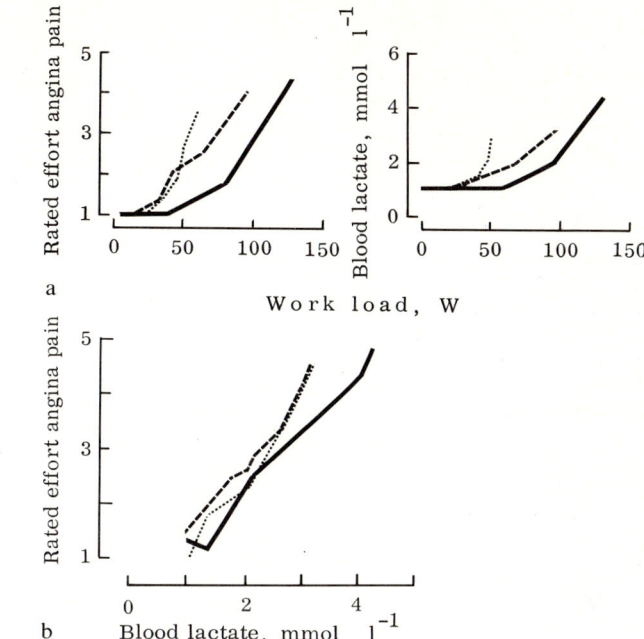

Fig. 17. a Rated anginal pain and blood lactate concentration during graded exercise for patients with confirmed ischaemic heart disease. The same patients are also presented in Fig. 14d. They are divided into three groups according to their symptom-limited exercise capacity (W_{max}), which averaged 53, 90, and 125 w, respectively. **b** Rated anginal pain and the corresponding blood lactates are interrelated and the functions seem to be identical

4. It has been found from methodological studies that by providing dynamic exercise with small intensity increments a blood lactate concentration corresponding to 4 mmol l^{-1} is practical to mark for *o*nset of *b*lood *l*actate *a*ccumulation over resting values or OBLA.

5. In sports on the leisure level as well as in top athletes, the OBLA exercise stress test protocol or related measures (anaerobic or lactate threshold, lactate breaking or turn point, respiratory compensation threshold, onset of plasma lactate accumulation or OPLA, etc.) have rapidly gained popularity, as it is a better predictor of endurance performance than maximal oxygen uptake determination.

6. The OBLA protocol has also been applied in clinical practice for evaluation of exercise capacity, in addition to or as an alternative to symptom-limited exercise stress tests. It has also been used to achieve a metabolic standard during exercise to evaluate blood pressure and heart rate responses. The test has also been used to "titrate" the drug regimen in relation to hypertension to achieve an optimal effect on blood pressure with as small a decrease in exercise performance as possible. It has also been used to evaluate the chest pain in patients with effort angina.

References

1. Christensen EH, Hansen O (1939) Zur Methodik der Respiratorischen Quotient-Bestimmungen in Ruhe und bei Arbeit. Skand Arch Physiol 137–151
2. Christensen EH, Hansen O (1939) Respiratorischer Quotient und O_2-Aufnahme. Skand Arch Physiol 180–189
3. Åstrand PO (1952) Experimental studies of physical working capacity in relation to sex and age. Munksgaard, Copenhagen
4. Hollmann W (1961) Zur Frage der Dauerleistungsfähigkeit. Fortschr Med 79: 439–453
5. Hollmann W, Hettinger T (1980) Sportmedizin. Arbeits- und Trainingsgrundlagen. Schattauer, Stuttgart
6. Holmgren A, Ström G (1959) Blood lactate concentration in relation to absolute and relative work load in normal men and in mitral stenosis, atrial septal defect and vasoregulatory asthenia. Acta Med Scand 163: 185–191
7. Wassermann K, Whipp BJ, Davis A (1981) Respiratory physiology of exercise: metabolism, gas exchange and ventilatory control. Int Rev Physiol 111 (23): 149–211
8. Wassermann K, Whipp BJ, Koyal SN, Beaver WL (1973) Anaerobic threshold and respiratory gas exchange during exercise. J Appl Physiol 35: 236–243
9. Karlsson J (1971) Lactate and phosphagen concentrations in working muscle of man. Acta Physiol Scand [Suppl] 358: 1–72
10. Karlsson J (1979) Localized muscular fatigue: role of muscle metabolism and substrate depletion. In: Hutton R, Miller D (eds) Exercise and sports sciences reviews, vol 7. Franklin, Philadelphia, pp 1–42
11. Karlsson J, Jacobs I (1982) Onset of blood lactate accumulation during muscular exercise as a threshold concept. I. Theoretical considerations. Int J Sports Med 3: 190–201
12. Karlsson J, Dlin R, Kaiser P, Tesch PA, Kaijser C (1983) Muscle metabolism, regulation of circulation and β-blockade. J Card Rehab 3: 404–420
13. Shepherd JT, Blomqvist CG, Lind AR, Mitchell JH, Saltin B (1981) Static (isometric) exercise: retrospection and introspection. Circ Res 48: 179–188
14. Bockman EL (1983) Possible role of K^+ as a mediator of active hyperemia in gracilis (white) muscles of cats. Proc of IUPS 15. 572.09
15. Mellander S (1983) Blood flow through fast and slow twitch skeletal muscles. Are they governed by different regulatory mechanisms? Proc of IUPS 15. 416.04
16. Karlsson J, Smith P, (1983) Effect of chronic sympathectomy on muscle fiber composition, isomyosin pattern, protein synthesis and calcium content in canine gracilis muscle. Acta physiol Scand (in press)
17. Karlsson J (1983) Muscle fibre composition, short term $\beta_1 + \beta_2$- and β_1-blockade and endurance exercise performance in healthy young men. Drugs 25: 241–246
18. Karlsson J (1983) β-blockade and exercise performance. Drugs 25: 257–261
19. Saltin B, Gollnick PD, Piehl K, Eriksson B (1971) Metabolic and circulatory adjustments at onset of exercise. In: Gilvert A, Guille P (eds) Onset of exercise. Toulouse, Univ. of Toulouse pp 63–76
20. Jorfeldt L, Juhlin-Dannfelt A, Karlsson J (1978) Lactate release in relation to tissue lactate in human skeletal muscle during exercise. J Appl Physiol 44 (3): 350–352
21. Liesen H, Dufaux B, Weber K, Lohman W, Fischer W, Hollmann W (1976) Das Verhalten von Serumimmunglobulinen bei Ausdauertraining und extremen Ausdauerbelastungen. Sportarzt Sportmed 6: 119–126
22. Mader A, Liessen H, Heck H, Philippi H, Rost R, Schurch P, Hollmann W (1976) Zur Beurteilung der sportartspezifischen Ausdauerleistungsfähigkeit im Labor. Sportarzt Sportmed 27: 80–88
23. Saltin B, Blomqvist CG, Mitchell JH, Johnson RL, Wildenthal H, Chapman CB (1968) Response to exercise after bed rest and after training. Circulation [Suppl 7]
24. Sjödin B, Linnarsson D, Wallensten R, Schéle R, Karlsson J (1982) The physiological background of onset of blood lactate accumulation (OBLA). In: Komi P. V. (ed): Exercise and Sport Biology. Human Kinetics Publisher, Inc., Champaign pp 43–55
25. Jacobs I (1981) Lactate, muscle glycogen and exercise performance in man. Acta Physiol Scand [Suppl] 495: 1–84

26. Jacobs I, Tesch P (1981) Short time, maximal muscular performance: relation to muscle lactate and fiber types in females. In: Borms J, Hebbelinck M, Venerando A (eds) Women and sport. Karger, Basel

27. Karlsson J, Jacobs I (1981) Is the significance of muscle fibre types to muscle metabolism different in females than in males? In: Borms J, Hebbelinck M, Venerando A (eds) Women and sport. Karger, Basel

28. Komi PV (1981) Fundamental performance characteristics in females and males. In: Borms J, Hebbelinck M, Venerando A (eds) Women and sport. Karger, Basel

29. Komi PV, Karlsson J (1979) Physical performance, skeletal muscle enzyme activities and fibre types in monozygous and dizygous twins of both sexes. Acta Physiol Scand [Suppl] 462: 1–35

30. Wilmore JH (1981) Women and sport: an introduction to the physiological aspects. In: Borms J, Hebbelinck M, Venerando A (eds) Women and sport. Karger, Basel

31. Eriksson BO, Gollnick PD, Saltin B (1973) Muscle metabolism and enzyme activities after training in boys 11–13 years old. Acta Physiol Scand 87: 485–497

32. Holmgren A (1956) Circulatory changes during muscular work in man. Scand J Clin Lab Invest [Suppl] 24: 1–97

33. Åström H, Jonsson B (1976) Design of exercise test with special reference to heart patients. Br Heart J 38: 289–296

34. Åström H, Holmgren A, Karlsson J, Orinius E (1983) Rated effort angina, perceived leg fatigue and blood lactate during graded exercise. In: Knuttgen HG, Vogel JA, Poortmans J (eds) Biochemistry of exercise. Human Kinetics, Champaign

35. Sannerstedt R (1966) Hemodynamic response to exercise in patients with arterial hypertension. Acta Med Scand [Suppl] 458: 1–76

36. Carlens P (1979) Effort angina. Doctoral thesis. Karolinska Institute, Stockholm

37. Kannel WB, Sorlie P, Gordon T (1980) Labile hypertension: a faulty concept? The Framingham study. Circulation 61: 1183–1190

38. Dlin RA, Hanne N, Silverberg DS, Bar-Or O (1983) Follow up of normotensive men with exaggerated blood pressure response to exercise. Am Heart J 106: 316–320

39. Dlin RA, Dotan R, Inbar O, Torstein A, Jacobs I, Karlsson J (1983) Exaggerated SBP response to exercise in a water polo team. Med Sci Sports Exerc (to be published)

40. Criqui MH, Haskell WL, Heiss G, Tyroler HA, Green P, Rubenstein CJ (1983) Predictors of systolic blood pressure response to treadmill exercise: the lipid research clinics program prevalence study. Circulation 68: 225–233

41. Karlsson J, Dlin R, Wahlberg F, Sannerstedt R, Kaijser C (1983) Blood pressure response in relation to blood lactate during exercise. In: Knuttgen HG, Vogel JA, Poortmans J (eds) Biochemistry of exercise. Human Kinetics, Champaign

42. Karlsson J, Kaijser C, Wahlberg F, Sannerstedt R (1983) Blood pressure response to the OBLA test work load in patients with essential hypertension. Int J Sports Med (to be published)

43. Karlsson J, Holmgren A, Åström H, Orinius E, Kaijser C (to be published) Angina pectoris and blood lactate concentration during graded exercise

44. Borg G, Holmgren A, Lindblad I (1981) Quantitative evaluation of chest pain. Acta Med Scand [Suppl] 644: 43–45

45. Wasserman K, McIlroy MB (1964) Detecting the threshold of anaerobic metabolism in cardiac patients during exercise. Am J Cardiol 14: 844–852

Specificity and Test Precision of the Anaerobic Threshold

N. Bachl

Introduction

The reference points of the aerobic-anaerobic transition are currently regarded in performance diagnostics as specific parameters for characterization of endurance capacity as well as for determination of the optimal intensity for extensive and intensive long-distance training. This is the case, although the physiological mutual effects of the determining factors are not adequately clarified, nor is the validity of these parameters histochemically or biochemically confirmed with reference to sports-type specificity, age, or pathological conditions. The models for individual threshold determination which have been worked out in recent years [1, 7, 11, 12, 15–17] indicate the efforts made to validate the endurance performance capacity.

To accomplish test precision for these reference points, which are so relevant to performance diagnostics, the reliability must also be taken into consideration [5]. Independent of the test protocols, the influence of which has been clarified by many groups working on the reference points of the aerobic-anaerobic transition, reproducibility must be possible for every ergometric protocol that arises. This, according to the definition of reliability, is related to the short-term variability [5], but must also be true of the long-term variability, as there would otherwise be no reliable statements about intraindividual changes.

Several investigations on the variability of ergometric measurements with maximum loading have been carried out, particularly for oxygen intake. Values between 6.4% and 11.2% [6], 3.5% and 5.7% [22], 4.9% and 6.8% [21], and around 5% [14] are mentioned as coefficients of variation for the $\dot{V}O_{2\,max}$. For the maximum lactate concentrations, there are values such as 6.6%–37.5% from venous blood [6] and 17.6% and 17.8% from arterial blood [21]. There are only a few studies [4, 8] available about the variability of submaximum lactate concentrations.

In the study presented here, the short- and long-term variabilities of the arterial lactate concentrations are analyzed at rest, with submaximal loads, and at reference points at lactate concentrations of 2 and 4 mmol/l, which are obtained from the lactate dynamics during incremental test protocol.

Finally, an attempt is made to make a statement about the validity of these performance-diagnostic parameters for the investigated groups and individuals from the comparison of the variabilities of the reference points of the aerobic-anaerobic transition at lactate concentrations of 2 and 4 mmol/l with individually determined reference points (ventilatory-respiratory determination).

Methods

A group of sports students and sports teachers were tested (*n*, 12; age, 25.8 ± 2.6 years; height, 180 ± 5.5 cm; weight, 77.1 ± 9.5 kg). Several other individuals were tested as well: a sports student (age, 29 years; height, 167 cm; weight, 59 kg) and a 400 m hurdler (age, 20 years; height, 186 cm; weight, 79 kg). Colleagues and technical assistants (MTAs) from the Institute of Sports Sciences (*n*, 7; age, 22–40) were available for the determination of the resting lactate daily rhythm.

The ergometric tests on the two individuals took place at the same time daily under standardized conditions. The group of sports students was investigated four times in the academic year, 1982/83, at the beginning and end of November, as well as at the beginning and end of May. The test half-day (8:30–12:00 and 13:00–16:00), but not the exact investigation time, was kept constant, according to the realities of outpatient and practice routine.

All ergometric tests were carried out on a bicycle or treadmill ergometer from the Jaeger Company (Würzburg). The spiroergometric tests were done with the Ergo-Pneumotest open circuit system from Jaeger (Würzburg). The lactate determination from capillary blood (20 µl) obtained in a standardized manner from the hyperemic earlobe was performed by means of the lactate analyzer 640 from the Roche Company. The loading tests on the bicycle ergometer were done in a standardized manner, beginning with a time increment of 3 min and a loading increment of 50 W, up to a loading level of 250 W. In the treadmill ergometer tests, the speed increases amounted to 2 km/hr, starting at 8 km/hr. The time increment was 3 min at a 5% treadmill inclination (test A) and 6 min at a 1.5% treadmill inclination (test B).

Aerobic threshold at 2 mmol/l and anaerobic threshold at 4 mmol/l lactate concentration were interpolated out of the lactate dynamics during incremental testing. The determination of the individual anaerobic threshold and the individual compensation threshold was done out of the plotted ventilatory parameters at those points where minute ventilation increased out of proportion to $\dot{V}O_2$ and $\dot{V}CO_2$, respectively.

Results

In Fig. 1 the daily variabilities of seven individual arterial resting lactate concentrations are shown. The thin lines show the intraindividual means from three experiments. The intraindividual coefficients of variation at all measuring points on the three test days lie between 20.2% and 47.5%, and the group mean value for all subjects is 28.9%. The group means of the interindividual coefficients of variation at the actual measuring times are noted at the respective points on the *x* axis and lie between 14.3% and 36.3%. The thick line shows the group mean values of the intraindividual mean lactate concentrations at the measuring times between 8:30 and 14:30 at a level between 0.95 and 1.19 mmol/l.

A further example of the short-term variability of resting lactate concentrations in individuals can be observed in Table 1. For the competitive sportsman, KG, a very low coefficient of variation of under 10% was found in tests on three consecutive days. For the sports student, WW, the submaximal test A, which took place in the

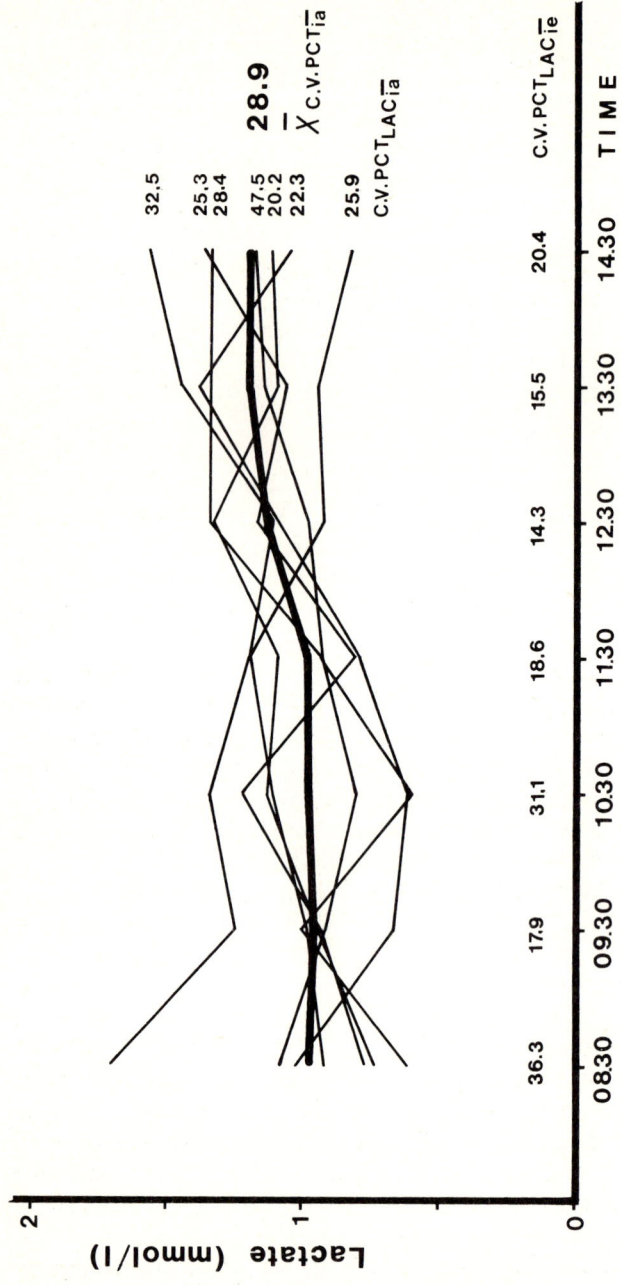

Fig. 1. Daily variabilities of 7 individual resting lactate concentrations. *Thin lines* show the intraindividual means from 3 test days with the intraindividual coefficients of variation over all measuring points marked at the *right* edge *(C. V. PCT$_{L_{ia}}$)*. The *thick line* shows the group mean values of the intraindividual mean lactate concentrations with the group mean of intraindividual coefficients of variation marked with capital characters *($\bar{x}_{C.V.PCT_{ia}}$)*. Furthermore the coefficient of variation of the interindividual means of lactate concentrations of all measuring points are noted at the respective points of the *x* axis *(C. V. PCT$_{LAC_{ie}}$)* (*PCT*, percent)

Table 1. Mean values $(\bar{x})$, standard deviation (s), and coefficient of variation (C. V. PCT) of intraindividual resting lactate concentrations for subjects, KG and WW, on 3 or 4 consecutive days

Subject	KG	WW	WW	WW	WW
Test		A	B	A	B
n	3		3		4
Lactate (mmol/l)					
$\bar{x}$	1.81	1.44	1.80	1.66	1.88
s	0.18	0.79	0.30	0.78	0.29
C. V. PCT	9.98	54.70	16.90	46.80	15.60

For WW, resting lactate was obtained before treadmill test A and test B.

Table 2. Mean values $(\bar{x})$, standard deviation (s), and coefficient of variation (C. V. PCT) of interindividual concentrations for the group of sports students and sports teachers

$n=12$	Test 1	Test 2	Test 3	Test 4
Lactate (mmol/l)				
$\bar{x}$	1.22	1.13	1.24	0.97
s	0.23	0.29	0.49	0.52
C. V. PCT	18.40	25.30	39.30	53.40

morning (9:00), produced very high (46.8% and 54.7%) values and results about two-thirds lower (15.6% and 16.9%) for the submaximal test B, which was carried out 2 h later. Calculation of the long-term variability of the resting lactate concentrations of a group of 12 sports teachers and students (four tests within 7 months) brought no significantly different results (Table 2).

The group mean value of all intraindividual resting lactate concentrations comes to 1.14 ± 0.2 mmol/l, with a coefficient of variation of 17.2%.

The group mean value of all intraindividual coefficients of variation was between 16.9% and 57.5%, with a mean value of $31.4 \pm 11.1\%$. The means and coefficients of variation for the interindividual resting lactate concentrations of that group are shown at Table 2 at the four test times.

The short-term variabilities in the changes in the submaximal lactate concentrations, as well as the aerobic and anaerobic threshold of the two subjects, KG and WW, are shown in such a way in Figs. 2 and 3 that the single measurements in the areas of the simple and double standard deviation of the intraindividual means are recorded. Additionally, the anaerobic threshold at 4 mmol/l lactate concentration is marked, together with the coefficient of variation of the running speed at the point.

In both test subjects, a reduction in the coefficient of variation is noticeable when loads and lactate concentrations are increased.

The long-term variabilities in the changes in the submaximal lactate concentrations, the interpolated aerobic and anaerobic thresholds resulting from them, and the individual thresholds can be seen from the results from the group of sports teachers. Tables 3 and 4 characterize the mean values for interindividual lactate concentrations and threshold reference points on the four test times. There was no significant difference in the t test in any of the parameters in any of the test/date combinations.

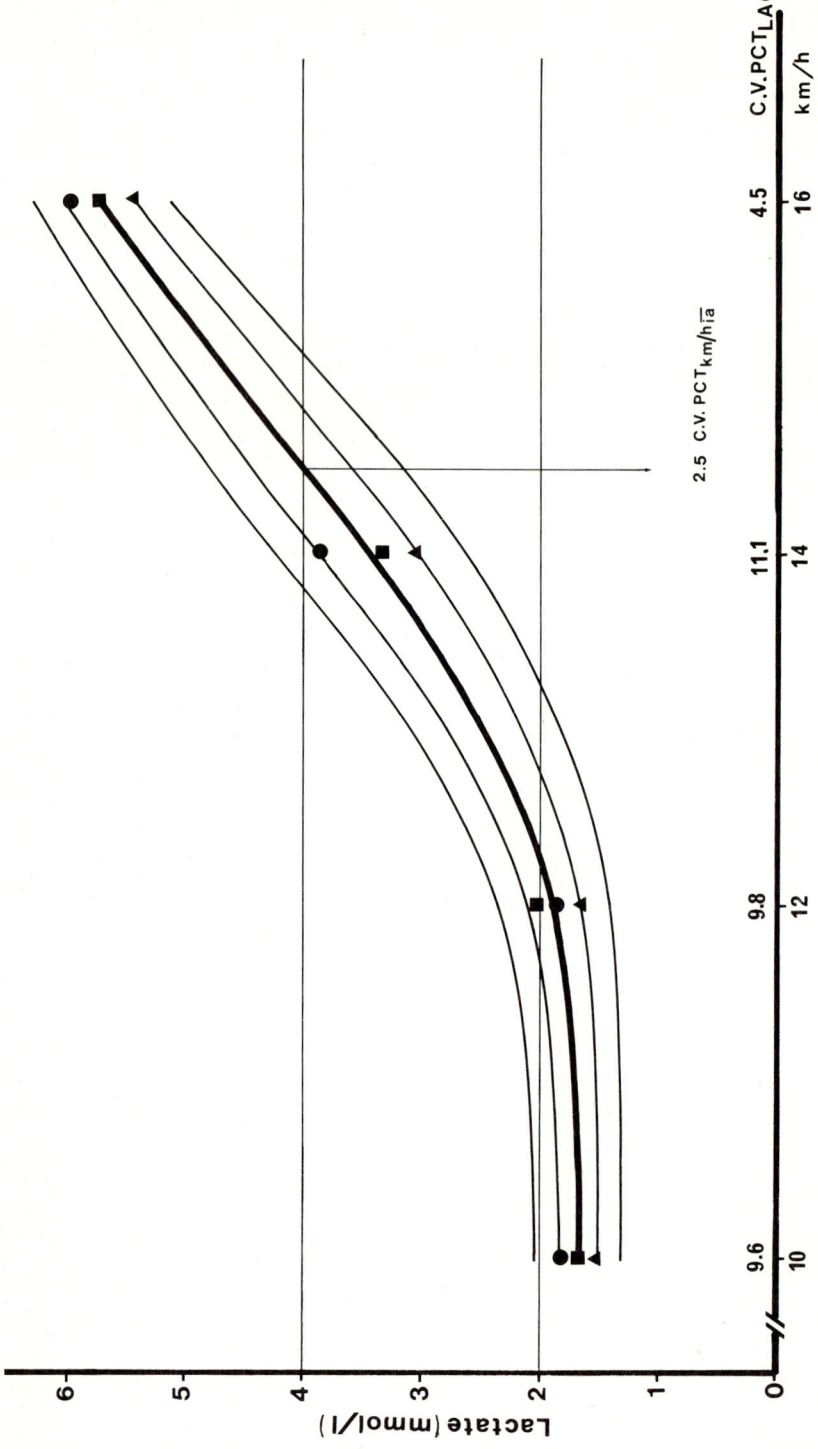

Fig.2. Dynamic response of arterial lactate concentrations for subject, KG, during incremental treadmill ergometry. The single measurements are recorded in the area of simple and double standard deviation of the mean intraindividual lactate concentration. The mean running speed of the anaerobic threshold (4 mmol/l lactate) is marked, together with its coefficient of variation. Coefficients of variation of the intraindividual lactate concentrations are noted at the respective points on the *x* axis

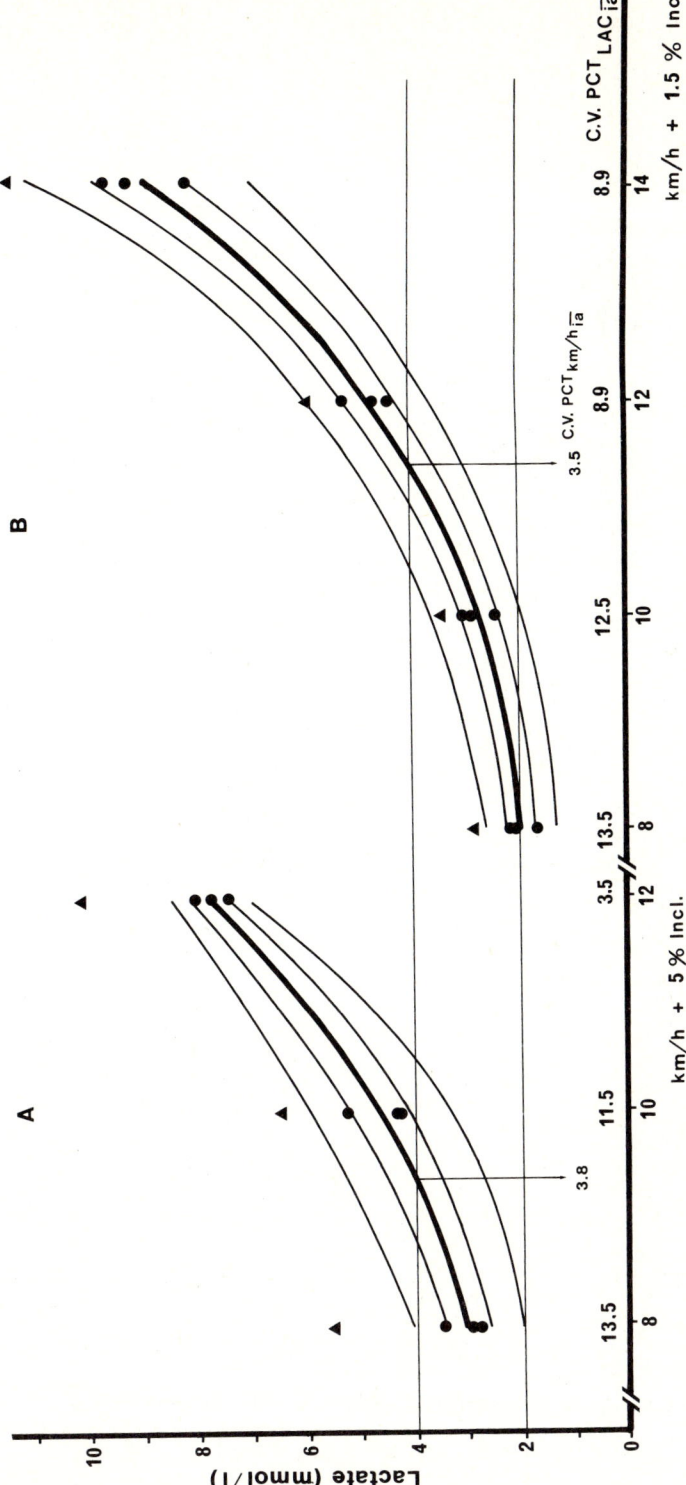

Fig. 3. Dynamic response of arterial lactate concentration for subject, WW, during incremental treadmill ergometry (*A*, *B*). The single measurements of "good days" (●) and a "bad day" (▲) are recorded in the areas of single and double standard deviation of the mean intraindividual lactate concentration of good days

Table 3. Mean values *(x̄)*, standard deviation *(s)*, and coefficient of variation (C. V. PCT) of interindividual lactate concentrations at 100, 150, 200, and 250 w for the group of sports students and sports teachers (*n* = 12) at the four test times

Lactate (mmol/l)		100 watts	150 watts	200 watts	250 watts
Test 1	$\bar{x}$	1.6	2.2	3.4	6.2
	s	0.3	0.5	0.95	1.9
	C. V. PCT	21.8	21.1	27.9	29.8
Test 2	$\bar{x}$	1.3	1.8	2.96	5.8
	s	0.4	0.6	0.96	1.98
	C. V.PCT	28.6	31.5	32.3	34.3
Test 3	$\bar{x}$	1.3	1.9	3.1	6.2
	s	0.7	0.8	1.3	2.96
	C. V. PCT	43.8	37.2	40.7	47.6
Test 4	$\bar{x}$	1.3	1.9	3.1	6.2
	s	0.4	0.6	1.2	2.1
	C. V. PCT	33.3	31.2	37.6	34.6

t Test showed no significant difference in any of the parameters in any of the test/date combinations.

Table 4. Mean values *(x̄)*, standard deviation *(s)*, and coefficient of variation (C. V. PCT) of interindividual lactate concentrations at fixed (2 and 4 mmol/l lactate) and individually determined threshold reference points for the group of sports students and sports teachers (*n* = 12) at the four test times

Watts		Aerobic threshold (2 mmol/l)	Anaerobic threshold (4 mmol/l)	Individual anaerobic threshold	Individual compensation threshold
Test 1	$\bar{x}$	139.2	210.7	135.6	227.4
	s	34.1	21.4	34.7	19.6
	C. V. PCT	24.5	10.2	25.6	8.6
Test 2	$\bar{x}$	161.3	223.3	145.5	228.5
	s	34.1	30.8	19.7	24.6
	C. V. PCT	21.1	13.8	13.6	10.8
Test 3	$\bar{x}$	149.9	220.8	148.3	222.2
	s	41.8	36.9	27.4	21.5
	C. V. PCT	27.9	16.7	18.5	9.7
Test 4	$\bar{x}$	156.6	221.3	147.8	229.4
	s	35.2	34.2	29.1	29.1
	C. V. PCT	22.4	15.5	19.7	12.7

t Test showed no significant difference in any of the parameters in any of the test/date combinations.

Tables 5 and 6 show the group mean values of the intraindividual lactate concentrations ($\bar{x}_{LAC_{i\bar{a}}}$) and watts ($\bar{x}_{WATT_{i\bar{a}}}$), the standard deviations and coefficients of variation, and the group mean values and range of these intraindividual coefficients of variation ($\bar{x}_{C. V. PCT_{i\bar{a}}}$) for submaximal loading steps, the aerobic and anaerobic threshold, and the individual thresholds. As with the individual subjects, a decrease in the mean values of the intraindividual coefficients of variation at increasing lac-

Table 5. Group mean values ($\bar{x}_{LAC_{i\bar{a}}}$), standard deviation *(s)*, and coefficients of variation (C. V. PCT) of the mean intraindividual lactate concentrations and the group mean values and range of these intraindividual coefficients of variation ($\bar{x}_{C.V.PCT_{i\bar{a}}}$) at 100, 150, 200, and 250 w for the group of sports teachers and sports students ($n=12$)

	100 watts	150 watts	200 watts	250 watts
Lactate				
$\bar{x}_{Lac_{i\bar{a}}}$	1.41	1.97	3.19	6.1
s	0.3	0.5	1.0	2.1
C. V. PCT	24.4	26.1	31.5	34.2
$\bar{x}_{C.V.PCT_{i\bar{a}}}$	25.3	18.7	18.3	16.7
s	11.1	8.8	7.7	5.1
Range				
C. V. PCT$_{i\bar{a}}$	11.2–43.6	8.4–37.3	7.6–30.3	8.1–25.8

Table 6. Group mean values ($\bar{x}_{WATT_{i\bar{a}}}$), standard deviation *(s)*, and coefficient of variation (C. V. PCT) of the intraindividual watt load and the group mean values and range of these intraindividual coefficients of variation ($\bar{x}_{C.V.PCT_{i\bar{a}}}$) at fixed (2 and 4 mmol/l lactate) and individually determined threshold reference points for the group of sports teachers and sports students ($n=12$)

	Aerobic threshold (2 mmol/l)	Anaerobic threshold (4 mmol/l)	Individual anaerobic threshold	Individual compensation threshold
$\bar{x}_{WATT_{i\bar{a}}}$	151.8	219.0	144.3	227.0
s	29.0	26.8	21.8	20.2
C. V. PCT	19.1	12.2	15.2	8.9
$\bar{x}_{C.V.PCT_{i\bar{a}}}$	15.6	7.7	12.7	5.3
s	7.8	3.8	7.1	3.4
Range				
C. V. PCT$_{i\bar{a}}$	6.0–34.1	3.1–18.4	3.6–25.3	2.2–14.1

There was no significant difference in *t* test between fixed and individual threshold reference points.

tate concentrations can be seen (group mean values of 25.3% to 16.7%), while the coefficients of variation of the group mean value of the absolute intraindividual lactate concentration ($\bar{x}_{LAC_{i\bar{a}}}$) do not show this trend. All the coefficients of variation mentioned show lower, however not statistically significant, values for the individual, in comparison with the "fixed threshold" (Table 6). Figure 4 finally documents the behavior of all lactate single measurements in comparison with the group mean values of the mean intraindividual lactate concentrations ($\bar{x}_{LAC_{i\bar{a}}}$) and their single and double standard deviation.

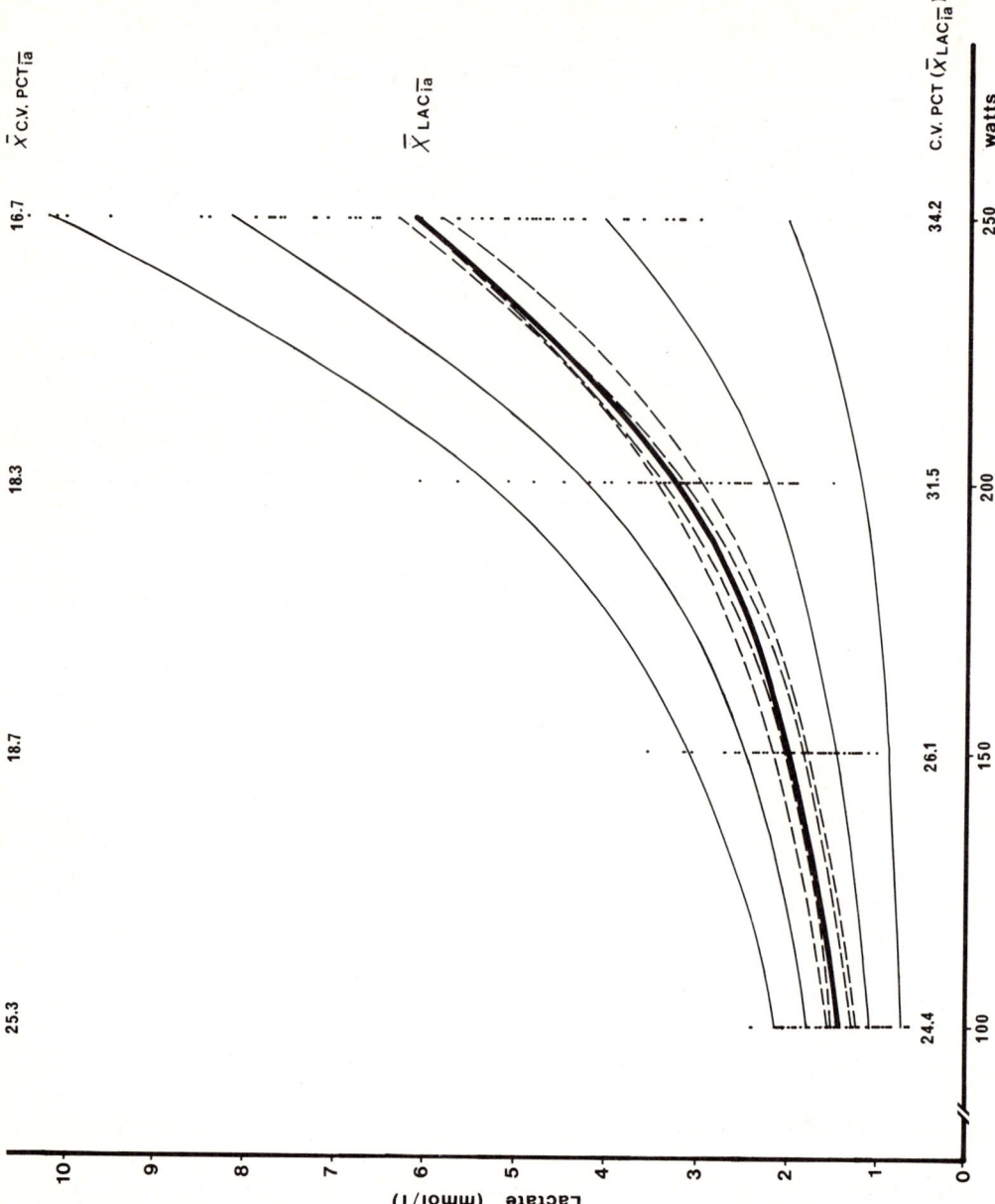

Fig. 4. Dynamic response of arterial lactate concentration for the group of sports teachers and students ($n=12$) during incremental bicycle ergometry. The single measurements of all subjects for all test times ($n=4$) are recorded in the areas of single and double standard deviation *(thin lines)* of the group mean values of the mean intraindividual lactate concentrations *(thick line)*. In contrast, mean interindividual lactate concentrations at test times 1, 2, 3, and 4 are noted with *broken lines*. The group mean coefficients of variation of the intraindividual mean lactate concentrations $(C. V. PCT_{\overline{X}LAC_{ia}})$ are noted at the respective points on the x axis, the group mean values of the intraindividual mean coefficients of variation $(\bar{x}_{C.V.PCT_{ia}})$ at the respective points on the *upper edge* of the figure

Discussion

Even the resting lactate concentrations in blood as an expression of the production and utilization [3] are subject to daily rhythmical fluctuations, which are closely connected with carbohydrate metabolism and adrenaline concentrations.

The results presented here, with lower values in the late morning and higher ones in the early afternoon, correspond to a great extent to those of Pannier [10] and Neumann [9]. The mean intraindividual coefficients of variation of the arterial lactate concentrations are at about the same level of 28.9% or 31.4% for both the daily rhythm and the short-term and long-term variability, i.e., a little higher than the values of 24.8% from venous blood from the arm in tests by Graham [6]. Nonetheless, there are considerable interindividual variabilities which are encountered in Graham's [6] study – between 7.2% and 37.8% – and in ours – 20.2% and 47.5%, and 16.9% and 57.5%. The intraindividual coefficients of variation are lower (Table 1) in the case of the competitive sportsman, a fact which is also confirmed from other observations. The explanation for the coefficients of variation for subject, WW, before test B (Table 1) – around one-third of the previous value, independent of test number – is that test A, which was carried out approximately 2.5 h beforehand, also had a certain regulating effect on longer-term recovery.

No differences could be found between the intra- and interindividual coefficients of variation for the resting lactate concentrations (Tables 1 and 2).

From the comparison of the statistical results of resting lactate concentration with changes in the loading lactate concentrations, it may be concluded that resting values up to 1.8–1.9 mmol/l need have no influence on the level of the loading values, especially when the duration of loading of one step is longer than 3 min.

The prerequisite for evaluation of the variability and reproducibility of the anaerobic threshold is the analysis of the lactate dynamics during incremental testing.

A decrease in the intraindividual variation coefficients of submaximal lactate concentrations with increasing loads is common to all loading tests. In precisely planned, repeated investigations at the same testing time, the coefficients of variation are found to be between 3.5% and 13% (Figs. 2 and 3). When the investigation time is within half a day, values are higher (16.7%–25.3%), although here, too, single values of under 10% are encountered (Table 5). A corresponding reference is that of Löllgen [8], who found coefficients of variation of 12%–18% for base excess at 150 W. Similar values may be derived from studies by Verstappen [19].

The results presented here show how important not only is a precise methodology corresponding to all standardization conditions, but also how necessary strict observance of the correct time of day seems to be, in order to achieve the highest possible reproducibility with variabilities under 5%, even at submaximal lactate concentrations.

This view is reinforced by a study by van Dam [4], who found a difference of 50 W at a lactate concentration of 4 mmol/l between tests at 11:00 in the morning and 19:00 in the evening, to the disadvantage of the morning test, in bicycle-ergometer-tested, endurance-trained sportsmen. Also studies from Baxter [2] and Reilly [13] confirm these results.

Large errors in performance diagnosis from submaximal parameters can, moreover, come into existence through disregard of the "daily form." This can be avoided to a great extent by exact exploration of all influencing factors with competition athletes (e. g., testing too soon after races, etc.), but probably never completely excluded. This source of error must be reckoned with much more often in the routine sports medicine field, as performance influencing factors are frequently concealed due to of ignorance. Figure 2 shows three investigations of a first-class, competitive sportsman on "good days," with the individual data marked into the single and double standard deviation of the intraindividual lactate mean values. The coefficients of variation of under 10% speak for good, if not optimal, reproducibility of the submaximal lactate concentrations. At the interpolated aerobic (v = 12.3 ± 0.38 km/hr or 3.42 ± 0.11 m/s) and anaerobic threshold (v = 14.4 ± 0.36 km/hr or 4 ± 0.1 m/s), the coefficients of variation lie in the seemingly desirable optimal area, at 3.1% and 2.5%. However, based on the absolute differences in speed, the standard deviations of 0.11 and 0.10 m/s are, just over the often-required limit of 0.05 m/s.

A "bad day" is contrasted with three good days for a sports student in Fig. 3 in the same way as in Fig. 2. The lactate to speed relation for the bad day lies on or above the double standard deviation, particularly for test A. The displacement to the left into an area twice the standard deviation exceeded gives the misleading appearance of a restriction in function which did not actually exist. The lactate concentration of over 5 mmol/l at 8 km/hr would not, moreover, permit an evaluation of the anaerobic threshold. The intraindividual coefficients of variation of the lactate concentrations for good days are slightly higher than with the competitive sportsman, KG, similar to those of the anaerobic threshold. The above-mentioned displacement to the left of the lactate to performance relation on bad days, also observable with the heart rate, points to a deterioration in the degree of efficiency on bad days in such a way that there is a higher energy requirement at the same loading steps. This is doubtlessly caused to a great extent by changes of coordination.

Verstappen [18, 19] also arrived at similar results, finding a definite increase in lactate concentrations, oxygen uptake, and heart and respiratory rate at the same loading steps on bad days, with no significantly different maximum values.

The same kinds of problems as with individuals are encountered even more in studies of apparently homogenous groups [20]. This applies all the more when reference values for certain performance-diagnostic parameters are obtained from such studies. The individual data seen in Fig. 4, in comparison with the intraindividual group mean, supports even more the demand to take individual fluctuations into consideration in the setting up of reference values.

According to expectations, the group mean coefficients of variation of the intraindividual mean lactate concentrations show no differences at various loading levels, while the group mean values of the intraindividual mean coefficients of variation of 25.3% at 100 w go down to 16.7% at 250 w.

Contrasting both these facts shows that the relative differences in performance within the group remain the same, but that reproducibility improves at higher loading levels, as the influence through greater resting variabilities becomes increasingly less.

The coefficients of variation for the group mean values of the intraindividual power output (19.1% for the aerobic and 12.2% for the anaerobic threshold) are distinctly higher than with the individual subjects (Table 6). This also applies to the group mean values of the intraindividual mean coefficients of variation of the aerobic (15.6%) and anaerobic (7.7%) threshold, although in individual cases similarly low values are found as with the test subjects for short-term variability. The order of magnitude of these coefficients of variation indicates that the long-term variability of the aerobic-anaerobic transition can be subject to greater fluctuations. If it is assumed that lactate determinations are influenced by few sources of error under condition of precise methodology, the significance of other important influencing factors, especially daily constitution, daily rhythm, and smaller changes in performance abilities is apparent. These factors should be taken into consideration in all long-term studies, as well as in the interpretation of performance improvements and impairments and in drug tests.

Finally, a statement about validity of these parameters for the subjects can be made from comparison of the variabilities of reference points of the aerobic-anaerobic transition at 2 and 4 mmol/l lactate with individually determined reference values.

From Tables 6 and 7 it can be seen that neither for the intraindividual mean value of the running speed nor for the group mean of the intraindividual mean power output at the aerobic-anaerobic threshold, any significant differences exist. With subject, KG, the coefficients of variation of the running speed are with one exception (1.5% at the individual aerobic threshold, Table 7) the same for both methods of determination. Therefore, the conclusion may be drawn that firstly, the parameters obtained are valid for the athlete at the time of the investigation and secondly, both methods appear to be reliable.

In principle, this statement is also true for the long-term investigation carried out on the sports teachers. However, for the individual determination, somewhat lower coefficients of variation are found for the group mean value of the intraindividual mean power output (Table 6) and the group mean value of the intraindividual mean coefficients of variation.

Whether better validity results of this individual method of determination for long-term studies on this group of people because of this trend can not be definitely deduced from the results presented here. Long-term investigation of extreme variants of various performance characteristics (short- and long-term endurance) would be necessary to clarify the validity of all individual methods of determination and the validity of parameters to be determined.

Table 7. Mean values $(\bar{x})$, standard deviation (s), and coefficient of variation (C. V. PCT) of the running speed of three consecutive test times for subject, KG, at fixed (2 and 4 mmol/l) lactate and individually determined threshold reference points

km/h	Aerobic threshold (2 mmol/l)	Anaerobic threshold (4 mmol/l)	Individual anaerobic threshold	Individual compensation threshold
$\bar{x}$	12.3	14.4	11.9	14.5
s	0.38	0.36	0.17	0.4
C. V. PCT	3.1	2.5	1.5	2.7

No relevant differences were found, either in the absolute values nor the coefficients of variation, for the two methods of determining the aerobic-anaerobic transition for the heart rates, which are often used for training control.

To summarize, it can be stated that only precise consideration of daily rhythm guarantees reproducible results for endurance performance limits determined from submaximal parameters. The variability of these results lies between 3% and 5%. This permits evaluation concerning improvement for impairment of physical performance. A further condition is the exact exploration and standardization of all other secondary influencing factors in order to exclude bad days. The greater stability of competitive sportsmen renders possible a reliable reference value tabulation without intraindividual comparative tests. This does not apply without restriction to untrained persons or to the comparison of group mean values in long-term experiments. Consideration should be taken particularly in the interpretation of stress tests with untrained individuals on a long-term basis.

References

1. Bachl N (1981) Möglichkeiten zur Bestimmung individueller Ausdauerleistungsgrenzen anhand spiroergometrischer Parameter. Österr J Sportmed Suppl 1
2. Baxter C, Reilly T (1983) Influence of time of day on all-out swimming. Br J Sports Med 17: 122–127
3. Chohen RD (1978) The production and removal of lactate. In: Bossart H, Perret C (eds) Lactate in acute conditions. Karger, Basel, pp 10–19
4. Van Dam B, Waterloh E, Knörzer H (1983) Der Einfluß von Tagesperiodik und Schweißproduktion auf das Laktatverhalten unter Ergometerbelastung. Lehre der Leichtathletik 13: 24–26
5. Drexler H, Löllgen H, Bodemann H, Just H (1983) Zur Reproduzierbarkeit ergometrischer und hämodynamischer Meßgrößen. In: Mellerowicz H, Franz I-W (eds) Standardisierung, Kalibrierung und Methodik in der Ergometrie. Fachbuch-Verlagsgesellschaft, Erlangen, pp 88–94
6. Graham TE, Andrew GM (1973) The variability of repeated measurements of oxygen debt in man following a maximal treadmill exercise. Med Sci Sports 5: 73–78
7. Keul J, Simon G, Berg A, Dickhuth H-H, Kubel R, Goertler I (1979) Bestimmung der individuellen anaeroben Schwelle zur Leistungsbewertung und Trainingsgestaltung. Dtsch Z Sportmed 7: 212–218
8. Löllgen H, Haninger B, Just H (1980) Langzeitvariabilität ergometrischer Meßgrößen. In: Kindermann W, Hort W (eds) Sportmedizin für Breiten- und Leistungssport. Berichtsband Deutscher Sportärztekongreß Saarbrücken. Demeter, Gräfelfing, pp 273–285
9. Neumann S, Pannier R (1973) Tagesprofile der Blutkonzentrationen von Lactat, Pyruvat und Glucose, des Säuren-Basen-Status sowie der Pulsfrequenz bei gesunden Kleinkindern. Ärzt Jugdkde 64: 453–460
10. Pannier R, Klimt F (1971) Blutmilchsäure – Tagesperiodik. Pädiatr Pedol 6: 113–116
11. Pessenhofer H, Schwaberger G, Sauseng N, Schmid P (1983) Methodische Grundlagen zur Bestimmung des individuellen aerob-anaeroben Übergangs. In: Mellerowicz H, Franz I-W (eds) Standardisierung, Kalibrierung und Methodik in der Ergometrie. 4th Int Sem Ergometry, Berlin. Fachbuch-Verlagsgesellschaft, Erlangen, pp 193–199
12. Pessenhofer H, Schwaberger G, Schmid P (1981) Zur Bestimmung des individuellen aerob-anaeroben Übergangs. Dtsch Z Sportmed 1: 15–17
13. Reilly T, Baxter C (1983) Influence of time of day on reactions to cycling at a fixed high intensity. Br J Sports Med 17: 128–130
14. Shephard R (1980) What can the applied physiologist predict from his data? J Sports Med Phys Fitness 20: 297–308
15. Simon G, Berg A, Dickhuth H-H, Simon-Alt A, Keul J (1981) Bestimmung der anaeroben Schwelle in Abhängigkeit vom Alter und von der Leistungsfähigkeit. Dtsch Z Sportmed 1: 7–14

16. Skinner JS, Mc Lellan T (1980) The transition from aerobic to anaerobic metabolism. Res Q 51: 234–248
17. Stegmann H, Kindermann W (1980) Modell zur Bestimmung der individuellen anaeroben Schwelle. In: Kindermann W, Hort W (eds) Deutscher Sportärztekongreß, Saarbrücken. Sportmedizin für Breiten- und Leistungssport. Demeter, Gräfelfing, pp 227–233
18. Verstappen FTJ, Kuipers H, Keizer HA (1981) Reproducibility of aerobic power and related physiological variables in women. In: Borms I, Hebbelinck M, Venerando A (eds) Women and sport. Karger, Basel, pp 133–140 (Medicine and Sport, vol 14)
19. Verstappen FTJ, Kuipers H, Keizer HA (1983) Daily variation in physical performance and its physiological correlates. In: Bachl M, Prokop L, Suckert R (eds) Current topics in sports medicine. Proceedings of the XXII world congress on sportsmedicine. Urban and Schwarzenberg (to be published)
20. Vogelaere P, De Meyer F, De Schrijver J (1983) Variations de repos de divers paramétres cardio-circulatoires, enregistrés journellement pendant une période de 9 mois. Med Sport 57: 226–229
21. Wright GR, Sidney K, Shephard RJ (1978) Variance of direct and indirect measurements of aerobic power. J Sports Med 18: 33–42
22. Wyndham CH, Strydom NB, Maritz JS, Morrison JF, Peter J, Potgieter ZU (1959) Maximum oxygen intake and maximum heart rate during stenuous work. J Appl Physiol 14: 927–936

Anaerobic Threshold and Oxygen Pulse as Fitness Criteria in Submaximal and in Peak Exercise Testing

B. van den Eynde and M. Ostyn

It has been well known since the studies of Daniels [2, 3] that in man, performance in endurance events can be improved by training without a concomitant increase in $\dot{V}_{O_2 max}$. This emphasizes the usefulness of testing not only the aerobic capacity but also submaximal fitness in athletes. Therefore, it seemed interesting to study various criteria of submaximal fitness.

Two points were particularly studied:

1. The role of the incremental profile when submaximal fitness is measured in continuous, gradual exercise tests.
2. The relation between submaximal fitness items and aerobic capacity. This part of the study yields several submaximal fitness levels together with real $\dot{V}_{O_2 max}$ determination on bicycle ergometer and treadmill.

Materials and Methods

To study the role of the incremental profile, 34 male students in physical education (21–23 years of age) volunteered to perform at random order seven different incremental test profiles on a bicycle ergometer until complete exhaustion. Oxygen uptake ($\dot{V}_{O_2}$) and ventilation ($\dot{V}_E$) were continuously recorded by closed circuit spirometry. Heart rate was registered minute by minute. The first anaerobic threshold (AT_1)[1] was determined according to Wasserman et al. [11]. The seven incremental profiles used in our study are given in Table 1.

Results

Mean values $(\bar{r})$ and standard errors of the mean (SE) of $\dot{V}_{O_2}$ at respective levels of AT_1 and of $\dot{V}_{O_2 max}$ are given in Table 2, together with AT_1 (l/min) as a percentage of $\dot{V}_{O_2 max}$.

Statistical tests (covariance analysis for repeated measurements) of the variables in Table 2 revealed that vita maxima tests resulted in respectively lower values for $\dot{V}_{O_2 max}$, $\dot{V}_{O_2}AT_1$, and $\dot{V}_{O_2}AT_1\%$ than the longer duration tests.

An example of statistical analysis is given for $\dot{V}_{O_2 max}$ in Table 3.

---

1 AT_1 is called the first anaerobic threshold because it is obtained by continous, incremental exercise testing, whereas AT_2 is detected using rectangular loads

Table 1. Schematic overview of seven incremental test profiles

Profile and author	Initial load (W)	Duration of exercise level (min)	Work load increment (W)	Mean working time (min)
1. Vita maxima, (variant of the authors)	1 × body weight times watts	1	1 × body weight times watts	4.9
2. Vita maxima	50	1	50	7.1
3. Vita maxima	70	1	40	8.0
4. Franz [5]	1 × body weight times watts	3	1 × body weight times watts	12.5
5. Mellerowicz [7]	50	3	50	17.7
6. Hollmann [6]	70	3	40	19.6
7. Franz [5]	70	1	10	22.8

Table 2. Means $(\bar{r})$ and standard error of mean (SE) of $\dot{V}O_2AT_1$, $\dot{V}O_{2\,max}$, and $\dot{V}O_2AT_1\%$

Test	$\dot{V}O_{2\,max}$ (l/min)	$\dot{V}O_2AT_1$ (l/min)	$\dot{V}O_2AT_1\%$ (percent)
1	3.07 ± 0.09	1.81 ± 0.07	59 ± 2
2	3.17 ± 0.10	1.77 ± 0.09	56 ± 2
3	3.33 ± 0.10	2.09 ± 0.09	63 ± 2
4	3.46 ± 0.10	2.33 ± 0.06	68 ± 2
5	3.54 ± 0.10	2.31 ± 0.08	66 ± 2
6	3.53 ± 0.08	2.58 ± 0.06	73 ± 1
7	3.38 ± 0.07	2.56 ± 0.06	76 ± 1

Table 3. Covariance analysis for repeated measurements and mean $\dot{V}O_{2\,max}$ levels (l/min) reached at the end of seven incremental bicycle ergometer tests

	T_7	T_6	T_5	T_4	T_3	T_2	T_1
T_1	***	***	***	***	**	NS	
T_2	**	***	***	***	**	shorter	
T_3	NS	**	**	**		tests	
T_4	NS	NS	NS				
T_5	*	NS	longer				
T_6	**		tests				
$\bar{x}\pm$	3.38	3.53	3.54	3.46	3.33	3.17	3.07
SE	± 0.07	± 0.08	± 0.10	± 0.10	± 0.10	± 0.10	± 0.09

$*p < 0.1$; $**p < 0.05$; $***p < 0.01$

Throughout the seven test protocols, $\dot{V}O_2AT_1$ values (in l/min) intercorrelated from 0.13 to 0.65, whereas oxygen pulse (OP) intercorrelations ranging at AT_1 from 0.34 to 0.81, seemed to be less profile-dependent than $\dot{V}O_2$.

In order to examine if OP is also more stable than $\dot{V}O_2$, as far as the relationship between $\dot{V}O_2AT_1$ (l/min) and $\dot{V}O_{2\,max}$ is concerned, a second study was undertaken.

Materials and Methods

To study the relation between submaximal fitness items and aerobic capacity, 23 male students (19–24 years of age) exercised at random order on a bicycle ergometer and a treadmill by continuously loading and by working discontinuously at different work load intensities. Treadmill speed was kept constant at 11.3 km/h for all sessions, and on the bicycle ergometer a constant pedaling rate of 60 rpm was required. Oxygen uptake ($\dot{V}O_2$), carbon dioxide production ($\dot{V}CO_2$), and ventilation ($\dot{V}_E$) were recorded every 30 s using an open circuit system. Heart rate was recorded every 30 s.

Using incremental or continuously graded exercise testing, PWC_{150} and PWC_{170} were determined for both bicycle ergometry and treadmill running [10]; AT_1 was also determined according to recent indications of Caiozzo et al. [1]. Further indicators of submaximum fitness, determined by rectangular loading (bicycle ergometer as well as treadmill), were:

– AT_2 (10 min), that is, the highest load at which $\dot{V}_E$ does not increase rectilinearly between the 5th and the 10th minute of exercise, as described by Reybrouck et al. [9]

– AT_2^+ (10 min), that is, one intensity level higher than AT_2 with an accuracy of 20 W (bicycle ergometer) or 2.5% slope (treadmill)

– *Régime stable limite* (RSL), that is, the highest 20-min lasting rectangular load which does not result in a ventilation increase between the 10th and the 20th min of more than 5% [8]

– *Régime stable approché* (RSA), that is, one level higher than RSL with the same accuracy as mentioned for AT_2^+

Table 4. Bicycle ergometry: correlations (Spearman) between $\dot{V}O_{2\,max}$ and respiratory, oxygen uptake ($\dot{V}O_2$) and oxygen pulse (OP) at different submaximal intensity levels

Intensity level	n	Correlation between $\dot{V}O_{2\,max}$ and		$\dot{V}O_2$ in % of $\dot{V}O_{2\,max}$
		$\dot{V}O_2$	OP	
$\dot{W}_{150}$	17	0.83***	0.83***	60
$\dot{W}_{170}$	20	0.82***	0.82***	77
AT_1	19	0.47*	0.70***	61
AT_2	19	0.67***	0.80***	78
AT_2^+	17	0.72***	0.86***	86
RSL	17	0.86***	0.93***	68
RSA	17	0.76***	0.89***	75

$n=20$ students in physical education. Significance of asterisks as in Table 3

Table 5. Treadmill running: correlations (Spearman) between respiratory, oxygen uptake ($\dot{V}O_2$) and oxygen pulse (OP) obtained in submaximal work and $\dot{V}O_{2\,max}$

Intensity level	n	Correlation between $\dot{V}O_{2\,max}$ and		$\dot{V}O_2$ in % of $\dot{V}O_{2\,max}$
		$\dot{V}O_2$	OP	
$\dot{W}_{150}$	15	0.83***	0.83***	60
$\dot{W}_{170}$	19	0.59**	0.59**	72
AT_1	21	0.46*	0.79***	82
AT_2	19	0.74***	0.87***	84
AT_2^+	17	0.54*	0.74**	93
RSL	19	0.68***	0.92***	76
RSA	14	0.84***	0.83***	83

$n=21$ students in physical education. Significance of asterisks as in Table 3

Results

From Tables 4 and 5 it can be seen that for working both on treadmill and bicycle ergometer:

1. $\dot{V}O_2AT_1$ (l/min) correlated with $\dot{V}O_{2\,max}$ to a lesser degree than did oxygen pulse at AT_1
2. The same but weaker tendency occurred for ventilatory criteria of submaximal fitness: AT_2, AT_2^+, RSL, RSA
3. The highest correlation between submaximal $\dot{V}O_2$ or OP with $\dot{V}O_{2\,max}$ was not a fortiori bound to the highest intensity level (in percent of $\dot{V}O_{2\,max}$)

   Of course at $\dot{W}_{150}$ and $\dot{W}_{170}$ no differences were found for $\dot{V}O_2$ and OP in their relationship to $\dot{V}O_{2\,max}$ (150 and 170 are constants in determination of OP).

Discussion

It is clear that in stepwise incremental loading (on a bicycle ergometer), AT_1 (in l/min $\dot{V}O_2$) differs considerably between vita maxima test profiles (tests 1, 2, and 3) and other tests (tests 4–7, lasting more than 12 min). Within vita maxima tests, at load increments of 50 W/min or the body weight (in kg) times watt, $\dot{V}O_2AT_1$ was nearly identical.

The 40 W increment per minute yielded significant differences with the two higher minute increments.

Davis et al. [4], using ramped load ranging from 20 to 100 W/min, did not find differences either for $\dot{V}O_2AT_1$ or for $\dot{V}O_{2\,max}$, which averaged in their study 41 ml kg^{-1} min^{-1}. The near-to-sedentary values together with different modes of intensity loading could be responsible for different findings. Time constants for $\dot{V}O_2$ by ramped load were the same within a range of loads from 20 to 100 W/min [4].

From the low $\dot{V}O_2AT_1$ (l/min) correlations with $\dot{V}O_{2\,max}$, it is reasonable to conclude, as did Weltman et al. [12], that AT_1 should only be used to evaluate submaximal fitness. However, relating oxygen pulse at AT_1 with $\dot{V}O_{2\,max}$ (l/min), oxygen pulse seems to be a more accurate predictor of $\dot{V}O_{2\,max}$ than $\dot{V}O_2$ itself.

References

1. Caiozzo VJ, Davis JA, Ellis JF, Azus JL, Vandagriff R, Prietto CA, McMaster W (1982) A comparison of gas exchange indices used to detect the anaerobic threshold. *J Appl Physiol* 53: 1184–1189
2. Daniels J, Oldridge N (1971) Changes in oxygen consumption of young boys during growth and running training. *Med Sci Sports* 3: 161–165
3. Daniels J, Oldridge N, Nagle F, White B (1978) Differences and changes in $\dot{V}O_2$ among young runners 10 to 18 years of age. *Med Sci Sports* 10: 200–203
4. Davis JA, Whipp BJ, Lamarra NL, Huntsman DJ, Frank MH, Wasserman K (1982) Effect of ramp slope on determination of aerobic parameters from the ramp exercise test. *Med Sci Sports Exerc* 14: 339–343
5. Franz I (1972) Vergleichende Untersuchungen zur Messung der PWC_{170}. In: Mellerowicz H, Jokl E, Hebbelinck M (eds) 3rd Int Seminar für Ergometrie, Ergon, Berlin, pp 136–142
6. Hollmann W (1963) Höchst- und Dauerleistungsfähigkeit des Sportlers. Barth, München
7. Mellerowicz H (1975) Ergometrie, 2nd ed. Urban und Schwarzenberg, München
8. Sadoul P, Durand D, Aubertin N (1958) Systèm sur l'exploration fonctionelle primaire. Etude des paramètres respiratoire au cours du travail de durée moyenne. *Pathol Biol* 6: 559–568
9. Reybrouck T, Ghesquiere J, Cattaert A, Fagard F, Amery A (1979–80) De relatie tussen anaerobe drempel en de maximale prestatiecapaciteit. Werken Belg Geneesk Ver L O Sport 28: 62–73
10. Wahlund H (1948) Determination of the physical working capacity. *Acta Med Scand* [Suppl] 215: 9–75
11. Wasserman K, Whipp BJ, Royal SN, Beaver WL (1973) Anaerobic threshold and respiratory gas exchange during exercise. *J Appl Physiol* 35: 236–243
12. Weltman A, Katch V, Sady S, Freedson F (1978) Onset of metabolic acidosis (anaerobic threshold) as a criterion measure of submaximum fitness. Res Q 49: 218–227

Ergometry: A Method for the Adjusted Common Functional and Metabolic Response Testing

B. Semiginovský, L. Havlíčková, and J. Vránová

Introduction

Recently the submaximal exercise protocol has been frequently used for endurance performance capacity analysis in sports medicine. Submaximal metabolic parameters estimated by both invasive as well as noninvasive techniques may be used for the anaerobic threshold determination [18, 20].

The concept of anaerobic threshold was introduced by Wasserman et al. [16, 18] and the anaerobic threshold was shown to be the last linear point in the work rate-ventilation curve during incremental exercise [18, 19].

According to current theory, the anaerobic threshold represents a transition between the predominantly aerobic and anaerobic energy yield, evoked by a physical load of graded intensity [1, 11]. It thus represents the shift in economy of energy liberation followed by blood plasma lactic acid accumulation. This specific variable is considered as an individually characteristic regulatory level for triggering of supplementary functional, metabolic, and endocrine homeostatic mechanisms in situations of progressive muscle tissue hypoxia, evoked by progressive incremental exercise [11, 12].

Phenomenologically, the anaerobic threshold is defined as the metabolic and functional response to the determined level of load or oxygen uptake, just below the point where the rapid development of metabolic acidosis and associated changes in respiratory gas exchange occur [18], or level of load for which a steady rate of the lactate production and removal is characteristic. The anaerobic threshold could be identified noninvasively by the point of nonlinear increase in ventilation, in carbon dioxide production, and an increase in end-tidal O_2 without a corresponding decrease in end-tidal CO_2 [18]. The invasive identification of anaerobic threshold is based on estimation of the blood lactate accumulation junction point [9].

These submaximal parameters can be measured independently of motivation and with significantly lower risk in patients with coronary heart disease, to determine the *individual* endurance performance capacity.

It can be hypothesized from the above that during progressive incremental exercise, each individual may be adjusted to the load with similar biological significance, corresponding to the starting point of curvilinear increase of the anaerobic glycolysis participation, in order to meet the actual energy demands.

In the present study, the load level, corresponding to the anaerobic threshold, was taken as a common denominator of the general metabolic response to the adjusted exercise intensity. Dynamics of cardiorespiratory parameters and blood plasma constituents were evaluated during a long-term exercise test at a load corresponding to the anaerobic threshold.

Table 1. Anthropometric and functional characteristics of subjects (mean $\pm$ SD, $n = 10$)

Age (years)	Height (cm)	Weight (kg)	Body fat (%)	W_{max} (W)	$\dot{V}_{E\,max}$ (l, BTPS)	$\dot{V}_{O_2max}$ (l, STPD)
21.5 ± 0.7	176 ± 5	68.5 ± 6.8	10 ± 3	298.8 ± 24.4	99.6 ± 13.5	3.06 ± 0.3

$\dot{V}_{O_2\,max}$ (ml kg^{-1})		HR$_{max}$ (beats min^{-1})		LA $\dot{V}_{O_2max}$ (mmol l^{-1})		
44.5 ± 2.95		191.04 ± 9.5		9.73 ± 2.76		

Subjects and Methods

Physical characteristics of ten male volunteer subjects are given in Table 1. None of them participated regularly in special training programs. Each subject completed four tests in 2 weeks.

The maximal oxygen uptake test was performed on the first experimental day. The discontinuous cycle ergometer standard International Biologic Program maximal oxygen uptake test was used. Test protocol consisted of three submaximal 6-min work bouts, alternated with 1-min rest periods. Initial load was 1 W kg^{-1} and was increased to 1.5 W kg^{-1} for the second step and to 2 W kg^{-1} for the third step. To determine $\dot{V}_{O_2\,max}$ a continuous increase of work load, starting approximately 15–30 W below PVC$_{170}$, with increases of 30 W each successive minute, was used until exhaustion. A plateau or drop-off in $\dot{V}_{O_2}$ with increasing work load was evidence that $\dot{V}_{O_2\,max}$ had been achieved [14]. Each subject was required to pedal at a constant rate of 1 Hz, paced by an auditory-visual metronome and total pedaling score was controlled. Open circuit spirometry methods were utilized for determination of $\dot{V}_{O_2}$. Expired gas was collected in meteorological balloons and its volume measured using a Tissot gasometer. Concentrations of O_2 and CO_2 in expired air were determined with a Zeiss interferometer. $\dot{V}_{O_2}$ (STPD) was calculated based on $\dot{V}_E$ (STPD) and true O_2 and CO_2 concentrations. Heart rate was taken from electrocardiograms. Blood samples were obtained from an antecubital vein at the 3rd min of recovery and analyzed for venous blood lactate concentration by an enzymatic method [4] using Boehringer kits.

The noninvasive determination of anaerobic threshold was realized on the 3rd successive day. The test protocol of Weltman and Katch [19] was used. The initial work was 30 W. This approximative value of anaerobic threshold was predicted from ventilation and gas exchange parameters monitored during the last 2 min of each submaximal step using the equation [11]

$$y = a + bx + c \log (x + 1)$$

where y is ventilation (l min^{-1}); x is the work rate (W); a, b, and c are individually characteristic constants.

The work was increased 30 W every 3 min until the seventh step (210 W). The work rate-ventilation curve and the work rate-CO_2 output curve or starting point of the curvilinear dissociation between ventilation and oxygen uptake were expressed graphically and the junction point was evaluated as the anaerobic threshold.

The invasive determination of anaerobic threshold was realized in the 5th successive day. This test consisted of pedaling on the cycle ergometer at three different work loads. The first load was $1 \, W^{-1} \, kg$; the second was adjusted to the anaerobic threshold; and the third corresponded to PWC_{170} for 10 min and alternated with 1-min rest periods during which venous blood samples were taken for blood lactate estimation. Ventilatory gas exchange parameters were evaluated in addition, and the noninvasive anaerobic threshold value was validated.

The anaerobic threshold long-term test was realized on the 8th successive day of experiments. The work load of a 90-min test was adjusted to the anaerobic threshold level predetermined by noninvasive measurements and validated by the lactate method. Actual work load level was just below that of the lactate junction point.

Analysis of acid-base equilibrium and arterial blood gases. Capillary blood from the ear-lobe was drawn into heparinized glass microtubes for measurement of pH, PO_2, and PCO_2 by electrodes (Radiometer AMT 1). Hemoglobin content was estimated photometrically, and plasma bicarbonate concentration and base excess were calculated.

Analysis of blood plasma constituents. Blood was sampled intermittently for determination of glucose, lactate, free fatty acids, triglycerides, cholesterol, urea, creatinine, and inorganic phosphate content; the blood plasma was examined for alanine aminotransferase (E. C. 2.6.1.2.), aspartate aminotransferase (E. C.2.6.1.1.), η-glutamyltransferase (E. C.2.3.2.1.), alkaline phosphatase (E. C.3.1.3.1.) activity and the immunoreactive human growth hormone and insulin concentration at rest and at the end of each 30-min time interval of the anaerobic threshold-adjusted 90-min test, as well as at 30th and 60th min of recovery. The proved sets for clinical chemistry estimations were used in the system of a Greiner analyzer.

Statistical analysis was performed using Student's t test for matched pairs. Differences were considered statistically significant with $p < 0.05$.

Results

The mean anthropometric and functional characteristics of subjects who volunteered in this study (presented in Table 1) were not significantly different from those of the Czechoslovakian population [10].

During incremental work of "short-term" test protocols used in our experiments, the minute ventilation and related parameters of the gas exchange increased disproportionally to work rate and oxygen consumption at approximately 60% of $\dot{V}O_{2 \, max}$ (see Table 2). The noninvasive anaerobic threshold ($AT_{\eta1}$), estimated through the analysis of ventilatory transients, preceded the invasive anaerobic threshold (AT_i) in all subjects examined (see Table 2 and Fig. 1). Thus the onset of the "decompensated" metabolic acidosis occurred at a relatively higher level of work load than the onset of respiratory compensation.

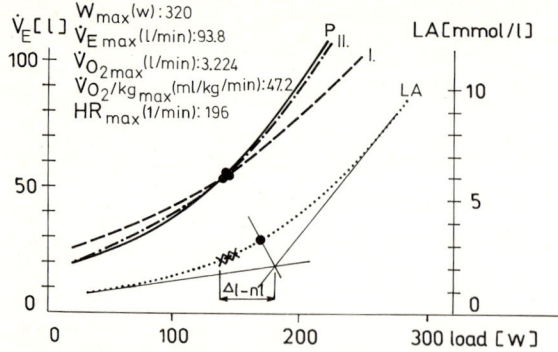

Fig. 1. Individual value of dissociation between the load-dependent start point of the ventilation and the blood lactate *(LA)* curvilinear increase. *P*, prediction of AT_{n1} from IBP $\dot{V}O_{2\,max}$ test ventilation variables using equation above; *I*, method of Weltman and Katch [19]; *II*, evaluation of gas exchange variables during AT_1 test protocol; *HR*, heart rate

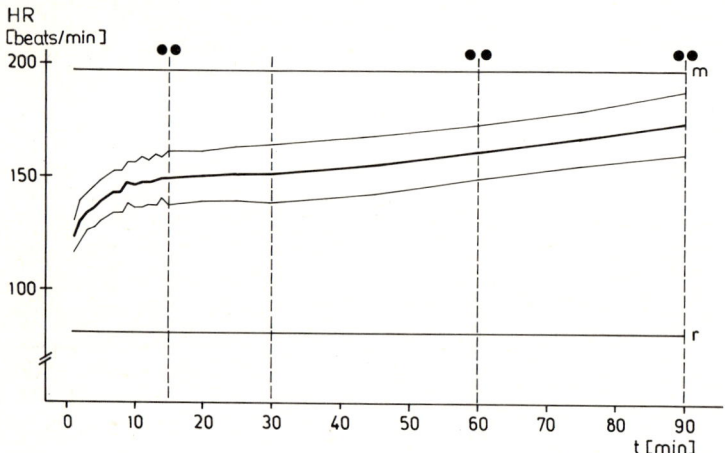

Fig. 2. Heart rate response to the AT_{n1} individually adjusted, long-term exercise test. Mean value, *thick line;* ±SD, *thin line; r,* resting level; *m,* maximal rate in $\dot{V}O_{2\,max}$ test; *interrupted vertical lines,* sections of sequential statistical evaluation; $p < 0.05$ ● ; $p < 0.01$ ● ●

Table 2. The noninvasive threshold-adjusted work load and related functional parameters

	Noninvasive	Invasive	Difference (Δi-ni)
Work load (W)	152.0 ±16 (51%)	168.0±22.0 (57%)	11%
$\dot{V}_E$ (l, BTPS)	52.0 ± 4.6 (53%)	61.3± 5.3 (62%)	17%
$\dot{V}O_2$ (l, BTPS)	1.9 ± 0.13 (62%)	2.2± 0.24 (72%)	16%
LA (mmol l^{-1})	2.85± 0.7 (30%)	3.4± 0.6 (35%)	19%

Description in absolute values (mean ± SD) and in % of $\dot{V}O_{2\,max}$ test

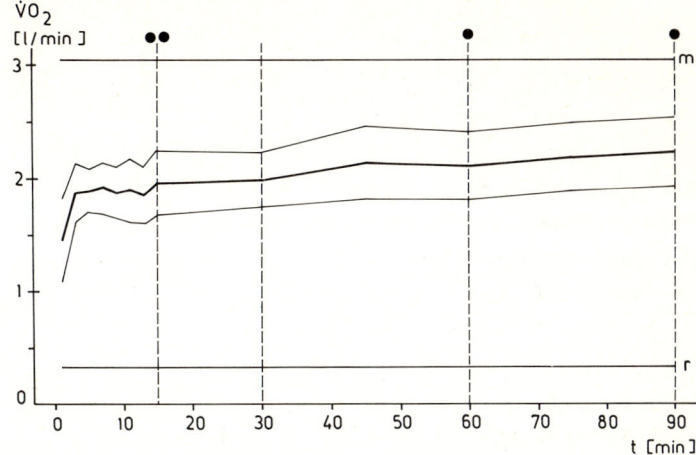

Fig. 3. Time course of oxygen uptake during long-term exercise test individually adjusted to AT_{n1}work rate. For explanation of symbols see Fig. 2

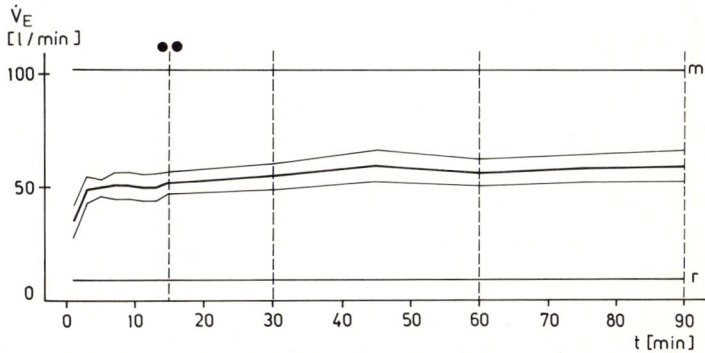

Fig. 4. Time course of pulmonary ventilation during long-term exercise test individually adjusted to AT_{n1} work rate. For explanation of symbols see Fig. 2

The work load corresponding to the short-term anaerobic threshold, predicted from the measurements in ventilatory gas exchange during the IBP $\dot{V}O_{2\,max}$ test and validated during specific test protocols for the $AT\eta_1$ estimation (see Methods), was used during the long-term test adjusted to the individual $AT\eta_1$ value. If heart rate (HR) and oxygen consumption are compared in the course of exercise, a "temporary" steady state was reached after 10–15 min (up to 30 min) of the long-term test, while a further slight increase of both values was obtained (HR: 15th min$\Delta + 67 \pm$ 12, 30th min$\Delta + 69 \pm 12$, 60th min$\Delta + 81 \pm 10$, and 90th$\Delta + 93 \pm 12$ beats min^{-1}; $\dot{V}O_2$ 15th min$\Delta + 1.62 \pm 0.27$, 30th min$\Delta + 65 \pm 0.23$, 60th min$\Delta 1.78 \pm 0.28$, 90th min$\Delta + 1.89 \pm 0.28$ l min^{-1}). Time course of both parameters of functional response to individually adjusted load is expressed in Figs. 2 and 3. However, in contrast with heart rate and oxygen uptake dynamics, a steady state for the pulmonary ventilation was maintained starting at the 15th min of the AT_{n1}-adjusted, long-term test (see Fig. 4). After a temporary increase at the 15th min of work load, the respiration quo-

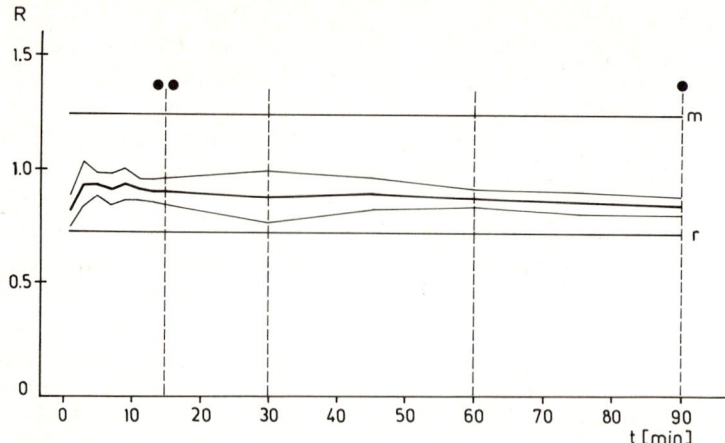

Fig. 5. Time course of respiration quotient changes during long-term exercise test individually adjusted to AT_{n1} work rate. For explanation of symbols see Fig. 2

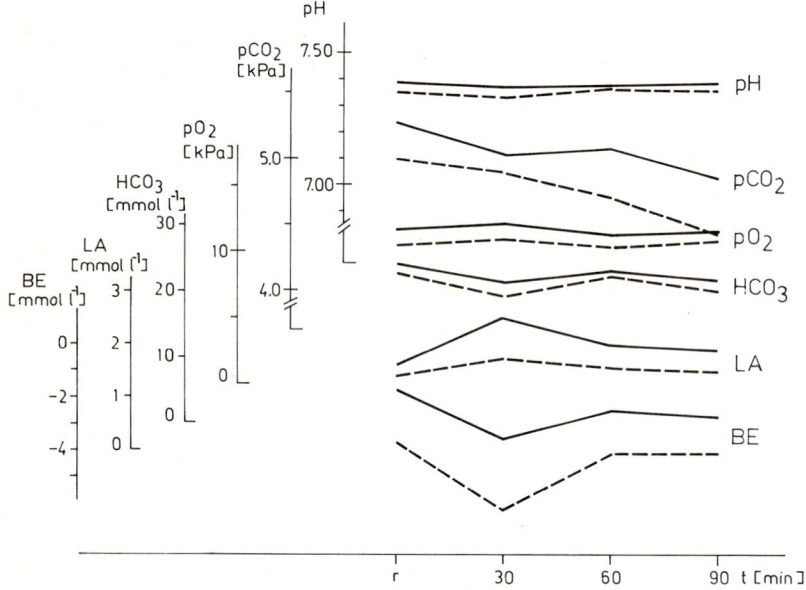

Fig. 6. Dynamics of acid-base equilibrium in response to the onset of metabolic acidosis during long-term exercise test individually adjusted to AT_{n1} work rate. Mean values, *solid lines;* SD, *broken lines; BE,* base excess; *LA,* lactate

tient (R) value steadily decreased and the difference was statistically significant at the last minute of long-term test (see Fig. 5).

Correlates of the internal milieu were evaluated as well as the functional response (see Fig. 6). Arterial blood homeostasis was well maintained with little change in pH. The most distinct changes were associated with a peak level of blood lactate (LA) accumulation, part of which was still present at the 30th min of adjusted exercise. An opposite feature of the base excess (BE) and actual bicarbonate load-dependent dynamics was found. Significant decrease of arterial blood PCO_2 during all work-time of the long-term test denoted respiratory compensation of metabolic acidosis by "hypocapnic buffering."

The responses of blood plasma constituents to the $AT\eta_1$-adjusted, long-term bicycle exercise are summarized in Table 3. Metabolic homeostasis is well maintained during the $AT\eta_i$ adjusted, long-term exercise with regard to blood glucose level. Any significant oscillations of blood glucose concentration and similarly of blood plasma triglycerides and β-lipoprotein content were not found. On the other hand, an outstanding increase of blood plasma level of inorganic phosphate (P_i) was found during the whole period of adjusted exercise ($p < 0.001$), with a return to resting values during recovery. While the load-dependent increase of the free fatty acids (FFA) in blood plasma started at the 30th min of adjusted exercise, a significant increase was found during recovery as well.

The blood plasma concentration of urea increased at the end of exercise; however, differences reached significance during recovery. The blood plasma creatinine content increased significantly at the 90th min of the long-term test and quickly recovered.

The intravascular fluid compartment changes were confirmed by plasma protein concentration analysis. Values were not corrected to hematocrit to express water balance dynamics.

Activity of blood plasma alanine aminotransferase, aspartate aminotransferase, and γ-glutamyltransferase were very discrete. The alkaline phosphatase activity showed a load-dependent feature.

Dynamics of hormonal release evoked by the $AT\eta_1$ individually adjusted load was studied in human growth hormone (IR-HGH) and insulin (IR-INS). For both, the protein-anabolic resultant effect and load-dependent changes were described. More than a fivefold exercise-induced increase of mean blood plasma IR-HGH content was found during the long-term test. This typical pattern of IR-HGH sequential changes induced by adjusted load was found in approximately 60% of examined persons; however, a negative or reciprocal effect of exercise was encountered as well. The time course of plasma IR-INS changes was closely related to the changes of blood glucose levels and is reminiscent of feature.

Discussion

The onset of metabolic acidosis, as estimated from both the gas exchange variables and blood lactate evaluation, is thought to be near the level where nonoxidative metabolism begins to play a more predominant role in energy production. In the range of this level of exercise intensity, the oxygen demand is greater than the oxygen supply. This interpretation of anaerobic threshold put forth by Wasserman et al. [18] seems to be in conflict with the finding of other authors [2, 8], which showed that increased lactate production may not be due to muscle hypoxia, but due to a discrepancy between glycolysis and pyruvate utilization [8]. It should be also noted that there is a functional preference of fast twitch fibers recruited primarily at higher intensities of short-term exercise [3] and their preference is for glycolytic glycogen de-

Table 3. Selected parameters of metabolic response to the noninvasive anaerobic threshold-adjusted long-term exercise and partial dynamics during recovery

Min	Rest	Exercise			Recovery		Significance				
	(1) 0	(2) 30	(3) 60	(4) 90	(5) 30	(6) 60	1-2	1-3	1-4	1-5	1-6
Blood plasma											
Glucose (mmol l^{-1})	4.94±0.87	4.48±0.68	4.61±0.59	4.55±0.78	4.22±1.03	4.76±0.57	NS	NS	NS	NS	NS
Inorganic phosphate (mmol l^{-1})	0.96±0.15	1.15±0.17	1.23±0.18	21.3±0.22	1.03±0.22	1.02±0.18	***	***	***	NS	NS
Free fatty acids (µmol l^{-1})	322±142	363±126	560±217	912±308	953±406	1127±249	NS	**	***	***	***
Triglycerides (mmol l^{-1})	1.12±0.46	1.04±0.36	1.13±0.37	1.30±0.34	1.06±0.31	0.93±0.15	NS	NS	NS	NS	NS
Cholesterol (mmol l^{-1})	4.27±1.07	4.48±0.99	4.63±1.14	4.77±1.13	4.57±1.00	4.50±0.84	*	*	**	NS	*
β-lipoprotein (g l^{-1})	2.16±0.39	2.14±0.35	2.12±0.38	2.22±0.37	2.04±0.40	2.13±0.37	NS	NS	NS	NS	NS
Urea (mmol l^{-1})	5.67±0.50	5.41±0.67	5.41±0.72	5.92±0.86	5.90±0.68	6.13±0.68	NS	NS	NS	**	***
Creatinine (µmol l^{-1})	80.0±14.9	82.7±10.8	85.0±9.4	92.1±10.2	84.6±11.7	88.3±9.4	NS	**	***	NS	NS
Proteins (g l^{-1})	77.81±4.25	80.75±4.15	81.12±5.50	81.15±4.32	76.93±5.65	78.28±2.60	**	**	***	NS	NS
Alanine aminotransferase (µkat l^{-1})	0.2±0.08	0.22±0.1	0.22±0.12	0.22±0.11	0.20±0.11	0.18±0.05	*	NS	NS	NS	NS
Aspartate aminotransferase (µkat l^{-1})	0.18±0.07	0.20±0.08	0.20±0.08	0.21±0.10	0.20±0.09	0.19±0.07	NS	*	*	NS	*
γ-glutamyltransferase (µkat l^{-1})	249±85	294±82	246±79	300±159	305±118	241±88	*	NS	NS	*	NS
Alkaline phosphatase (µkat l^{-1})	2.17±0.62	2.25±0.59	2.27±0.61	2.30±0.61	2.26±0.55	2.18±0.56	NS	*	**	NS	**
Immunoreactive human growth hormone (µg l^{-1})	5.8±4.9	30.4±14.5	31.2±9.3	29.8±13.9	16.6±8.6	8.1±3.7	**	**	**	*	NS
Immunoreactive insulin (pmol l^{-1})	40.2±22.2	30.9±17.2	23.7±13.6	30.1±13.6	25.1±10.6	24.4±10.8	NS	**	**	**	**

Mean±SD, $n=10$; *$p<0.05$; **$p<0.01$; ***$p<0.001$. All data except proteins were corrected for hematocrit values

pletion. Thus, the onset of metabolic acidosis during graded exercise may not always be caused primarily by muscle hypoxia.

The anaerobic threshold concept has recently received much attention, including comparison of its invasive and noninvasive determination [for review see 1]. In the present study we demonstrated that the nonlinear increase in ventilatory gas exchange preceded the junction point of blood lactate accumulation. Similar findings were also reported elsewhere [7, 13]. A tendency to the acid-base equilibrium disturbances during incremental work switch-on ventilatory control, which is a more sensitive indicator than indication of onset of metabolic acidosis. Lactate concentration did not adequately reflect the point of the gas exchange anaerobic threshold.

Since lactate production is associated with hydrogen ion release in equimolar concentrations, the H^+ have to be buffered. Buffering, minimizing the reduction in pH, takes place within skeletal muscle cells, in erythrocytes, and in plasma by combination with protein interaction as well as with the bicarbonate or phosphate buffer systems [7]. The reaction of H^+ with bicarbonate results in the formation of excess CO_2, which is eliminated by increased ventilation. An increase in ventilation causes reduction in arterial PCO_2 and bicarbonate and provides respiratory compensation for nonrespiratory acidosis [6, 17]. During AT_{n1}-adjusted, long-term exercise, respiratory compensation resulted in the onset of hypocapnia. It was assumed that decrease of arterial PCO_2 speeds lactate uptake from blood [13].

In spite of an unsteady state for heart rate and oxygen consumption, a steady state for pulmonary ventilation was maintained during long-term exercise adjusted to the "short-term" estimated anaerobic threshold [5].

Characteristic dynamics of venous blood lactate were observed [15]. An initial rise in heart rate subsequently decreased but was further maintained at higher than resting values, and represents a balance between the sustained production and the utilization of lactate. The graded prevalence of slow twitch fiber recruitment in the course of prolonged exercise is supposed as background for that dynamic [5].

Metabolic response to the individually adjusted anaerobic threshold, long-term exercise revealed the load-dependent changes of blood plasma Pi, FFA, urea, creatinine, insulin, and HGH content as well as alkaline phosphatase activity. Typical load-dependent pattern of human growth hormone release as well as dissociation in load evoked a response to standardized intensity of exercise and could be utilized in diagnosis of training adequacy, especially during ontogeny, and evaluated from the point of view of triggering or "lack of triggering" (triggering over the individual capacity) of general regulatory and homeostatic processes.

Conclusions

During an incremental, short-term exercise test, curvilinear increase of gas exchange parameters (AT_{n1}) preceded similar characteristics of venous blood lactate accumulation. A close regulatory relationship between respiratory and acid-base variables was confirmed. A well-maintained blood glucose level during the AT_{n1}-adjusted, long-term test was found.

Findings of load-dependent dynamics of inorganic phosphate content and alkaline phosphatase activity during long-term exercise suggested investigation of calcium balance in endurance load.

The importance of individual evaluation of the HGH load-dependent release from the point of view of the adequate volume and intensity of physical activity during ontogeny is reported.

References

1. Bachl N (1981) Möglichkeiten zur Bestimmung individueller Ausdauerleistungsgrenzen anhand spiroergometrischer Parameter. Österreich Sportmed Suppl 1
2. Doll E, Keul J, Maiwald C (1968) Oxygen tension and acid-base equilibria in venous blood of working muscle. Am J Physiol 215: 23–29
3. Gollnick RD, Piehl K, Saltin B (1974) Selective glycogen depletion pattern in human muscle fibers after exercise of varying intensity and at varying pedal rates. J Physiol 241: 45–47
4. Gutmann I, Wahlefeld AW (1974) L-(+)-Lactate. Determination with lactate dehydrogenase and NAD. In: Bergemeyer HU (ed) Methods of enzymatic analysis, second english edn. Academic, New York, pp 1464–1468
5. Havlíčková L, Macková E, Melichnà J, Semiginovský B, Bartůňková S, Bass A, Bunc V, Vránová J (1982) The anaerobic threshold adjusted physical load I. Cardiorespiratory and muscle biopsy characteristics. Physiol Bohemoslov 31: 443–444
6. Hugson LR, Green HJ (1982) Blood acid-base and lactate relationship studied by ramp work test. Med Sci Sports 14: 297–302
7. Jones NL (1980) Hydrogen ion balance during exercise. Clin Sci 59: 85–91
8. Jöbsis FF, Staineby W (1968) Oxidation of NADH during contraction of circulated mammalian skeletal muscle. Respir Physiol 4: 292–300
9. Kindermann W, Simon G, Keul J (1979) The significance of the aerobic-anaerobic transition for the determination of work load intensities during endurance training. Eur J Appl Physiol 42: 25–34
10. Seliger V, Bartůněk Z (ed) (1976) Mean values of various indices of physical fitness in the investigation of Czechoslovak population aged 12–55 years. ÚV ČSTV, Prague
11. Semiginovský B, Bunc V, Havlíčková L, Vránová J, Bartůňková S (1981) Anaerobic threshold – its physiology, estimation, prediction, medical implications. Physiol Bohemoslov 30: 465
12. Semiginovský B, Doležal V, Havlíčková L, Bartůňková S, Vránová J (1982) The anaerobic threshold adjusted physical load III. Human growth hormone and insulin response. Physiol Bohemoslov 31: 468
13. Simon J, Young JL, Gutin B, Blood DK, Case RB (1983) Lactate accumulation relative to the anaerobic and respiratory compensation thresholds. J Appl Physiol 54: 13–17
14. Taylor HL, Buskirk ER, Henschel A (1955) Maximal oxygen intake as an objective measure of cardio-respiratory performance. J Appl Physiol 8: 73–80
15. Vránová J, Semiginovský B, Havlíčková L, Bartůňková S (1982) Anaerobic threshold physical load II. Blood plasma constituents. Physiol Bohemoslov 31: 479
16. Wasserman K, McIlroy MB (1964) Detecting the threshold of anaerobic metabolism. Am J Cardiol 14: 844–852
17. Wasserman K, Whipp BJ, Davis JA (1981) Respiratory physiology of exercise: metabolism, gas exchange and ventilatory control. Int Rev Physiol 23: 149–211
18. Wasserman K, Whipp BJ, Koyal SN, Beawer WL (1973) Anaerobic threshold and respiratory gas exchange during exercise. J Appl Physiol 35: 236–243
19. Weltman A, Katch VL (1979) Relationship between the onset of metabolic acidosis (anaerobic threshold) and maximal oxygen uptake. J Sports Med 19: 135–142
20. Yoshida T, Nagata A, Muro M, Takeuchi N, Suda Y (1981) The validity of anaerobic threshold determination by a Douglas bag method compared with arterial blood lactate concentration. Eur J Appl Physiol 46: 423–430

Factorial Structure of Measures Assessing the Energetic Capacities of Trained Individuals

S. Heimer, R. Medved, and K. Komirović

Introduction

The assessment of the value of aerobic and anaerobic energetic capacity has at present become an inevitable, routine procedure in various medical fields, primarily within the field of sports medicine, occupational medicine, cardiology, and rehabilitation.

To what extent do routinely obtained determinant parameters of the energetic chain give an insight into the real structure and magnitude of energetic capacity?

The first procedure commonly applied for the assessment of the validity of variables deriving from some pattern of measuring processes is the determination of the object of measurement of this pattern of variables.

Purpose of the Study

The purpose of the study was to determine the latent structure of energetic capacity and the participation of different ergometric parameters in the architectonics of isolated factors as a basis for the determination and validity of different measures for the assessment of energetic capacity.

Sample of Subjects

The study covered a sample of 126 3rd-year students of the faculty of physical culture in Zagreb. Their mean age was 21.5 ± 1.5 years, body weight 72.9 ± 7.2 kg, and body weight 177.7 ± 5.7 cm.

Sample of Variables

The parameters chosen for inclusion in the sample were those which can justifiably be characterized as segmentary or complex measures of aerobic and anaerobic capacity. Theoretically, they cover part of the structural and functional magnitudes of the oxygen transport system or express processes and extent of anaerobic metabolism. The following 19 variables were considered in this study:

1. Concentration of hemoglobin in the blood (HEMOG) – g/%
2. Forced vital capacity (FVC) – ml

3. Forced expired volume in a second (FEV_1) – ml
4. Steady state heart rate in the Åstrand test at the loading of 100 W (ASTFSR) l/min
5. Heart rate at rest (PULSMM) – l/min
6. Minute ventilation at rest (MIRMVD) – l/min
7. Oxygen intake at rest ($MIRVO_2$) – l/min
8. Maximum load in the test (MAXW) – W
9. Maximum minute ventilation in aerobic capacity test (MAXMVD) l/min
10. Maximum oxygen intake in aerobic capacity testing ($MAXVO_2$) – l/min
11. Maximum carbon dioxide output in aerobic capacity testing (MAXVCO) – l/min
12. Maximum heart rate in aerobic capacity test (MAXP) – l/min
13. Maximum excess carbon dioxide in aerobic capacity testing (MAXEXC) – l/min
14. Heart rate at the aerobic-anaerobic threshold (PRAGP) – l/min
15. Load intensity at the aerobic-anaerobic threshold (PRAGW) – l/min
16. Maximum minute ventilation volume in anaerobic capacity test (ANTMVD) – l/min
17. Maximum heart rate in anaerobic capacity test (ANTP) – l/min
18. Maximum oxygen intake in anaerobic capacity test ($ANTVO_2$) – l/min
19. Oxygen debt in anaerobic capacity test (O_2DUG) – l

Treatment of Results

The results were analyzed by the algorithm CLIMAX. The CLIMAX is a program for the factor analysis of a pattern of variables. It is based on the relations of the factorial model [1], image model [2], and a model with universal metrics [3]. The algorithm CLIMAX is described elsewhere [4–6]. Details of the statistical procedure have been performed according to pertinent literature [7–11].

Results and Discussion

By using the procedure for the isolation of pseudocanonic factors, four dimensions were isolated: they have 57%, 19%, 17%, and 8%, respectively, of the common variation.

Table 1 shows the pattern of pseudocanonic factors. The first factor, which behaves like the principal object of measurement, explains the highest percentage of the variance and contains, in the system observed, spiroergometric variables at maximum (and submaximum) loading. In view of this, there is no doubt that the general measure of functional capacity defined in this way is represented by energetic capacity, the more so because the markedly dominant variables are maximum oxygen intake and maximum CO_2 output.

The second factor is mostly defined by the measures of the aerobic-anaerobic

Table 1. Pseudocanonic factors

	FAC 1	FAC 2	FAC 3	FAC 4
HEMOG	−0.18	0.26	0.12	0.12
FVC	0.31	0.23	−0.14	0.06
FEV$_1$	0.28	0.23	−0.26	0.08
ASTFSR	−0.35	−0.34	(0.54)	−0.04
PULSMM	−0.12	0.13	(0.47)	0.07
MIRMVD	−0.03	0.21	−0.25	(0.54)
MIRVO$_2$	0.17	0.32	−0.21	(0.49)
MAX W	(0.78)	0.10	−0.18	0.05
MAXMVD	(0.78)	−0.22	−0.03	−0.09
MAXVO$_2$	(0.88)	0.12	0.02	−0.05
MAXVCO	(0.92)	−0.17	0.02	−0.06
MAX P	0.30	−0.18	(0.72)	0.07
MAXEXC	(0.65)	−0.38	0.10	−0.02
PRAG P	0.02	(0.66)	0.53	−0.18
PRAG W	0.40	(0.75)	0.03	−0.16
ANTMVD	0.51	0.05	−0.05	0.27
ANT P	0.09	−0.17	(0.65)	0.41
ANTVO$_2$	0.51	0.29	0.02	0.33
O$_2$ DUG	0.23	0.16	−0.02	(0.51)
λ	16.50	5.39	4.84	2.41
α	0.78	0.46	0.49	0.20

threshold. Of interest is the pattern of the third factor, on which the variables of heart rate are projected in all observed phases (rest, submaximum, maximum, and supramaximum loading).

On to the fourth factor there is a dominant projection of spiroergometric measures observed before and after physical activity (MIRMVD, MIRVO$_2$, and O$_2$DUG). As these measures are in a certain way the expression of the homeostatic mechanism (maintenance of the basal energetic level and postactivity restitution), this factor could be ascribed to homeostatic regulation processes.

The first factor in the promax position (Table 2) is defined by all classical measures of aerobic capacity. For this reason, this latent dimension could probably be ascribed to the functional capacity which depends on the efficacy of the convergent functioning of all the processes affecting energy transport and energy utilization.

The second factor in the promax position practically depends entirely on the level of the aerobic-anaerobic threshold, which expresses the ability of the maintenance of aerobic metabolism at increasing work intensity, i. e., the resistance to increase in lactate concentration.

Also in the promax position there was a special latent dimension which entirely depends on different heart rate measures. It appears that the system for regulation of the cardiovascular function is dominant over other regulation systems to such an extent that all the representatives of the function of the cardiovascular system form only one latent dimension.

The fourth factor in the promax position can be ascribed to the capacity of the maintenance and reestablishment of homeostasis at rest, intensive activity, and re-

Table 2. The pattern of promax factors

	VRX 1	VRX 2	VEX 3	VEX 4
HEMOG	−0.32	0.20	0.09	0.18
FVC	0.18	0.14	−0.12	0.17
FEV$_1$	0.14	0.09	−0.22	0.21
ASTFSR	−0.18	−0.09	(0.50)	0.25
PULSMM	−0.18	0.27	(0.48)	0.01
MIRMVD	−0.33	−0.18	0.03	(0.76)
MIRVO$_2$	−0.14	−0.02	0.03	(0.71)
MAX W	(0.70)	0.06	−0.09	0.13
MAXMVD	(0.88)	−0.10	0.04	−0.14
MAXVO$_2$	(0.83)	0.22	0.02	−0.03
MAXVCO	(0.99)	−0.03	0.08	−0.11
MAX P	0.37	0.13	(0.73)	−0.11
MAXEXC	(0.80)	−0.22	0.22	−0.14
PRAG P	−0.16	(0.88)	0.20	−0.17
PRAG W	0.16	(0.78)	−0.22	−0.02
ANTMVD	0.38	−0.06	0.13	0.35
ANT P	0.03	−0.07	(0.85)	0.31
ANTVO$_2$	0.27	0.15	0.17	(0.46)
O$_2$ DUG	−0.03	−0.09	0.25	(0.65)
α	0.77	0.51	0.51	0.64

Table 3. Correlations of promax factors

	VRX 1	VRX 2	VRX 3	VRX 4
VRX 1	1.00	0.22	−0.14	0.45
VRX 2	0.22	1.00	−0.03	0.37
VRX 3	−0.14	−0.03	1.00	−0.33
VRX 4	0.45	0.37	−0.33	1.00

covery, connected with the efficacy of the oxygen consumption in tissue metabolism.

The association of aerobic capacity with the remaining three observed latent dimensions is different for each of them (Table 3). A relatively low correlation with the anaerobic threshold factor is understandable, because aerobic capacity is only one of a great many suppressors of the increase of the lactic acid concentration, along which a significant role is played by the receptive capacity of the myocardium, liver, kidneys, and skeletal muscles to bind or remove this metabolite from body fluids, and also by the buffer capacity which for its part slows down homeostatic disturbances. Numerous studies have confirmed the necessity to determine the threshold when assessing energetic capacity just because of its relatively high reliability in relation to the magnitude of aerobic capacity.

The inversely proportional association between heart frequency and oxygen intake is the basis for an indirect assessment of maximum oxygen intake. The same relationship is maintained also between the aerobic capacity factor and the heart frequency regulation. A low correlation between both these factors corroborates their high independence, at least within the framework of the observed selected sample, yet calls, of course, for caution if trying to generalize the finding.

The maintenance of homeostasis at rest, during intense activity, and at recovery depends, as has already been mentioned, to a great extent on the efficacy of the supply of tissues with oxygen, that is, on the efficacy of aerobic tissue metabolism. This is indicated also by the highest correlation coefficient between the aerobic capacity factor and homeostasis ($r = 0.45$).

The dimension of the aerobic-anaerobic threshold practically shows no association with heart rate, while the correlation with the homeostasis factor is moderately high. In the above discussion the significance of homeostatic mechanisms in the manifestation of the increase of the concentration of anaerobic metabolites has already been pointed out.

A negative association between heart rate and homeostatic capacity gives grounds to conclude that in trained persons there is a more efficacious activity of corresponding regulation mechanisms of the transport system (a lower heart frequency at a defined energetic level), that is, a higher stability of homeostasis.

References

1. Rao CR (1955) Estimation and tests of significance in factor analysis. Psychometrika 20: 93
2. Guttman L (1953) Image theory for the structure of quantitative variates. Psychometrika 18: 277
3. Harris CW (1962) Some Rao-Guttman relationships. Psychometrika 27: 247
4. Bosnar K, Prot F, Momirović K, Lužar V, Dobrić V (1982) Algoritam za procjenu pseudokanoničkih faktora. Kineziologija 12: 1–2, 29–34
5. Štalec J, Momirović K (1971) Ukupna količina valjane varijance kao osnov kriterija za odredjivanje broja značajnih glavnih komponenata. Kineziologija 2: 78
6. Momirović K, Štalec J (1973) DMEAN i DMAX kriteriji za odredjivanje broja značajnih image faktora pri analizi zadataka u psihologijskim testovima. Stručni skupovi psihologa "Dani Ramira Bujasa", 1970 i 1972. Društvo psihologa Hrvatske, Zagreb, p 95
7. Kaiser HF (1958) The varimax criterion for analytic rotations in factor analysis. Psychometrika 23: 187
8. Hendrickson AE, White PO (1964) PROMAX: a quick method for rotation to oblique simple structure. Br J Statist Psychol 17: 65
9. Kaiser HF, Caffrey J (1965) Alpha factor analysis. Psychometrika 30: 1
10. Thurstone LL (1947) Multiple factor analysis. University of Chicago Press, Chicago
11. Bartlett MS (1937) The statistical conception of mental factors. Br J Psychol 28: 97

R Wave Response and Exercise ECG

A. Tahy*

Introduction

Analysis of R wave amplitude changes that occur during exercise has been proposed for the diagnosis of coronary heart disease in recent years [2, 5, 6–8, 18–22].

A reduction of R wave amplitude during exercise has been defined as a sign of normal coronary arteries and normal left ventricular function, whereas its increase or the absence of R wave changes are interpreted as an abnormal response due to coronary artery disease or left ventricular dysfunction [6–8, 12].

Other studies, however, demonstrated different results [3, 11, 13, 14, 17, 29, 30].

The reasons behind this controversy might be related to many factors, such as the changes in left and right ventricular function, lung volume in respect to electrophysiologic properties, depolarization sequence, and rotation of the heart reflected by QRS axis [4, 9, 10, 16, 23–26, 28].

The purpose of this study is first to investigate the effect of the physical training program on exercise-induced R amplitude changes of patients after an acute myocardial infarction (AMI); secondly, to determine whether a correlation exists between the R wave and RS complex amplitude changes and the later attainable physical working capacity (PWC) after AMI; thirdly, to determine predictive value of R wave amplitude changes for angina pectoris (AP) after AMI; and, finally, to study the correlation between R wave amplitude changes and lung function.

Patients and Methods

The investigations were carried out on five patient groups. The group AMI_1 was composed of 60 males, whose ages ranged from 29 to 59 (mean 46 ± 7.3). The group AMI_2 was composed of 77 males, ranging in age from 32 to 60 years (49.5 ± 7.1). The patients in groups AMI_1 and AMI_2 were hospitalized in our Institute for 4–8 weeks after a documented AMI for a 3- to 4-week training program.

The group $COPD_1$ was composed of 41 patients with mild chronic obstructive lung disease (COPD). There were 35 men and 6 women, aged 47.4 ± 10.4 years. The ratio of residual lung volume to total lung capacity (RV/TLC) in this group was $< 50\%$. The group $COPD_2$ was composed of 63 patients with severe COPD. There were 52 men and 11 women, aged 49.5 ± 8 years. In this group the RV/TLC was

* I thank Edit Korom and Bogdán Lászlóné for assistance in the preparation of this manuscript and Sándor Pátzay and László Rigó for the statistical analysis

> 50%. The control group was composed of 24 persons (21 men and 3 women, aged 46 ± 12 years). All were examined in our Institute.

Subjects with QS pattern in V_5 or V_6 ECG leads were not included in this study.

The examination of the *impact of a training program on the exercise-induced R amplitude changes* and the study of *correlation of R amplitude changes with the later PWC* have been carried out on the patients of the group AMI_1.

The predictive value of exercise-induced R wave changes for AP were studied on patients of the group AMI_2.

The *correlation between lung function and exercise-induced R wave amplitude changes* was investigated on patients of the groups AMI_2, $COPD_1$, and $COPD_2$ and on control subjects.

Exercise Protocol

Bicycle ergometry was used according to a modified version of the WHO protocol [31]. The initial work load was 25 W during 3 min; thereafter, work load was increased every 3 min by 25 W until one of the stop criteria was fulfilled. An electrocardiogram of the chest leads V_2, V_5, and V_6 was monitored continuously and recorded.

The recordings were made before the start of the exercise test with the patient in the sitting position and subsequently every 3 min at the end of each exercise stage and at the end of the stress test. Twelve-lead ECG were recorded before and immediately after the stress testing and in the 1st and 3rd min of recovery. At the admission to the training program the patients in the groups AMI_1 and AMI_2 underwent a preliminary stress test in order to provide guidelines for planning of the program. At the end of the program and at the control investigation, i.e., 6 months after the onset of AMI (3–4 months after the end of the program), the exercise test was stopped when the patient achieved 85% of age-predicted maximal heart rate or at the onset of other criteria (symptoms, signs, blood pressure and ECG changes) [1].

In this study, PWC means the maximal (heart rate-predicted or symptom-limited) work load (in watts) at the termination of the exercise test. The R wave and RS complex amplitudes were measured in lead V_5 or V_6. The same lead was always used in the same patients. R wave amplitude was measured from the isoelectric line to the peak of the R wave in mm. The summation of R amplitude and S amplitude gives the RS amplitude in mm. The average value of at least ten consecutive R waves and RS complexes was used in order to minimize the respiratory variation. The ΔRST was formed from the algebraic summation of the R wave changes in V_5 or V_6 (that is, ΔR) and S-T segment changes (in any of the 12 leads). S-T depression was taken as a positive value. The R wave was taken as positive if there was an increase, and negative if there was a decrease during the stress testing [8]. The ΔRS indicates the amplitude changes of the RS complex during an exercise. The ΔRS was formed from the algebraic summation of the R and S amplitude changes (increase, positive; decrease, negative) (Fig. 1).

The R wave and RS complex amplitude changes and the ΔR, ΔRS, and ΔRST values were measured and calculated at every work load level. In the group AMI_1, the amplitude changes and the ΔR, ΔRS, and ΔRST values of the same patient al-

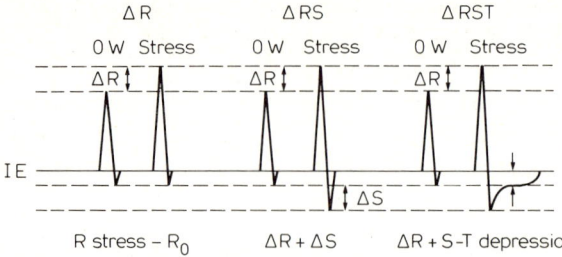

Fig. 1. Measuring of ΔR, ΔRS, and ΔRST. *OW*, at rest; *stress,* during exercise; *R stress-Ro,* the difference between R wave amplitudes during exercise and at rest; IE, isoelectric line

Table 1. Effect of a 4-week training program on the ΔR, ΔRS, and ΔRST in patients after an acute myocardial infarction ($n = 60$)

		ΔR		ΔRS		ΔRST	
		BT	ET	BT	ET	BT	ET
	$\bar{x}$	+1.86	+0.38	+2.60	+1.06	+2.35	+0.75
	$\pm SD$	2.81	2.32	2.81	3.15	2.79	2.64
Differences	$\bar{x}$		−1.48		−1.53		−1.60
	$\pm SD$		3.56		4.17		3.24
	p		<0.005		<0.01		<0.001

BT, before training; *ET,* end of the training

Table 2. ΔRST and ΔRS values at the end of the training program and the physical working capacity 3–4 months later

	ΔRST				ΔRS			
	≤0	+1+2	≥+3	≥+6	≤0	+1+2	≥+3	≥+6
	$n=26$	$n=16$	$n=18$	–	$n=30$	$n=12$	$n=18$	$n=8$
PWC	75.0	73.9	49.9[b]	–	80.9	65.0	46.6[b]	45.0[a]
Watt	±16.2	±34.9	±14.5	–	±28.8	±21.5	±15.3	±16.0

Delta values, mean ± SD; n, number of patients in each group; PWC, physical working capacity
[a] $p<0.05$ vs group ≤0; [b] $p<0.001$ vs group ≤0
ΔRST vs watts, $r = -0.527$, $y = -4.833\,x + 69.125$, $p < 0.01$
ΔRS vs watts, $r = -0.740$, $y = -6.678\,x + 75.789$, $p < 0.01$

ways refer to the same load level attained as maximum load during the preliminary test testing (Tables 1 and 2).

In group AMI_2, the results (that is, the attained stress levels, R amplitude changes, and ventilatory function) refer to the investigation carried out 6 months after the onset of AMI (Tables 3–6). The patients of the group AMI_2 were followed for 1–4 years after AMI (mean 2 years). The study on the predictive value of ΔR values for AP refers to this period.

Table 3. R wave amplitude changes during stress testing in control subjects, in patients with chronic obstructive pulmonary disease, and in patients after an acute myocardial infarction

	n	Age (years)	Max stress (W)	Ro (mm)	R 50 W (mm)	R_{max} (mm)	RV/TLC %
Control	24	46 ±12	102 ± 41.6	12.54 ± 3.6	12.25 ± 3.4 $(-2.3\%)^{+•••}$	11.29^+ ± 3.8 $(-9.9\%)^{+++}$	$31.4^{+}_{+}{}^{+++}$ ± 6.7
$COPD_1$	41	47.4 ±10.4	90.2 ± 40.6	12.51 ± 4.8	12.61 ± 4.4 $(+0.8\%)$	12.19 ± 4.7 $(-2.5\%)^{++}$	41.4^{+++} ±10.8
$COPD_2$	63	49.5 ± 8.0	81.8 ± 34.5	13.4 ± 5.8	12.14 ± 4.9 $(-9.4\%)^{+}_{+}{}^{++}$	$11.37^•$ ± 5.0 $(-15.1\%)^{•••}$	$61.8^{+}_{•••}{}^{++}$ ± 7.6
AMI	77	49.5 ± 7.1	94.8 ±33.2	11.91 ± 4.29	13.36 ± 4.93 $(+12.2\%)^{+}_{•••}{}^{++}$	$13.57^+_•$ ± 4.78 $(+13.9\%)^{+}_{•••}{}^{++}$	$36.9^{•}_{•••}$ ± 9.6

Control, patients with no pulmonary or cardiovascular disease
COPD, patients with chronic obstructive pulmonary disease. In group $COPD_1$, RV/TLC < 50%; in group $COPD_2$, RV/TLC > 50%; RV/TLC, residual volume per total lung capacity
AMI, patients after an acute myocardial infarction
Max stress, highest "submaximal" stress level (see text)
R0, R50W, and R_{max}, R wave amplitudes (R means) in mm in ECG lead V_5 or V_6 at rest (R0), and at the end of 50-W (R50W) and "submaximal" (R_{max}) stress testing, respectively (for explanation see text)
$\Delta R\%50W$ and $\Delta R\%max$ (in parentheses in columns 2 and 3), ΔR in percentage of R0; $\Delta R\%50W = \Delta R50W/R0\ 100$; $\Delta R\%_{max} = \Delta R_{max}/R0\ 100$
Statistical analysis (Student's t test),
differences in vertical direction (R amplitudes and $\Delta R\%$), $+,• = p<0.05$, $++ = p<0.01$, $+++,••• = p<0.001$. Differences in horizontal direction (R amplitudes), control: 1–3 $p<0.05$; $COPD_1$: nonsignificant; $COPD_2$: 1–3 $p<0.05$; AMI: 1–2, 1–3 $p<0.05$

Table 4. R wave amplitude changes during stress testing in patients after an acute myocardial infarction

	R0 (mm)	R50W (mm)	R_{max} (mm)	$\Delta R50W$ (mm)	ΔR_{max} (mm)
AP after AMI ($\bar{x}\pm SD$)	11.9 ±4.31	14.26^a ±5.34	14.75^a ±5.20	$+2.36^b$ ±2.29	$+2.85^b$ ±2.22
NO AP after AMI ($\bar{x}\pm SD$)	11.9 ±4.32	12.37^a ±4.34	12.26^a ±4.34	$+0.47^b$ ±2.47	$+0.36^b$ ±3.5

AP, patients who had angina pectoris after the acute phase of myocardial infarction (AMI)
NO AP, patients who had no angina pectoris after the acute phase of myocardial infarction
R0, R50W, and R_{max}, R wave amplitudes (R means) in mm in ECG leads V_5 or V_6 at rest (R0), and at the end of 50-W (R50W) and "submaximal" (R_{max}) stress testing, respectively (for explanation see text)
ΔR, changes of R wave amplitudes (R means) during stress testing (positive, increase; negative, decrease); $\Delta R50W = R50W - R0$; $\Delta R_{max} = R_{max} - R0$
Statistical analysis (Student's t test): differences in vertical direction, a = $p<0.05$, b = $p<0.01$
Differences in horizontal direction, AP after AMI: 1–2, 1–3 $p<0.05$

Table 5. Sensitivity, predictive value, and specificity of ΔR for angina pectoris after an acute myocardial infarction

	$\Delta R \geqslant 0$ (%)	$\Delta R > 0$ (%)	$\Delta R \geqslant 1.5$ mm (%)
Sensitivity	92.6	82	66
Predictive value	62.3	65	79
Specificity	36	50	81

$\geqslant 0$, > 0, and $\geqslant 1.5$ mm, ΔR levels for predicting angina pectoris, that is, an abnormal response

Table 6. Ventilation-dependent R wave amplitude changes at rest and during stress testing in control subjects, in patients with chronic obstructive pulmonary disease, and in patients after an acute myocardial infarction

	n	Rvd 0 W mm	Rvd 50 W mm	Rvd max mm	RV/TLC
Control	24	$0.87^{+++}_{\bullet\bullet\bullet}$ ± 0.74 $(6.9\%)^{+}$	2.12^{++}_{++} ± 1.26 $(17.3\%)^{+++}_{+}$	3.45 ± 2.20 (30.5%)	31.4^{+++}_{+} ± 6.7
COPD$_1$	41	1.14 ± 0.82 (9.1%)	2.34 ± 1.08 (18.5%)	2.61 ± 1.26 (21.4%)	41.4^{+++} ± 10.8
COPD$_2$	62	1.61^{+++} ± 1.04 $(12.0\%)^{+}$	3.13^{+++} ± 2.17 $(25.8\%)^{+++}$	3.38 ± 2.19 (29.7%)	$61.8^{++}_{\bullet\bullet\bullet}$ ± 7.6
AMI	77	$1.63^{\bullet\bullet\bullet}$ ± 1.03 $(13.7\%)^{\bullet}$	3.0^{++} ± 1.39 $(22.4\%)^{+}$	3.20 ± 1.58 (23.6%)	$36.9^{\bullet\bullet\bullet}_{+}$ ± 9.6

Control, patients with no pulmonary or cardiovascular disease

COPD, patients with chronic obstructive pulmonary disease. In group COPD$_1$, RV/TLC $< 50\%$; in group COPD$_2$, RV/TLC $> 50\%$. RV/TLC, residual volume per total lung capacity

AMI, patients after an acute myocardial infarction

Rvd, ventilation-dependent R wave amplitude changes at rest (Rvd 0 W), at the end of 50-w (Rvd 50 W), and at "submaximal" (Rvd max) stress testing

Rvd% (in parentheses in columns 1, 2, and 3), Rvd in percentage of the R amplitude (R means) at the same stress level. Rvd% $0\,W = \dfrac{\text{Rvd } 0\,W}{R0} \times 100$; Rvd% $50\,W = \dfrac{\text{Rvd } 50\,W}{R50W} \times 100$; Rvd% max $= \dfrac{\text{Rvd max}}{Rmax} \times 100$

Statistical analysis (Student's t test), differences in vertical direction (Rvd and Rvd%, respectively, $+ = p < 0.05$, $++ = p < 0.01$, $+++$, $\bullet\bullet\bullet\, p < 0.001$

Differences in horizontal direction, control (Rvd, Rvd%): 1–2, 1–3, 2–3 $p < 0.001$; COPD$_1$ (Rvd, Rvd%): 1–2, 1–3 $p < 0.001$; COPD$_2$ (Rvd, Rvd%): 1–2, 1–3 $p < 0.001$; AMI (Rvd): 1–2, 1–3 $p < 0.001$, AMI (Rvd%): 1–2, 1–3 $p < 0.05$

The lung function of patients in the groups AMI_2, $COPD_1$, and $COPD_2$ and the control group was investigated at rest with a body plethysmograph. The inspiration and expiration phases were recorded intermittently using a thermosensor inserted in the nostril of the patient. The respiratory changes of the R wave amplitudes were calculated by measuring the variation of the absolute magnitudes of the R waves during ventilation ($R_{max} - R_{min}$ during respiratory phases) at rest and during stress testing (Rvd).

The sensitivity, predictive value, and specificity of ΔR for AP were calculated as follows: sensitivity (%) is all patients who had AP who showed a defined abnormal response ÷ all patients who had AP 100; predictive value is all patients who had AP who showed an abnormal response ÷ total number of patients with a defined abnormal response 100; specificity is true negatives ÷ true negatives + false positives × 100.

Statistical Methods

Analysis of differences was made by Student's *t* test and regression equations were obtained by standard formulas. Probability less than 0.05 was considered significant.

Results

In the course of the training program of the patients in group AMI_1, the ΔR, ΔRS, and RST values significantly decreased ($p < 0.005$, $p < 0.01$, and $p < 0.001$ respectively) (Table 1).

There was a negative correlation between the ΔRST and ΔRS values measured at the end of the training program and the PWC 3–4 months later ($r = -0.527$ and -0.740 respectively) (Table 2).

The group ($\leqslant 0$) attained a higher PWC than the group ($\geqslant 3$) and the group ($\geqslant 6$).

At rest the mean R wave amplitude of the patients after AMI was lower than that of the $COPD_2$ patients and that of the control patients. These differences were not significant (Table 3).

During stress testing the R wave amplitude in control patients and in the group $COPD_2$ significantly decreased and in patients after AMI (group AMI_2), it significantly increased ($p < 0.05$) (Table 3).

There was statistically significant difference in the change of R amplitude (ΔR in mm and $\Delta R\%$ as percent change) between the control group and group $COPD_2$ ($p < 0.05$), and the group of patients after AMI ($p < 0.001$). Also the difference between the $COPD_2$ and AMI groups was significant ($p < 0.01$). According to the findings, there was a correlation between the lung function of control patients, patients with COPD, and the exercise-induced R wave amplitude changes (ΔR 50 W% vs RV/TLC, $r = -0.282$, $n = 127$, $p < 0.01$, $y = 0.311 \cdot x + 12.917$).

Separating the patients in the group AMI_2 on the basis of having typical AP or not during the follow-up period (before or after the stress testing), the figures show that the patients with AP after AMI had significantly greater R wave increase during stress testing than the patients without it (Table 4).

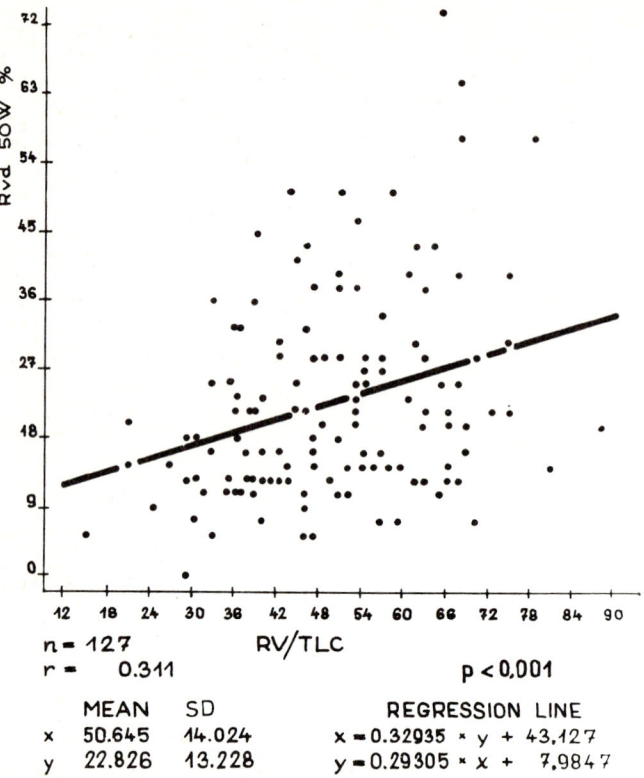

Fig. 2. Correlation between RV/TLC at rest and Rvd% during exercise ECG (*50-W* stress level)

The sensitivity, specificity, and predictive value of this investigation varies according to the ΔR level defined as an abnormal response, that is, according to the different criteria for positivity (Table 5). According to our findings, there were significant differences in ventilation-dependent R wave amplitude changes among the different groups (Table 6).

The $COPD_2$ and AMI groups have significantly higher Rvd and Rvd% (that is, the percent ventilatory change of the R wave amplitude at rest and 50-W stress level) than the control patients ($p < 0.001$). In the group of control patients and patients with COPD, there existed a positive correlation between the "severity" of COPD and Rvd. [In the case of RV/TLC vs Rvd% at 50-W stress level: $r = 0.31$, $n = 127$, $p < 0.001$ (Fig. 2).]

No similar correlation was found in the group of patients after AMI. A typical response of R wave amplitude to exercise testing in a patient with AP after AMI is illustrated in Fig. 3, and in a patient with COPD in Fig. 4.

Discussion

The diagnostic value of exercise-induced R wave amplitude changes is a subject of controversy.

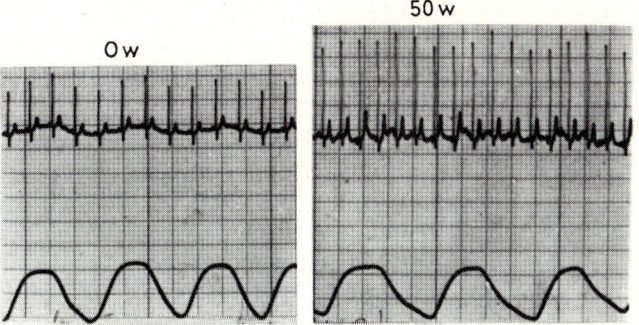

Fig. 3. Response to exercise in a patient after acute myocardial infarction. ECG lead V_5 and curve of ventilation. (Inspiration, upward movement; expiration, downward movement of the curve.) R wave amplitude is increased during stress testing. *0 W,* at rest; *50 W,* during *50-W* exercise)

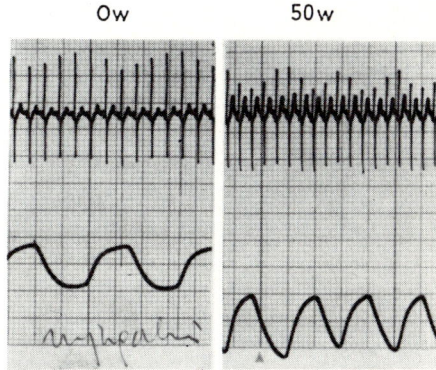

Fig. 4. Response to exercise in a patient with chronic obstructive lung disease (COPD). ECG lead V_5 and curve of ventilation. R wave amplitude is decreased during exercise. Note that during inspiration, R wave amplitude decreases. *0 W,* at rest; *50 W,* during *50-W* exercise

According to some investigators, the QRS complex changes are accurate in detecting coronary artery disease [6–8, 13]; others could not establish any useful correlation between R wave amplitude changes and presence of coronary artery disease or left ventricular dysfunction [3, 11, 13, 29]. Some investigators could prove only a correlation between R wave amplitude changes and left ventricular ejection fraction [18].

Our results may suggest that there is a correlation between R wave amplitude changes and left ventricular function.

Otherwise we could not explain the correlation between QRS changes and the later PWC, or the association of exercise-induced R wave increase with AP after AMI.

The results of our study indicate that the lung function is an important factor in determining the nature of the exercise-induced R wave amplitude changes. Probably much of the confusion in the literature regarding the significance and practical value of R amplitude changes during exercise is due to the fact that the lung plays an important role in exercise-induced R amplitude changes. The changes in R wave amplitude could be due in part to the increase of right ventricular end-diastolic volume. We assume that in COPD during exercise, the increased right ventricular afterload causes an increase in right ventricular volume. Along with this, the left ventricular volume decreases (ventricular interdependence) [15, 27, 30] and, in accordance with the "Brody effect", the left precordial R wave voltage decreases [9].

The significant decrease of R wave in severe COPD patients during exercise could be explained by the changing shape and size of the left ventricle.

The ventilation-dependent R wave amplitude changes during stress testing may indicate the magnitude of ventricular interdependence and probably may have predictive value concerning the later cardiovascular complication of the COPD patients.

References

1. Andersen KL, Shephard RJ, Denolin H, Varnauskas E, Masironi R (1971) Fundamentals of exercise testing. World Health Organization, Geneva
2. Baron DW, Ilsley C, Sheiban I, Poole-Wilson PA, Richards AF (1980) R-wave amplitude during exercise. Relation to left ventricular function and coronary artery disease. Br Heart J 44: 512–517
3. Battler A, Froelicher V, Slutsky R, Ashburn W (1979) Relationship of QRS amplitude changes during exercise to left ventricular function and volumes and the diagnosis of coronary artery disease. Circulation 60: 1004–1013
4. Bayley RH, Kalbfleisch JM, Berry PM (1969) Changes in the body's QRS surface potentials produced by alterations in certain compartments of the nonhomogeneous conducting model. Am Heart J 77: 517–528
5. Berman JL, Wynne J, Cohn PF (1980) Multiple-lead QRS changes with exercise testing. Diagnostic value and haemodynamic implications. Circulation 61: 53–61
6. Bonoris PE, Greenberg P, Castellanet M, Ellestad MH (1977) Predictive value of R-wave amplitude changes in treadmill stress testing. Circulation 56 [suppl III] (abstract)
7. Bonoris PE, Greenberg P, Christison G, Castellanet M, Ellestad M (1978) Evaluation of R-wave amplitude changes versus ST-segment depression in stress testing. Circulation 57: 904–910
8. Bonoris PE, Greenberg P, Castellanet M, Ellestad M (1978) Significance of changes in R-wave amplitude during treadmill stress testing: angiographic correlation. Am J Cardiol 41: 846–851
9. Brody DA (1956) A theoretical analysis of intracavitary blood mass influence on heart-lead relationship. Circ Res 4: 731–739
10. David D, Naito M, Chen CC, Mishelson EL, Morganroth J, Schaffenburg M (1981) R-wave amplitude variations during acute experimental myocardial ischemia: an inadequate index for changes in intracardiac volume. Circulation 63: 1364–1374
11. Deanfield JE, Davis G, Mongiadi F, Savage C, Selvin PA, Fox KM (1983) Factors influencing R-wave amplitude in patients with ischaemic heart disease. Br Heart J 49: 8–14
12. Ellestad MH, Cooke BM, Greenberg PS (1979) Stress testing. Clinical application and predictive capacity. Prog Cardiovasc Dis 21: 431–460
13. Eenige Van MJ, Feyter De PJ, Jong De JP, Roos JP (1982) Diagnostic incapacity of exercise-induced QRS wave amplitude changes to detect coronary artery disease and left ventricular dysfunction. Eur Heart J 3: 9–16
14. Harris FJ, Lee G, Mason DT (1980) Detection of treadmill exercise induced myocardial ischemia in coronary disease with conduction abnormalities at rest: comparison of ST and R-wave change. Clin Res 28: 178A
15. Hirschfeld SS (1982) Ventricular interdependence during exercise in cystic fibrosis. Chest 82: 524–525

16. Ilsley C, Canepa-Anson R, Langley G, Foale R, Poole-Wilson P, Rickards A (1981) R-wave amplitude and left ventricular volume: changes with nitroglycerine and atrial pacing. Cardiology 68 [Suppl 2]: 153–160

17. Iskandrian AS, Hamid Hakki H, Horowitz L, Mintz GS, Anderson GJ, Kane SA, Segal BL (1982) Changes in R-wave during exercise: correlation with left ventricular function and volumes. J Electrocardiol 15: 199–203

18. Janota M, Fabian J, Rohác J, Belan A (1983) The diagnostic value of R-wave amplitude changes during exercise testing. Cor Vasa 25: 168–176

19. Kaku R, Lee G, Lui H (1980) R-wave amplitude analysis during exercise testing reliably detects coronary disease in patients receiving digitalis. Clin Res 28: 185A

20. Kentala E, Luurila O (1975) Response of R-wave amplitude to postural changes and to exercise. Ann Clin Res 7: 258–263

21. Lee V, Lee G, Joye JA (1980) Effect of nitroglycerin on V_5 R-wave amplitude in coronary patients with normal versus elevated left ventricular preload. Clin Res 28: 190A

22. Lee G, Mason DT, Low R (1980) Evaluation of markedly increased R-wave amplitude during spontaneous myocardial ischemic episodes in patients with variant angina. Clin Res 28: 191A

23. Lekven J, Chatterjee K, Tyberg JV, Parmley WW (1979) Reduction in ventricular endocardial and epicardial potentials during acute increments in left ventricular dimensions. Am Heart J 98: 200–207

24. Lerman J, Mele E, Chiozza M, Swetlize H, Perosio AM (1981) Effects of nitrates on R-wave variations after exercise in coronary heart disease. Chest 80: 137–141

25. Roberts DE, Hersh LT, Scher AM (1979) Influence of cardiac fiber orientation on wavefront voltage, conduction velocity, and tissue resistivity in the dog. Circ Res 44: 701–714

26. Simoons ML, Hugenholtz PG (1975) Gradual changes of wave form during and after exercise in normal subjects. Circulation 52: 570–577

27. Stool E, Mullins C, Leshin S, Mitchell J (1974) Dimensional changes of the left ventricle during acute pulmonary arterial hypertension in dogs. Am J Cardiol 33: 868–872

28. Vine DL, Finchum RN, Dodge HT, Bancroft WH, Hurst DC (1971) Comparison of the vectorcardiogram with the electrocardiogram in the prediction of left ventricular size. Circulation 43: 547–558

29. Wagner S, Cohn K, Selzer A (1979) Unreliability of exercise induced R-wave changes as indexes of coronary artery disease. Am J Cardiol 44: 1241–1246

30. Wise RA, Robotham JL, Summer WR (1981) Effects of spontaneous ventilation on the circulation. Lung 159: 175–186

31. World Health Organisation (1973) Evaluation of comprehensive rehabilitation and preventive programmes for patients after acute myocardial infarction. Regional Office for Europe, Copenhagen/Euro 8206 (8)

Significance of Asymptomatic ST-Segment Depression During Exercise in Postinfarction Patients

L. Samek, P. Betz, and H. Roskamm

The existence of myocardial ischemia without angina pectoris is no longer doubted today. The phenomenon has been called asymptomatic or silent myocardial ischemia [3].

On the basis of postmortem examinations, asymptomatic ischemia is to be expected in 6.4% of males in the age group 30–60 years [1].

The reason why ischemic episodes are silent in some patients remains unknown. A reduced sensitivity to pain might play an important role, since it has been shown that patients in whom asymptomatic ischemia occurs have a higher threshold for external pain [2].

In an earlier study [4], we could demonstrate in 146 postinfarction patients with inferior myocardial infarction that the presence of ST-segment depression and angina pectoris was correlated with multivessel disease. The positive predictive value of these two parameters was 71%. A negative test, i. e., neither ST-segment depression nor angina pectoris after reaching the target heart rate, also correlated well with the absence of multivessel disease, the negative predictive value being 90%. ST-segment depression alone, however, showed only a poor correlation with multivessel disease, the positive predictive value being 41%.

Our present study was aimed at answering the following two questions:

1. Is it possible to improve the positive predictive value of ST-segment depression for three-vessel disease in patients who do not experience angina pectoris?
2. Is the prognosis of asymptomatic patients with ST-segment depression different from the prognosis of symptomatic patients?

Patients and Methods

From the total group of patients who had undergone coronary angiography between 1975 and 1979, we selected 102 consecutive patients who fulfilled the following criteria:

1. Proven transmural myocardial infarction (at least two of the following criteria had to be fulfilled: ischemic chest pain lasting for more than 20 min, typical rise and fall of enzymes, or evolving Q wave abnormalities)
2. ST-segment depression of 0.1 mV or more during an exercise test without subjective angina symptoms

Patients on digitalis were excluded from the study as well as patients with ST-segment depression at rest, left ventricular hypertrophy, bundle branch block, or any type of heart disease other than coronary artery disease. The clinical characteristics of 30 patients later treated surgically and 72 patients treated medically (follow-up group) are presented in Table 1. The follow-up group was correlated with 325 post-infarction patients with ST-segment depression and angina during the exercise test (Table 1).

Table 1. Clinical characteristics

	Partially asymptomatic			Symptomatic ST-depression and angina
	ST-depression medically and surgically treated	ST-depression surgically treated	ST-depression medically treated	
Patients (n)	102	30	72	325
Age (years, $\bar{x}\pm$SE)	50.2 ± 0.6	49.9 ± 1.2	50.4 ± 0.7	51.7 ± 0.3
Male/female	99/3	30/0	69/3	325/0
Height (cm, $\bar{x}\pm$SE)	170.7 ± 0.6	170.2 ± 1.2	170.9 ± 0.8	170.5 ± 0.3
Weight (kg, $\bar{x}\pm$SE)	72.5 ± 0.8	71.2 ± 1.6	73.1 ± 0.9	72.9 ± 0.3
Site of myocardial infarction				
anterior	22%		24%	23%
inferior	74%		75%	71%
anterior and inferior	4%		1%	6%
1-Vessel disease	25%	0	36%	22%
2-Vessel disease	32%	27%	33%	30%
3-Vessel disease	43%	73%	31%	48%
Coronary score (Kaltenbach)			30.6 ± 0.5	34.1 ± 1.1
Ventricular function[a]				
0	8%		8%	8%
1	68%		68%	56%
2	21%		21%	32%
3	3%		4%	4%
Heart volume[b] (ml, $\bar{x}\pm$SE)	823 ± 13	825 ± 21	822 ± 17	838 ± 7

[a] Ventricular function: 0, no or one mildly hypokinetic segment; 1, two or fewer hypo- or akinetic segments; 2, three or more hypo- or akinetic segments; 3, nearly all segments of the left ventricle with severe hypo- or akinesia

[b] Heart volume measured radiologically from two projections in the supine position

The exercise test was performed on an electrically braked bicycle ergometer in the supine position; 12 leads were recorded before and immediately after exercise; V_1–V_6 were recorded at the end of each minute during exercise. Blood pressure was measured indirectly. Exercise was started at 25–50 W and increased stepwise by 25–50 W every 6 min. ST-segment response was considered ischemic if in the absence of a preexisting ST-segment abnormality, a horizontal or downsloping depression occurred and reached 0.1 mV or more 60–80 ms after the J point, as compared with the PR-segment.

Coronary angiography and left ventriculography were performed using Sones' or Judkin's technique. Multiple projections, including caudocranial views, were per-

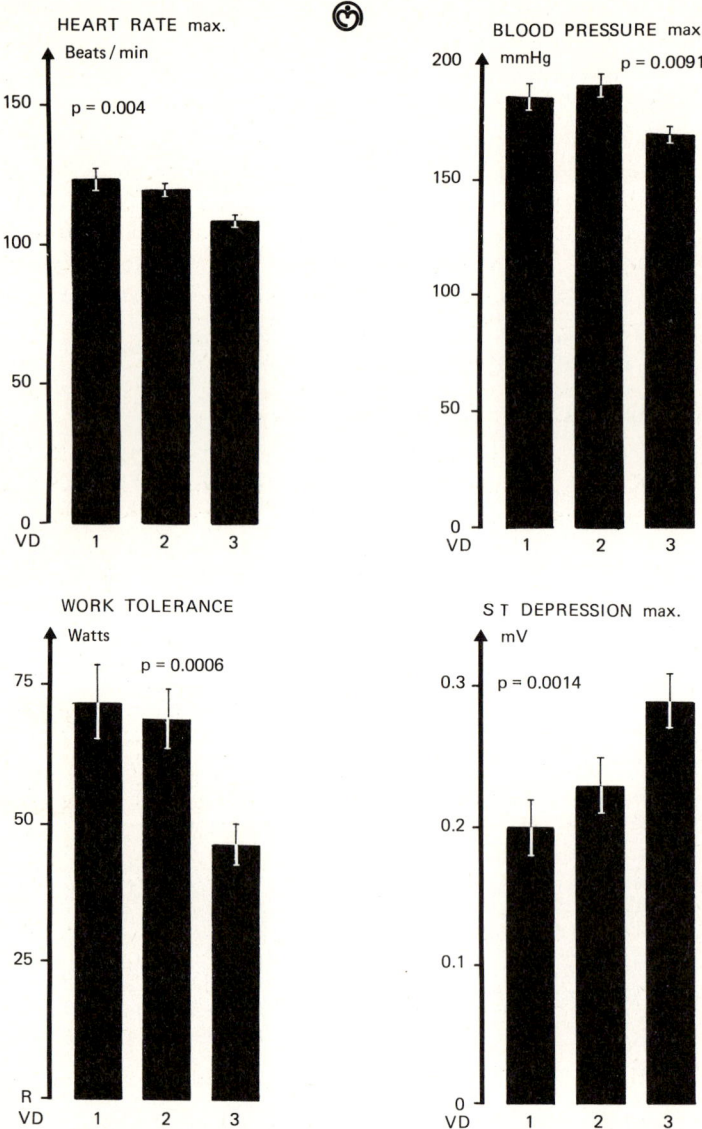

Fig. 1. Results of exercise testing in asymptomatic postinfarction patients with one-, two-, and three-vessel disease *(VD)*. ($n = 102$; $\bar{x} \pm SE$)

formed to ensure adequate visualization of all vessels. A more-than-50% narrowing of the luminal diameters of one of the three main vessels, as measured with a caliper, was considered a significant lesion.

Statistical differences between the groups were verified with the analysis of variance. For multivariate analysis, the Cox model was used.

Results

Predictive Value

Figure 1 and Table 2 show for the variables maximal heart rate, blood pressure, work tolerance, and maximal ST-segment depression only very slight differences between patients with one-vessel disease and two-vessel disease, while the group with three-vessel disease shows marked differences. Therefore, for further statistical analysis, we formed a joint group of the patients with one- and two-vessel disease. This joint group was discriminated from the group of patients with three-vessel disease by stepwise logistic regression according to the Cox model.

The univariate analysis revealed significant differences between both groups as far as the variables maximal work tolerance, maximal heart rate, maximal systolic blood pressure, maximal ST-segment depression, and the product of maximal heart rate and maximal blood pressure are concerned. In multivariate analysis with the Cox model, the maximal ST-segment depression and product of maximal heart rate and maximal blood pressure turned out to be the two variables best discriminating between both groups.

Table 2. Exercise test variables in asymptomatic patients with ST-segment depression of ≥ 0.1 mV in groups with one-, two-, and three-vessel disease

	1-Vessel disease	2-Vessel disease	3-Vessel disease	p value
Patient *(n)*	23	31	48	
Work tolerance (W)[a]	71.2 ± 6.9	68.4 ±5.7	46.1 ±3.8	0.000
Heart rate, max (beats/min)[a]	124.1 ± 4.0	120.2 ±3.2	109.3 ±2.8	0.004
Systolic blood pressure, max (mmHg)[a]	185.9 ± 6.5	190.7 ±5.5	170.2 ±4.1	0.009
ST-depression (mV)[a]	0.2 ± 0.02	0.23±0.02	0.29±0.02	0.010
Σ ST-depression (mV)[a]	0.67± 0.08	0.72±0.06	0.98±0.08	0.006
SBP × HR × 10^{-2}[a]	230.7 ±11.3	231.0 ±9.3	188.6 ±7.4	0.000

[a] $\bar{x}$±SE

SBP, systolic blood pressure; *HR*, heart rate; Σ *ST*, sum of all ST-depressions

Table 3. Stepwise logistic regression (Cox) discriminating one- and two-vessel disease from three-vessel disease

	Univariate		Multivariate	
	χ^2	p value	χ^2	p value
Age	0.28	0.5991		
Watt max.	14.05	0.0002		
HR max.	10.09	0.0015		
Syst. BP max.	10.14	0.0014		
ST $\downarrow$ max.	12.45	0.0004	7.58	0.0059
Σ ST $\downarrow$ max.	8.93	0.0028		
HR × SBP	17.28	0.0000	12.42	0.0004

102 postinfarction patients, asymptomatic ST-depression; *HR*, heart rate; *BP*, blood pressure; Σ *ST* $\downarrow$, sum of all ST-depressions; *SBP*, systolic blood pressure

The probability of the absence of a three-vessel disease was determined by the following formula:

$$\frac{1}{1+e^{-1.9+0.17\times D-5.8+ST}}$$

where D is the double product of heart rate and systolic blood pressure and ST is the maximal ST-segment depression in mV.

These two independent variables correctly classified a total of 79% of the patients as belonging to either group. An at random classification was only correct for 52% (Table 3).

Figure 2 illustrates how, with the extent of ST-segment depression alone, the number of patients with three-vessel disease increases. If ST-segment depression is between 0.10 and 0.19 mV, only 29% of the patients have three-vessel disease; if ST-segment depression is 0.4 mV or more, 78% of the patients have three-vessel disease, but one must keep in mind that this last group is only a small fraction of the whole group.

Prognosis

To find out whether the prognosis of patients with asymptomatic ST-segment depression differs from that of symptomatic patients, we compared 72 medically treated, asymptomatic postinfarction patients (30 patients had undergone aortocoronary bypass surgery before) with 325 postinfarction patients who had both ST-segment depression and angina during the exercise test.

The two groups were comparable with respect to age and left ventricular function but differed with respect to vessel involvement ($p < 0.01$, Table 1). The frequency of

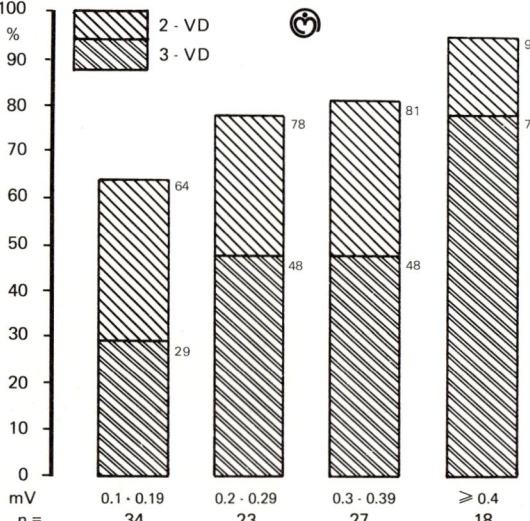

Fig. 2. Relation between the degree of ST-segment depression and multivessel disease *(VD)* in asymptomatic postinfarction patients

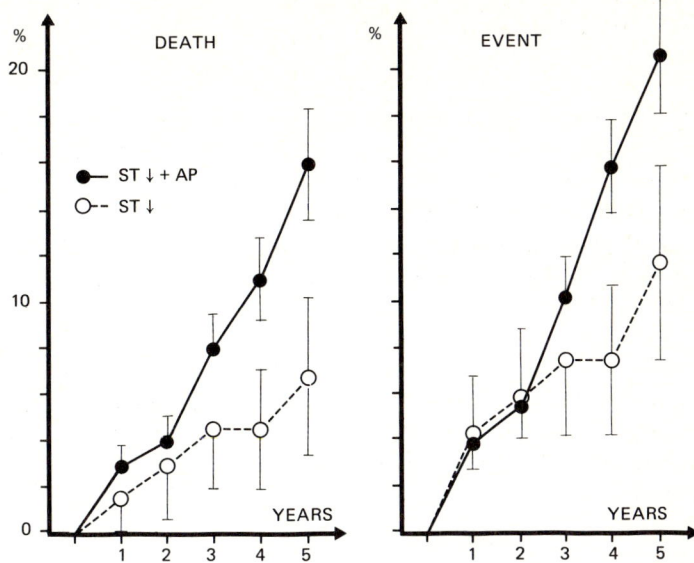

Fig. 3. Cumulative death and event rates (death and reinfarction) in 72 asymptomatic (○) and 325 symptomatic (●) postinfarction patients. *ST↓*, ST-depression; *AP*, angina pectoris

three-vessel disease was 31% in the asymptomatic patients and 48% in the symptomatic patients. Figure 3 shows the 5-year cumulative death rate for both groups. In the first 3 years, there is only a slight difference between the curves; the difference becomes increasingly significant after the 3rd year with a death rate of 7% (1.2%/year) for the asymptomatic patients and 16% (3.5%/year) for the symptomatic patients. Figure 3 also shows cardiac events like nonfatal reinfarctions, which were added to the cumulative death rates. The tendency is the same as for cumulative death rates; the difference after 5 years is 12% (2.0%/year) versus 21% (3.9%/year).

Perhaps the differences between the two groups would be more obvious also for the earlier postinfarction years if the highly symptomatic patients had not been treated surgically.

Conclusions

1. In asymptomatic patients after myocardial infarction, an exercise test with a low double product of heart rate and blood pressure and significant ST-segment depression is suggestive of three-vessel disease. In these patients coronary angiography is indicated.
2. The long-term prognosis of asymptomatic patients with significant ST-segment depression is – with respect to death and reinfarction – significantly better than in patients with both ST-segment depression and angina during exercise.

142   L. Samek et al.

References

1. Diamond GA, Forrester JS (1979) Analysis of probability as an aid in the clinical diagnosis of coronary artery disease. N Engl J Med 300: 1350–1358
2. Droste C, Roskamm H (1983) Experimental pain measurement in patients with asymptomatic myocardial ischemia. J Am Coll Cardiol 1: 940–945
3. Lindsey HE Jr, Cohn PF (1978) "Silent" myocardial ischemia during and after exercise testing in patients with coronary artery disease. Am Heart J 95: 441–447
4. Samek L, Roskamm H, Rentrop P, Kaiser P, Stürzenhofecker P, Schober B, Görnandt L, Velden R (1975) Belastungsprüfungen und Koronarangiogramm im chronischen Infarktstadium. Z Kardiol 64: 809–814

Significance of Ambient Air Temperature and Humidity in Tropical Ergometry: The Critical Temperature

O. Ketusinh, C. Chintanaseri, and S. Prasertsiripandh*

Introduction

As is well known, physical exertion is strongly influenced by the environment, especially in the tropics. Air temperature and humidity have been found to chiefly affect the recovery period [1], but the work capacity is also affected [2]. Further it has been shown that measures which favor heat dissipation favorably influence work [3, 5], while prewarming of the body exerts an adverse effect [4]. To evaluate the influence of heat and humidity on ergometric work in more detail, the following investigations were carried out.

Methods and Results

To provide the required environment, we made use of a climatic chamber in which the air temperature could be adjusted between 19 °C and 50 °C, and the relative humidity adjusted between 45% and 90%. All subjects were male and untrained, and three groups were employed: (a) youths, that is, school children aged between 13 and 16 years; (b) young adults, comprising technicians, students, and a clerk, aged 20 to 28 years; (c) old adults, made up of a teacher, a janitor, and a doctor, aged 55 to 70 years. All subjects wore wide-legged shorts and wide, open singlets. Before each experiment the subject was acclimatized to the particular atmosphere in the climate chamber for 1 h or longer. $\dot{V}o_{2max}$ was determined by Åstrand's method with the Monark bicycle [6].

Three experiments were performed: (a) the influence of rising temperature with humidity unchanged; (b) the influence of changing humidity while temperature was unchanged; and (c) the influence of age on the reaction to temperature changes. The first two experiments involved five young adults, while the third employed six youths and three old adults.

Experiment 1

The first experiment was performed to determine the influence of rising temperature with the humidity unchanged.

The young men worked on the ergometer in the climate chamber with the hu-

---

* We wish to record our thanks to Mr. Santana Montsa for technological assistance

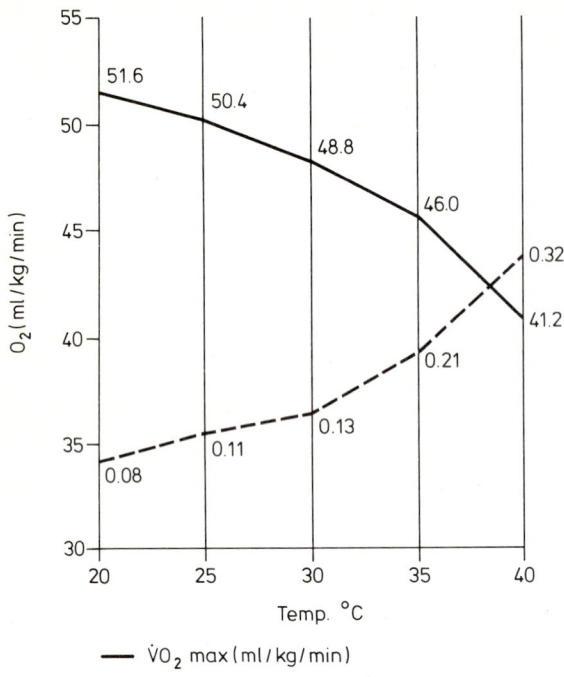

Fig. 1. Constant humidity, rising temperature

—— $\dot{V}O_2$ max (ml/kg/min)

-- - Weight loss (kg)

Table 1. Changes in $\dot{V}O_{2max}$ with temperature changes, humidity constant

Temperature °C	$\dot{V}O_{2max}$ (ml kg^{-1} min^{-1})
20	51.6
25	50.4
30	48.8
35	46.0
40	41.2

Relative humidity 55%. Means of 5 subjects

midity kept constant at 55% and air temperature adjusted to 20 °C. The $\dot{V}O_{2max}$ was determined in the usual way. After an interval of not less than 24 h, the experiment was repeated with the air temperature adjusted to 25 °C, then after 24 h, 30 °C, and so on. The highest temperature tried was 40 °C. Results are shown in Table 1 and Fig. 1.

As seen in Fig. 1, as the air temperature rose from 20 °C to 30 °C the $\dot{V}O_{2max}$ decreased gradually, the mean decrease being 0.32 ml per degree Celsius. Between 30 °C and 35 °C there was a remarkable dip in the fall, which became even more pronounced between 35 °C and 40 °C. The mean decrease in $\dot{V}O_{2max}$ per degree between 30 °C and 40 °C was 0.72 ml. Incidentally the loss in body weight, presumably sweat loss, showed corresponding changes but in the opposite direction.

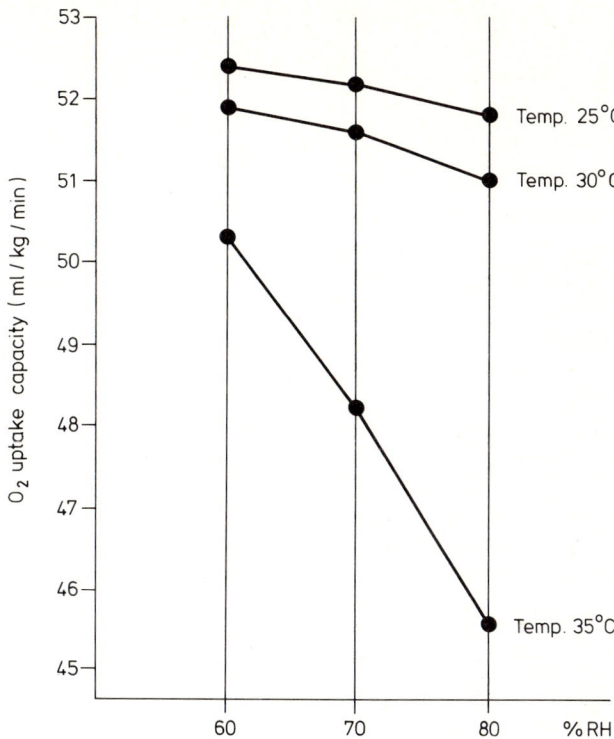

Fig. 2. Constant temperature, rising humidity

Table 2. Changes in $\dot{V}O_{2\,max}$ with relative humidity changes, temperature constant

Temperature 25 °C	RH %	$\dot{V}O_{2\,max}$ $(ml\,kg^{-1}\,min^{-1})$
	60	52.5
	70	52.5
	80	52.0
Temperature 30 °C		
	60	51.9
	70	51.7
	80	51.2
Temperature 35 °C		
	60	50.5
	70	48.1
	80	45.3

Means of 5 subjects; *RH*, relative humidity

Experiment 2

The second experiment was performed to determine the influence of changing humidity with the temperature unchanged.

The same subjects as in experiment 1 were employed. They performed on the Monark bicycle in the climate chamber, at first with the air temperature kept constant at 25 °C while the humidity was varied from 60% to 80% in steps of 10%, with 24-h intervals. The experiments were repeated with the temperature adjusted to 30 °C, then for the third time with the temperature at 35 °C. Results are summarized in Table 2 and shown graphically in Fig. 2.

A scrutiny of Fig. 2 brings out quite distinctly the significance of an environmental temperature of 35 °C. In the first two parts of the experiment, with the temperature at 25 °C and 30 °C respectively, there was very little change in the $\dot{V}_{O_2 max}$ when the relative humidity was increased from 60% to 80%. The mean fall of $\dot{V}_{O_2 max}$ per % rise in humidity was only 0.01 ml kg^{-1}min^{-1} for the temperature 25 °C and 0.023 ml kg^{-1}min^{-1} for 30 °C. But when 35 °C was reached, the $\dot{V}_{O_2 max}$ showed a sudden great fall, the mean decrease being 0.17 ml kg^{-1}min^{-1} for 1% rise in relative humidity.

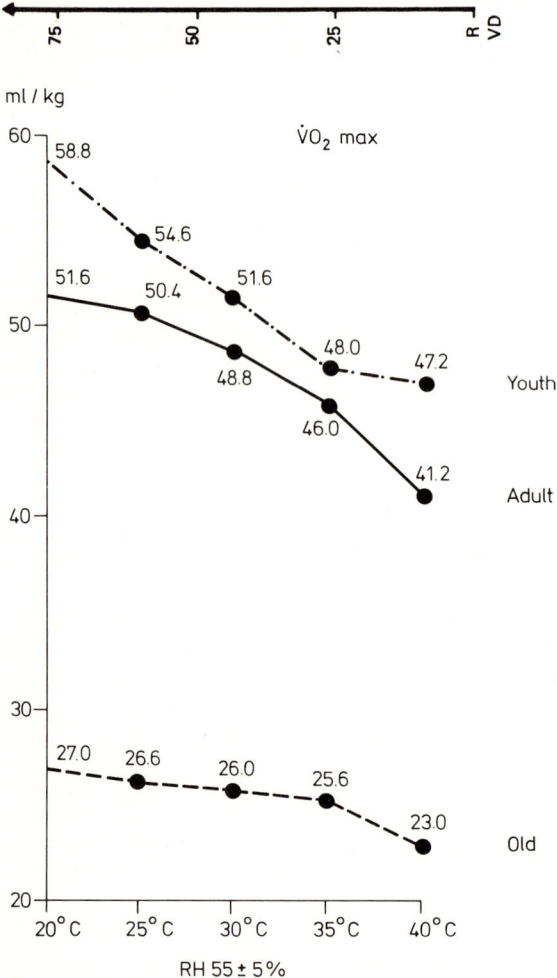

Fig. 3. Influence of temperature changes with RH constant

Experiment 3

The third experiment was performed to determine the influence of age on the reaction to temperature change.

Six youths, aged 13 to 16, served as subjects for procedures similar to experiment 1.

Three old adults, aged 55 to 70 years, worked on the Monark bicycle under surrounding conditions similar to those in experiment 1.

Results are shown in Tables 3 and 4 and Fig. 3, with results of young adults (from experiment 1) for comparison.

Referring to Fig. 3 it can easily seen that all ages are affected by the environmental temperature, but the reactions are not quite the same. The children show a more rapid and steady decline, without the "dip" at 35 °C strikingly noticeable in young adults and – less markedly – in old adults. The latter are marked by an "inactive" type of response: the slant in the line of changes is very slight and the dip is hardly noticeable. But on the whole, referring to Table 4, all ages are almost equally influenced. The decrements between 20 °C and 40 °C, calculated in terms of the starting values (20 °C), for old adults, young adults, and youths were 15%, 20%, and 19% respectively.

Table 3. Age and reaction to temperature rise

Temperature °C	Youths	Old adults	Young adults
20	58.8	27.0	51.6
25	54.0	26.6	50.4
30	51.6	26.0	48.8
35	48.0	25.6	46.0
40	47.2	23.0	41.2

Means of 6 youths, 3 old adults, 5 young adults
$\dot{V}_{O_2 max}$ (ml kg^{-1} min^{-1})
Relative humidity 55%

Table 4. Decrements in $\dot{V}_{O_2 max}$ with temperature changes in subjects of different ages

	Old adults	Young adults	Youths
Decrement between			
20 °C and 35 °C	0.09	0.37	0.72
35 °C and 40 °C	0.52	0.96	0.16
Total decrease in terms of starting value	15%	20%	19%

$\dot{V}_{O_2 max}$ (ml kg^{-1} min^{-1}) per degree Celsius temperature rise
Relative humidity 55%

Table 5. Influences of air temperature and humidity on $\dot{V}O_{2\,max}$ ($ml\,kg^{-1}min^{-1}$)

Temperature °C	Relative humidity (RH)			Difference between RH 60% and RH 80%
	60%	70%	80%	
25	52.5	52.5	52.0	0.5
30	51.9	51.7	51.2	0.7
35	50.5	48.1	45.9	4.6
Difference between 25 °C and 35 °C	2.0	4.4	6.1	

Discussion

The results of the experiments point to the particular significance of thermoregulation when performing ergometry in a tropical country.

Air temperature as well as humidity negatively affect the $\dot{V}O_{2\,max}$. The influences are best seen by scrutinizing Table 5, which is actually a rearrangement of Table 2.

Reading the table vertically, the influence of air temperature becomes evident. At any level of humidity, temperature rise always lowered the $\dot{V}O_{2\,max}$, the more the higher the humidity. Thus in RH 60% the difference between the $\dot{V}O_{2\,max}$ at 25 °C and 35 °C was 2.0, whereas in RH 80% it was 6.1.

Reading the table horizontally the influence of air humidity may be scanned. At temperatures of 25 °C and 30 °C, increases of RH were accompanied by only small changes in $\dot{V}O_{2\,max}$. But at 35 °C the effects became much more significant. Thus at 30 °C the difference in $\dot{V}O_{2\,max}$ between RH 60% and 80% was 0.7 $ml\,kg^{-1}min^{-1}$, while at 35 °C the difference was 4.6 $ml\,kg^{-1}min^{-1}$.

Age differences (between 13 and 70 years) do not seem to exert a very important influence on the reaction of the body to temperature changes during work. The youthful subjects showed a rapid fall in $\dot{V}O_{2\,max}$ as the temperature rose, while older subjects exhibited only a sluggish fall. The difference might be due to dissimilarity in the metabolic rate. But the end results were about the same: all groups – youthful, young, and old – had the $\dot{V}O_{2\,max}$ reduced by 15%–20% when the air temperature increased from 20 °C to 40 °C.

The influences of air temperature and humidity are interdependent, with humidity being less significant. When the temperature was low (30 °C and less), humidity had very slight effect. Only when the temperature was raised to 35 °C did changes in humidity bring about a marked reduction in the $\dot{V}O_{2\,max}$. The effects of air temperature, on its part, was also influenced by humidity. When this became higher, the $\dot{V}O_{2\,max}$ in any temperature became lower. The most marked decrease in $\dot{V}O_{2\,max}$ was seen when the temperature was maintained at 35 °C and the humidity raised to 80% (Table 5).

Going through our experimental results, one is struck by the particular significance of the temperature 35 °C. It seems to be a critical level at and above which the depressing effects of ambient temperature becomes markedly aggravated. Its role may be explained in the light of thermoregulation. When the air temperature reaches 35 °C, it exceeds the mean skin temperature [7] and the three channels of heat dissipation, which depend upon a thermogradient between the skin and ambient air, are switched off, leaving only evaporation in maintaining the heat balance. This, by the way, also explains the greater depression in higher humidities, which obstruct evaporation. The decline in $\dot{V}_{O_2 max}$ is probably the result of heat accumulation or "heat debt" [3], which interferes with the physiochemical processes of muscular contraction.

Conclusions

From our study of the effects on $\dot{V}_{O_2 max}$ of stepwise raising of the air temperature and air humidity, the following conclusions may be made:

1. Both temperature and humidity of the ambient air exert a deleterious influence on the $\dot{V}_{O_2 max}$; the two factors are interdependent, but humidity is of secondary importance, its effects depending in part on temperature level.
2. Youths, young adults, and old adults reacted to temperature rise in the same way. However, youths appear to be more sensitive, while old adults are less so.
3. Working in an environment with rising temperature, the $\dot{V}_{O_2 max}$ became much lower when 35 °C was reached or exceeded. The effects of rising humidity were also unusually augmented in an atmosphere at 35 °C.
4. In view of its special significance, the temperature 35 °C should be regarded as a critical temperature in connection with ergometric work: when this temperature is reached or exceeded the influences of air temperature and humidity will be augmented out of proportion to what has been at lower temperatures. Further, it is suggested that in any work involving ergometry the ambient temperature should be recorded.

References

1. Ketusinh O (1970) Pulsveränderungen, Blutdruckveränderungen und Gewichtsverlust beim Arbeiten in warmen Umgebungen. Arbeitsmed Sozialmed Arbeitshygiene 6: 80–81
2. Ketusinh O (1972) Heat and humidity as limiting factors of physical exertion. Proc Sci Conf Munich Olympics, Munich
3. Ketusinh O (1972) Tropical ergometry. In: Hansen G, Mellerowicz H (eds) III International Seminar on Ergometry. Ergon-Verlag, Berlin, pp 34–40
4. Charoenrajata T (1974) Influence of body temperature on work capacity. Dissertation, Faculty of Education, Chulalongkorn University, Bangkok
5. Udomsatya S (1981) Influence of air movement on work. Dissertation, Faculty of Education, Chulalongkorn University, Bangkok
6. Astrand PO (1956) Human physical fitness with special reference to sex and age. Physiol Rev 36: 307–312
7. Lehmann G (1962) Praktische Arbeitsphysiologie. Thieme, Stuttgart

Dependence of W_{170} Results on Ambient Temperature (16 °C–24 °C)

H. Reißmann and H.-V. Ulmer

Introduction

The shortened W_{170} test [7] is a modified W_{170} test procedure saving considerable time by using a rapid and almost continuous load increase of 5 W/10 s without having significant influence on the test criteria [1–3]. This time-saving method has special significance for large-scale testing in industrial medicine, for example, in the investigation of working capacity in accordance with the German recommendations "G 26, G 30, and G 31" of the Industrial Injuries Insurance Companies [4].

In particular, field experiments outside of laboratories cannot always guarantee a constant ambient temperature. The International Council of Sport and Physical Education [6] recommends as standard conditions for ergometry room temperature between +18 °C and +22 °C if possible, and it should not fall below 16 °C or exceed 24 °C. Contradictions in the literature suggest that this range of temperature may leave the margin too wide and could therefore influence shortened W_{170} test results. Concerning this question, Mellerowicz [5] suggests that more exact quantitative examinations of the influence of ambient temperature on the work load/heart rate relationship should take place.

Methods

Twenty-one subjects between the age of 18 and 27 were tested. The group comprised five female and male sportsmen each and six female and five male non-sportsmen. Each subject wore lightweight sportswear.

A Dynavit bicycle ergometer with a working range from 0 to 400 W was employed. It works with an eddy current brake and independently of the rotational speed. The ergometer system employs a unit for programming the work load. In our case a special control unit for the shortened W_{170} test was used, which increased the work load automatically by 5 W/10 s (corresponding to 30 W/min). The pedal rate was 75 rpm. The heart rate was measured by means of photoelectrical earclip linked to the ergometer system.

The tests were performed in a climatic room of the Physiological Institute of the Johannes Gutenberg University, Mainz.[1] At the beginning and at the end of the tests the actual dry temperature was measured near the ergometer using a gauged mercu-

---

1 We thank Prof. Dr. Thews very much for giving the permission to use the climatic room in the Physiological Institute of the Mainz University

ry thermometer, the bulb of which had been wrapped in a radiation-reflecting foil. Simultaneously the relative humidity was determined by means of a so-called Hygronom (manufactured by Weiss, Gießen).

Each subject pedaled once at the five temperatures (16 °C, 18 °C, 20 °C, 22 °C, and 24 °C) in a random order. In addition everyone took part in at least one repeated experiment. For about 15 min preceding the test, each subject adapted himself to the existing temperature conditions.

The test procedure is shown in Fig. 1.

Results

Table 1 gives a survey of the climatic conditions during the ergometer experiments. The actual mean dry temperatures only deviate little from the required figures; the standard deviations do not exceed 0.1 °C. The relative humidity depended upon the ambient temperature and was, according to the standard conditions of the ICSSPE, within the limits of 30%–60%. The effective temperatures were calculated on the basis of the mean dry temperatures and the mean relative humidities. There was no measurable wind velocity near the ergometer.

Table 1. Climatic conditions during the ergometer experiments

Required dry temperature (°C)	Actual dry temperature (°C)	Relative humidity (%)	Effective temperature (BET) (°C)	Number of experiments n
16	16.3 ± 0.1	45 ± 4	11.8	27
18	18.1 ± 0.1	45 ± 3	12.9	25
20	20.1 ± 0.1	42 ± 3	14.4	27
22	22.1 ± 0.1	39 ± 2	15.5	28
24	24.1 ± 0.1	36 ± 2	16.8	25

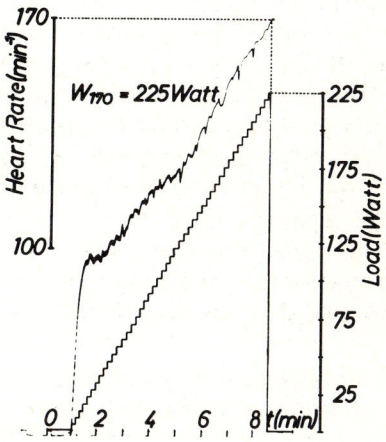

Fig. 1. Load procedure while performing the shortened W_{170} test with the special program control unit of the Dynavit ergometer system. After 1 min of warming up without any load, the program commenced. Beginning at 0 W, the load increased 5 W every 10 s. The corresponding increase of heart rate is shown. At the heart rate of 170 min^{-1}, the equivalent load was recorded, in this example, 225 W. Finally, the subjects had to pedal a further minute without any load

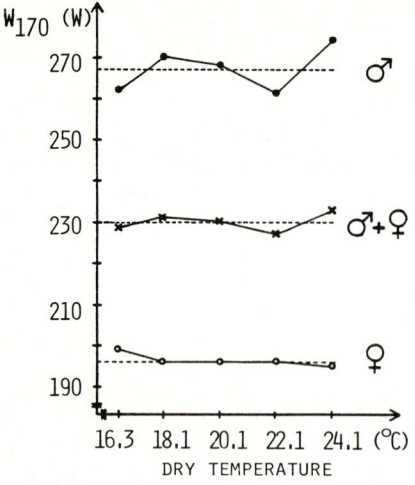

DRY TEMPERATURE

Fig. 2. Absolute W_{170} test results of the men *(top)*, women *(bottom)*, and the total group *(middle)* plotted against the 5 dry temperatures. Each mean is indicated by *dotted lines*

Table 2 see p. 153

Table 3. Comparison of the W_{170} test results (means in W) at 16.3 °C (M_1) and 24.1 °C (M_2)

		Men	Women	Total group
	n	10	11	21
$M_{16.3\,°C}$	(w)	262	199	229
$M_{24.1\,°C}$	(w)	274	195	233
$M_2 - M_1$	(w)	12	− 4	4
$M_2 - M_1$	(%)	4.5	− 2.0	1.7
	t	2.71	− 0.65	1.11
	$2p$	< 0.05	> 0.05	> 0.05

The duration of the shortened W_{170} test was 9 min 51 s on average. This included 1 min of pedaling without any load, both at the beginning and at the end of the test.

The absolute and relative W_{170} test results are shown in Table 2. The means of the total group do not indicate any dependence on ambient temperature. There is a difference of only 3% between the lowest value of 227 W and the highest of 233 W. Similar results are shown in Fig. 2. The men's average is 267 W, equal to 3.8 W/kg. The women's mean is 196 W, equal to 3.2 W/kg. Within these two subgroups the deviations from the means are small, the largest deviation being 7 W (equal to 3%) at 24.1 °C for the men. For the women it is only 3 W (equal to 2%) at 16.3 °C. When the temperatures of 18.1 °C, 20.1 °C, and 22.1 °C are considered, the mean test results for the women are the same. For reasons of clarity no standard deviations are shown in this figure, but they are substantial: 36–50 W for the men and 45–49 w for the women. A Student's t test was applied to prove a possible tendency of the W_{170} values due to the ambient temperature (Table 3). The differences of the means between the experiments at 16.3 °C and 24.1 °C are between 1.7% (total group) and 4.5% (men). For the men this Student's t test reveals a significantly increasing tendency of the W_{170} test results ($2p < 0.05$; see Fig. 2).

Table 2. Absolute and relative W_{170} test results of the 21 subjects at the 5 temperatures and the means with standard deviations and variation coefficient of the total group

Subject		16.3 °C W_{170}		18.1 °C W_{170}		20.1 °C W_{170}		22.1 °C W_{170}		24.1 °C W_{170}	
		Absolute (W)	Relative (W/kg)	Absolute (W)	Relative (W/kg)	Absolute (W)	Relative (W/kg)	Absolute (W)	Relative (W/kg)	Absolute (W)	Relative (W/kg)
Sportsmen											
	1	310	3.8	330	4.1	330	4.1	315	3.9	300	3.7
♂	2	315	4.3	320	4.4	335	4.6	315	4.3	320	4.4
	3	290	4.1	255	3.6	305	4.4	280	4.0	295	4.2
	4	255	3.7	270	4.0	250	3.7	265	3.9	300	4.4
	5	310	4.1	325	4.3	320	4.2	310	4.1	315	4.1
	6	230	3.3	235	3.3	220	3.1	240	3.4	220	3.1
	7	190	2.8	205	3.0	195	2.9	195	2.9	225	3.3
♀	8	225	3.9	225	3.9	230	4.0	230	4.0	220	3.8
	9	250	4.0	240	3.8	240	3.8	240	3.8	250	4.0
	10	270	4.5	275	4.6	280	4.7	275	4.6	275	4.6
Nonsportsmen											
	11	230	3.3	235	3.3	235	3.3	225	3.2	245	3.5
♂	12	225	3.8	230	3.9	220	3.7	215	3.6	240	4.1
	13	235	3.3	250	3.6	220	3.1	225	3.2	250	3.6
	14	205	2.8	240	3.3	205	2.8	210	2.9	220	3.0
	15	240	3.6	245	3.7	255	3.8	245	3.7	250	3.7
	16	155	2.7	130	2.2	130	2.2	135	2.3	150	2.6
	17	235	3.0	220	2.8	225	2.9	230	2.9	215	2.8
♀	18	170	3.2	170	3.2	170	3.2	170	3.2	160	3.0
	19	130	2.4	145	2.7	135	2.4	130	2.4	140	2.6
	20	160	2.7	155	2.6	165	2.7	160	2.7	150	2.5
	21	170	3.4	160	3.2	170	3.4	155	3.1	145	2.9
$\bar{r}$		229	3.5	231	3.5	230	3.5	227	3.4	233	3.5
± SD		53	0.6	57	0.6	60	0.7	56	0.6	58	0.7
V (%)		23	16	24	17	25	20	24	18	24	18

Discussion

These results show that the shortened W_{170} test values are not systematically influenced by an increase or decrease in ambient temperature within the range of 16 °C–24 °C. Therefore, varying temperatures within this range can be ignored.

Although there is a significant difference between the test results of the men at the lowest and at the highest temperature level, Fig. 2 shows, that this is not a systematic increase. Moreover, one should have expected a lower working capacity with an increase in ambient temperature, according to the knowledge of thermoregulation. Therefore it can be concluded that the above-mentioned significant difference is independent of the ambient temperature.

Obviously these statements only apply when the W_{170} test is used in conjunction with the Dynavit ergometer. As it cannot be excluded that an influence due to the physical dependence of the ergometer could compensate for any physiological temperature effect.

Further conceivable objections are that the time for acclimatization might be too short to achieve a new thermal balance. For a genuine steady state a longer period of time surely would be necessary. Considering the light clothes of our subjects, however, a time of 15 min for acclimatization before the experiments ought to have caused measurable differences in performance if a relevant influence existed.

The substantial standard deviations from the mean test results are not surprising when the inhomogeneous group of men, women, sportsmen, and nonsportsmen is taken into account. This inhomogeneity does not affect the intraindividual comparison; the standard deviations underline the ability of the test to differentiate subjects according to their physical fitness.

In conclusion, the experiments confirm the economy and validity of the shortened modification of the W_{170} test, which make the test very useful for serial experiments, especially.

Summary

The shortened W_{170} test is a modified W_{170} test procedure saving considerable time by using a rapid and almost continuous load increase of 5 w/10 s. Twenty-one subjects performed the test in repeated experiments on the Dynavit ergometer at the temperatures of 16 °C, 18 °C, 20 °C, 22 °C, and 24 °C. After an acclimatization period of about 15 min and a test duration of about 10 min on average, the mean W_{170} value was 196 W for the women and 267 W for the men. There was no systematic influence of ambient temperature, so that varying temperatures within the range of 16 °C–24 °C can be ignored. This is a further argument for the usefulness of this test modification.

References

1. Albrecht KL, Ulmer H-V (1979) W_{170}-Kurztest zur Prüfung der körperlichen Ausdauer-Leistungsfähigkeit bei Reihenuntersuchungen. Wehrmed Monatsschr 23: 336–339
2. Barth G, Ulmer H-V (1979) Comparison of 3 methods for W_{170} with different test duration. Eur J Physiol [Suppl] 382: 32
3. Goerz V, Ulmer H-V (1983) W_{170}-Kurztest (30 W/min) mit einem programmierten Ergometer. In: Mellerowicz H, Franz I-W (eds) Standardisierung, Kalibrierung und Methodik in der Ergometrie. Perimed, Erlangen, p 135–139
4. Hauptverband der gewerblichen Berufsgenossenschaften e. V. (ed) (1981) Berufsgenossenschaftliche Grundsätze für arbeitsmedizinische Vorsorgeuntersuchungen – G 26 Atemschutzgeräte, G 30 Hitzearbeiten, G 31 Überdruck, 2nd edn. Gentner, Stuttgart
5. Mellerowicz H (1975) Ergometrie. Urban and Schwarzenberg, München
6. Smodlaka V, Mellerowicz H, Horák J (1981) Revidierte Standardisierungsvorschläge für Ergometrie. Inform Arzt 10/17: 52–53
7. Ulmer H-V (1980) Experience in a shortened W_{170}-test using a continuous increase of load (30 W/min). In: Ostyn M, Beunen G, Simons J (eds) Kinanthropometrie II. University Park Press, Baltimore, p 464 (International Series of sports science, vol 9)

Factor Analysis of Physical Fitness: Comparison of Rowers with Cross Country Skiers

J. Horák, J. Pirič, and M. Jirásek

Introduction

Rowing and cross country skiing require different types of cardiopulmonary and skeletomuscular capacities. The present study was designed to investigate by means of factor analysis the different adaptive mechanisms relevant for these two different kinds of sports activities.

Material and Methods

Thirty-eight rowers and 29 skiers were studied by bicycle exercise testing. All subjects exercised three times for 6 min with work loads of 2.0, 2.5, and 3 W/kg. During the exercise test, acid-base status was determined by taking blood samples from the hyperemized earlobe. Lactate analysis was performed from an antecubital vein in the third minute after conclusion of the test. Cardiopulmonary function data were analyzed in a standardized way as has been described earlier [3]. Pertinent data are shown in Tables 1 and 2.

Table 1. Anthropometric characteristics of rowers and skiers

	Rowers		Skiers		$p<$
	$\bar{x}$	SD	$\bar{x}$	SD	
Age (years)	22.30	2.76	24.17	1.42	0.001
Body mass (kg)	86.20	5.66	72.29	5.12	0.001
Height (cm)	186.55	5.47	177.38	4.52	0.001
Lean body mass (kg)	77.56	6.48	68.12	5.07	0.001
Body fat (%)	11.08	3.18	5.42	1.99	0.001

Table 2. Main parameters of cardiorespiratory and strength-endurance performance and lactate

	Rowers		Skiers		$p<$
	$\bar{x}$	SD	$\bar{x}$	SD	
$\dot{V}_{O_2 max}$ (ml min^{-1})	5138.40	566.80	5020.00	509.20	N.S.
$\dot{V}_{O_2 max}$ kg^{-1} (ml min^{-1})	59.61	6.58	69.44	7.04	0.001
W_{max} (w)	425.39	47.43	448.45	34.50	N.S.
W_{max} kg^{-1} (w)	4.93	0.45	6.20	0.48	0.001
Total work (kJ)	124.02	38.45	210.63	28.85	0.001
Total work kg^{-1} (kJ)	1.44	0.45	2.91	0.40	0.001
Lactate (mmol l^{-1})	11.34	2.18	11.63	2.01	N.S.

Factor analysis was performed according to Jöreskog [5] with analytical rotation [2]. This gives a model based on oblique solution as well as on a correlation matrix.

Results

Age and anthropometric data (Tables 1 and 2) differed significantly between the two groups of athletes; especially weight and height were different. Skiers showed higher values for weight-related oxygen uptake and total amount of work load. Maximal lactate values were similar in the two groups (Table 2).

A total of eight factors were calculated by factor analysis. The various factors with their specific loading are listed in Table 3 (rowers) and Table 4 (skiers). Differences in loading (or saturation) can clearly be seen between the two lists, with a more pronounced loading body mass in skiers. On the other hand, the strength-endurance factor (II; with strength identical to muscle strength) contains far more parameters for rowers than for skiers. The remaining factors demonstrate similar loadings with some minor differences for the single values. With regard to the metabolic components, base excess is superior to lactate (the latter being analyzed *after* cessation of exercise).

A correlation matrix was also calculated for the groups of athletes. Rowers clearly show a close relationship between the factors for strength, cardiovascular fitness, and body composition.

In contrast, skiers demonstrate a closer interrelationship between anthropometric factors and cardiopulmonary values at maximal power output (Table 5).

Discussion

Factor analysis in ergometry is one possible approach to discriminating parameters which assist in predicting physical fitness [4]. Such parameters can be weighted according to the strength of the arithmetically developed correlation. In this study, some specific differences occurred for the components of the factors between two kinds of sportsmen, rowers and skiers. These differences can be described by a strength-endurance component and an endurance-cardiopulmonary function component.

In rowers, physical stature and strength play a greater role than in skiers, in whom endurance and body weight-related parameters predominate. This then reflects the quite different constitution of these two groups of athletes. Rowing requires athletes with a greater weight and height.

Heart rate measured at peak exercise is of less significance in discriminating the two groups. This can in part be explained by the methodological approach used in this study. The athletes tested here did not reach their real maximal heart rate during bicycle testing. Higher heart rate values would be achieved if specific testing (rowing or treadmill testing, respectively) is performed [4, 6–8].

Another difference between the two groups exists for maximal pulmonary ventilation.

Table 3. Factor structure of cardiorespiratory performance in top rowers

Factor	Variables	Loadings	Communalities
1	*Anthropometry*		
	1. Height	1.047	0.927
	2. Body surface	0.738	0.995
	3. Body mass	0.435	0.970
	4. Lean body mass	0.364	0.835
	5. HR rec.	-0.367	0.558
2	*Strength-endurance capacity*		
	1. Total work	0.958	0.763
	2. W_{max}	0.915	0.936
	3. WP_{max}	0.881	0.799
	4. Body mass	0.756	0.970
	5. Lean body mass	0.630	0.835
	6. WP 3W	0.611	0.983
	7. Body surface	0.431	0.995
	8. $\dot{V}_{E\,max}$	0.396	0.832
3	*Circulatory performance*		
	1. HR_{max}	0.935	0.809
	2. HR 3W	0.882	0.988
	3. HR rec.	0.681	0.556
	4. Body fat	0.474	0.343
	5. WP 3W	-0.443	0.985
4	*Cardiorespiratory performance under submaximal load*		
	1. $\dot{V}_{O_2}$ 3W	0.958	0.983
	2. $\dot{V}_{O_2}/HR$ 3W	0.870	0.986
	3. $F_{ECO_2} - F_{ICO_2}$ 3W	-0.415	0.397
5	*Cardiorespiratory performance under maximal load*		
	1. $\dot{V}_{O_2\,max}$	0.876	0.983
	2. $\dot{V}_{O_2}/HR_{max}$	0.875	0.938
	3. $F_{IO_2} - F_{EO_2}$ 3W (s)	-0.452	0.612
6	*Respiratory performance*		
	1. $F_{IO_2} - F_{EO_2\,max}$	0.798	0.664
	2. $\dot{V}_{E\,max}$	-0.767	0.832
	3. $F_{IO_2} - F_{EO_2}$ 3W	0.612	0.612
	4. $F_{ECO_2} - F_{ECO_2\,max}$	0.583	0.626
	5. $F_{ECO_2} - F_{ICO_2}$ 3W (s)	0.387	0.397
7	*Metabolic acidosis*		
	1. Base excess	0.993	0.976
	2. Standard bicarbonate	-0.989	0.984
	3. pH	-0.984	0.999
	4. Lactate	0.586	0.485
8	*Partial pressure of arterial CO_2*		
	1. P_{CO_2}	0.984	0.986
	2. Age	0.366	0.429

Table 4. Factor structure of cardiorespiratory performance in top skiers

Factor	Variables	Loadings	Communalities
1	*Anthropometry*		
	1. Body mass	0.991	0.997
	2. Lean body mass	0.918	0.927
	3. Body surface	0.844	0.999
	4. WP 3W	0.555	0.996
	5. Height (s)	0.447	0.960
	6. Body fat	0.439	0.230
2	*Strength-endurance capacity*		
	1. Total work	1.067	0.985
	2. W_{max}	0.840	0.995
	3. WP_{max}	0.703	0.987
3	*Circulatory performance under submaximal load*		
	1. HR 3W	0.941	0.998
	2. HR rec.	0.696	0.670
	3. WP 3W	-0.677	0.987
	4. $\dot{V}O_2/HR$ 3W (s)	-0.355	0.997
4	*Circulatory performance under maximal load*		
	1. HR_{max}	0.934	0.970
	2. WP_{max} (s)	-0.448	0.987
	3. PCO_2 (s)	-0.392	0.380
5	*Cardiorespiratory performance under submaximal load*		
	1. $\dot{V}O_2$ 3W	0.990	0.999
	2. $\dot{V}O_2/HR$ 3W	0.870	0.997
	3. HR rec. (s)	-0.434	0.670
6	*Cardiorespiratory performance under maximal load*		
	1. $\dot{V}O_{2max}$	0.925	0.999
	2. $\dot{V}O_2/HR_{max}$	0.892	0.998
	3. $\dot{V}E_{max}$	0.544	0.876
	4. PCO_2	-0.499	0.380
	5. $FIO_2 - FEO_2$ 3W	0.428	0.228
7	*Respiratory performance under maximal load*		
	1. $FIO_2 - FEO_{2max}$	0.615	0.733
	2. $FECO_2 - FICO_{2max}$	0.472	0.602
	3. Height	0.596	0.960
8	*Metabolic acidosis*		
	1. Base excess	0.999	0.937
	2. Standard bicarbonate	-0.998	0.993
	3. pH	-0.978	0.987
	4. Lactate	0.530	0.670

Table 5. Significant correlations of the factors

	r
I. Rowers	
1. *Anthropometric factor*	
– factor of strength-endurance capacity	0.62
– factor of cardiorespiratory performance under maximal load	0.31
2. *Strength-endurance capacity factor*	
– factor of circulatory performance	– 0.54
– factor of cardiorespiratory performance under maximal load	0.52
II. Ski-runners	
1. *Anthropometric factor*	
– factor of cardiorespiratory performance under maximal load	0.52
– factor of strength-endurance capacity	0.47
2. *Factor of strength-endurance capacity*	
– factor of cardiorespiratory performance under maximal load	0.40

This value is related to cardiopulmonary factor in skiers but to the anthropometric factor in rowers. This can be explained by the well-known fact that pulmonary function strongly correlates to physical stature even in athletes.

The significance of alveolar ventilation or respiratory gas exchange as one component of physical fitness is reflected in the high loading of alveolar concentration of carbon dioxide (P_{CO_2}). P_{CO_2} is known to be influenced by hyperventilation to compensate for the metabolic acidosis occurring beyond the anaerobic threshold. The role of P_{CO_2} in assessing physical fitness in athletes has been stressed by several authors [1, 8, 9], especially as a marker of ventilatory control during exercise.

It seems remarkable that in these lists of parameters of physical fitness lactate is less significant. This may be due to methodological reasons, namely that lactate was analyzed after the end of stress testing. The higher loading of base excess during exercise testing emphasizes the need of measuring metabolic parameters at peak stress rather than afterwards.

In summary, factor analysis proves useful to discriminate between different kinds of sports activities which differ in the anthropometric requirements but which are similar in cardiovascular demands. Factor analysis may also be helpful in reducing data in exercise testing and in elaborating those parameters which are of most significance. This can lead to test batteries yielding a diagnostic score as a measure of physical fitness in different types of sports.

References

1. Grodins FS (1981) Exercise hyperpnea. The ultra secret. In: Hutas I, Debreczeni LA (eds) Respiration. Pergamon, Budapest (Adv physiol, vol 10)
2. Harris ChW, Kaiser H (1964) Oblique analytic solution by orthogonal transformation. Psychometrika 24: 347–362
3. Horák J (1983) Faktorielle Struktur der kardiorespiratorischen Leistungsfähigkeit gemessen mit Hilfe der Spiroergometrie bei Sportlern und Nichtsportlern. In: Mellerowicz H, Franz I-W (eds) Standardisierung, Kalibrierung und Methodik in der Ergometrie. Perimed, Erlangen, pp 169–174

4. Israel S (1970) Zur Problematik der maximalen Herzschlagfrequenz bei Sportlern. Med Sport 9: 193–200
5. Jöreskog KG (1963) Statistical estimation in factor analysis. Almquist and Wiksell, Stockholm
6. Keul J, Huber G, Link K, Dickhut H-H, Simon G (1979) Untersuchungen und leistungsbeglei-tende Untersuchungen zur Beurteilung der Leistungsfähigkeit im Radsport. Leistungssport 9: 254–261
7. Mellerowicz H, Smodlaka VN (1981) Ergometry. Basics of medical exercise testing. Urban and Schwarzenberg, Baltimore
8. Nowacki PE (1977) Sportmedizinische und leistungsphysiologische Aspekte des Ruderns. In: Adam K, Lenk H, Nowacki PE, Rullfs M, Schröder W (eds) Rudertraining. Limpert, Bad Homburg
9. Tavastšerna NI (1958) Variation of respiratory center irritability of sportsmen and its dependence on speed and endurance. Teor Prakt fiz Kult 3: 213–215 (in Russian)

Quality Criteria and Power Calibration of Ergometers

K. Goffloo and W. Sontopski

Introduction

Among other details, ergometers should give vital information such as:

- Is there coronary heart disease?
- What safe activity range is left to the patient after a myocardial infarction?
- What pathological diagnosis can be made under defined physical stress and could this reduce the economic occupation of the patient?

These few aspects already confirm the paramount importance of reliable and precise ergometric results. High ergometer standards, however, have not been taken for granted and accepted by all manufacturers. This was revealed by tests made by the *Physikalisch-Technische Bundesanstalt* on several ergometers available on the market [1, 2].

Also it should not be forgotten that the practitioner can not evaluate the precision quality of the ergometer on site. He will suspect malfunction only when an indication is missing or when it gives unrealistic values incompatible with common diagnosis. Such proceedings are explained by Schnellbächer [3]. He suggests observing the heart rate of healthy middle-aged men who are apparently fit at a standard speed of 60 rpm. At 120 W the heart rate should be in a steady state within the normal range of 110–130/min.

The impossibility for the practitioner to take control of instrument precision is illustrated mildly ironically by Schnellbächer [3] as follows: "Just as little as the practitioner can be expected to have his own private test dog to sample the first pill or vial of each package of medicine to ensure that it will not be lethal, he also cannot be expected to erect his own calibration equipment in his practice. But he should be expected to use only instruments of high quality, as permitted by the authorities and regularly maintained."

In [4], reference is made to the fundamental problems and solutions of power measuring with ergometers. In the meantime, exact requirements for measuring precision [5] are available. In [5] the maximum permissible error for power measuring under 60 W is ±3 W. The following system analysis is based on these premises. It will demonstrate the appropriate graduation of the brake force measurements. The basic modules for brake force measuring are the force sensor and the evaluation electronics. Field experience with medical weighing units gives valuable information for feasible and economic error splitting to these two modules. From this splitting the required resolutions for the force sensor and the evaluation electronics can be made.

We shall explain the influence of these resolution requirements for the choice of the force sensor type and the design of the evaluation electronics.

Then follows a summary on designer's and manufacturer's efforts for making available ergometers with the required precision.

The demonstration is arranged in such a way that it will always be possible to refer to requirements and solutions for medical personal weighing units for reference and comparison.

Measurement of Braking Force

Physical power is proportional to the product of braking momentum and rotations. For a given ergometer model, the braking momentum is always proportional to the braking force. Thus the power measuring requires the measuring of the braking force and speed. In the following we shall focus on the requirements for measuring braking force.

Determination of the Required Resolution in the Measuring Instrument for the Braking Force

For determining the required resolution for force measuring we start with the following data:

Maximum pedal power	$P_{max} = 400\,W$
Admissible error in power measuring	$\Delta P = \pm 3\,W$
Minimum pedal revolutions	$n_{min} = 20/min$
Maximum pedal revolutions	$n_{max} = 90/min$
Maximum braking force	$F_{max} \sim \dfrac{P_{max}}{n_{min}}$
Minimum braking force to be resolved	$F_{min} \sim \dfrac{\Delta P}{n_{max}}$
Rearranged:	$\dfrac{F_{min}}{F_{max}} = \dfrac{\Delta P\ n_{min}}{n_{max}\ P_{max}} = \dfrac{1}{600}$

The example shows that the minimum braking force to be resolved by the measuring system is smaller than the maximum braking force by a factor of 600. That means that a measuring system with 600 steps will be necessary. For a given indicating precision in the measuring instrument, the internal precision must be higher. Thus it follows for the power measuring that the resolution must be made in at least 900 steps.

The power measuring instrument is composed of the force sensor and evaluation electronics systems. Now the question arises to what percentage the component systems of the force sensor and the evaluation electronics may participate in the total error. The splitting should be done in such way that the price of the whole system will be as low as possible. For medical personal weighing units, the error splitting of ⅔ for the force sensor and ⅓ for the evaluation electronics was a good solution. Tak-

ing this confirmed and satisfactory rule, the force sensor should get a minimum resolution of 1350 steps and the evaluation electronics of 2700 steps.

As with all measuring instrument, here also a temperature- and age-conditioned
drift of the zero point should be taken into consideration.

That means that the measuring range of the electronic system should be increased
for one drift range for preventing measuring errors. Based on a drift range of $\pm 25\%$
of the measuring range, it will be possible to ensure the necessary periods under
operating conditions of the ergometer.

This means for the electronic system that the minimum graduation of 2700 must
be increased to about 4000 steps. This corresponds to a medical personal weighing
unit with a 200-kg capacity and 50-g graduations. This example shows the high standard for braking force measurement in the ergometer.

For getting the correct weight with a weighing machine, it is necessary to set the
system to zero before stepping onto the platform. The same applies to braking force
measurement. One has to start from the actual zero point to get the correct force
value. The best procedure is an automatic and obligatory one, before starting the
next ergometric measurement.

Interferences for the Force Sensor

As mentioned before, the power sensor must have a resolution of at least 1350 steps.
The most economic force sensor with long stability will be the strain gauge load
cells. Such sensors are been used for many years in medical weighing machines under the Weight and Measures Act.

Interferences for the Evaluation Electronics

As mentioned above, the evaluation electronics should have a resolution of at least
4000 steps. For ensuring a constantly stable, high precision (independent from temperature), only high-quality components should be used.

Furthermore, the evaluation electronics should be tested for several days under
"burn-in" conditions for eliminating faulty components before shipping the unit
and for compensating the solder heat stress on the components (Fig. 1). During the
burn-in the electronic system should be activated at rapid intervals, operated under
extreme conditions, and exposed to high ambient temperature.

*Provisions for Recognizing Operational Faults, Autocalibration,
and Automatic Zero Point Return*

There are two kinds of provisions for recognizing operational faults:

– Those which are controlled by the microcomputer and which will not be noticed
  by the operator.
– Those which are started with the push button for controlling the pertaining oper-
  ations.

Fig. 1. "Burn-in" for tracing faculty components

In the first group is included the selftest of the microcomputer with test circuits provided in the normal operation program. When starting, all code signals are added and compared with a check number. This program test, however, does not yet warrant that all inputs – as for instance the start power or power steps – will be correctly processed. For this, the memories of these variables get check inputs, and when there is an incompatibility, the computer stops and signals an error.

The second group includes test numbers for checking the force-measuring circuit. A test value is put in at the head of the measuring circuit and the value put out at the end of the measuring circuit can then be compared, automatically or by pushing a button, with the input value. With this procedure you can easily find all creep faults. This test procedure is also included in the official test regulations of the German Polytechnical Institute for Ergometers [5].

As mentioned above, temperature and ambient conditions may provoke a zero point drift in most of the force sensors and evaluation electronics systems. Errors up to 50 W may occur [1]. Therefore, an automatic zero point compensation at the start of each ergometer examination will be necessary. For medical weighing machines, the zero-setting button or the automatic zero setting has been an obligatory component for many years.

Coupling Ergometers to "Intelligent" ECG Recorders

New ECG recorders not only register and evaluate the heart rate but they now also control and evaluate ergometric data. For instance, it is now possible to put in a step-up load program through the ECG recorder. Extra interfaces in these ECG recorders combined with extended software allow an easy telecontrol of the ergometer.

Care should be taken, however, that the extra interfaces do not create extra errors.

Standard procedure is the output of continuous power values for ergometer control. The ECG recorder provides a given power value for conducting the ergometer operation, while the power control will be made in the ergometer only. The continuous interface, however, will have the following inconveniences:
– Contact problems at the plugs
– Digital errors in the transmitter (for instance, the ECG recorder)
– Digital errors in the receiver (for instance the ergometer)
– Thermal and contact stress
– Sensitivity to disturbances
– Safety hazard through galvanic bridges

All these problems can be avoided with digital interfaces such as are now available for modern telesystems. Design standards for ergometers should be also established for the interfaces for ensuring safe operation and avoiding extreme model variation.

Assembly Line Production of Medical Ergometers

Precision and quality of the ergometer depend on correct design and very much on constant quality control along the whole assembly line.

Fig. 2. Computer control and adjustment for conductor chips

In order to avoid getting "Monday ergometers," such controls should be automatic. Automatic control can be total. Control facilities such as contact points for test computers should be provided for as early as the design stage.

The following summarizes the most important quality controls for assembly line ergometers:

- Computer control and adjustment of conductor chips (Fig. 2).
- Prior to the end test the mechanical and electronic components should operate for several hours under full load.
- Comparative output diagram to be established on each ergometer on a calibrated test stand (Fig. 3).
- The printed diagram (Fig. 4) should be shipped with the ergometer for testifying to its quality. From this diagram the operator will see the individual precision of the unit.

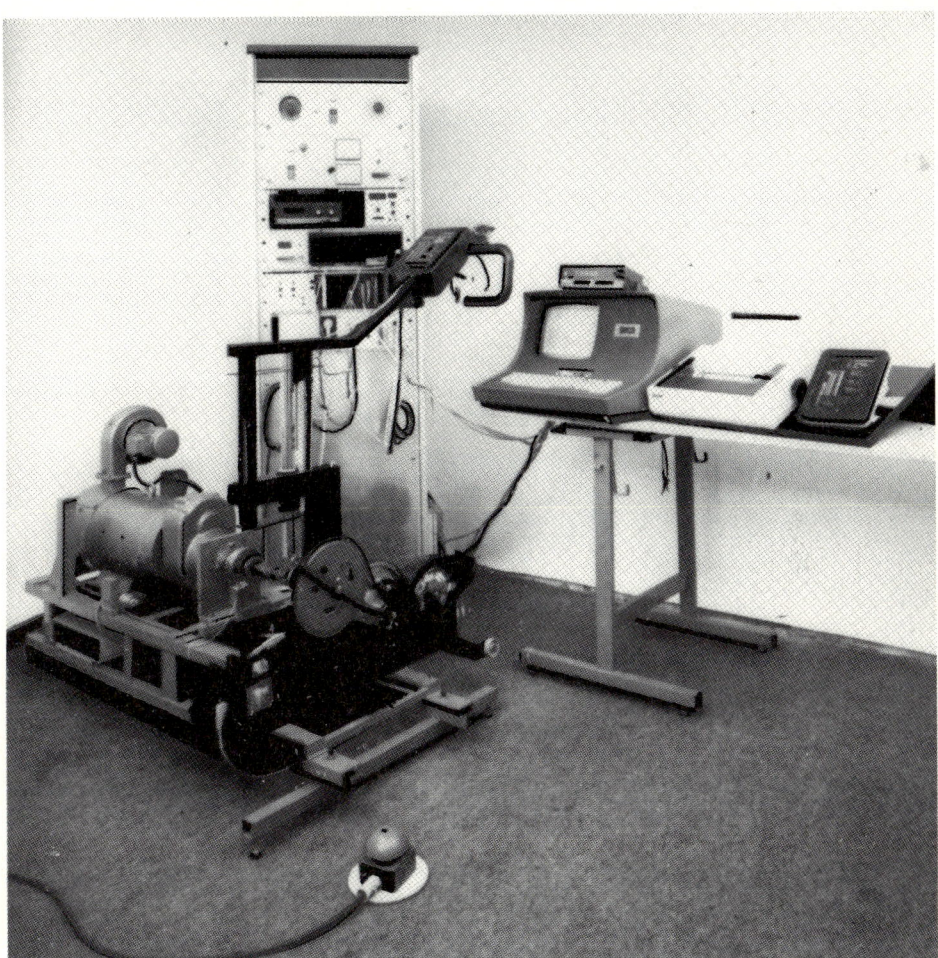

Fig. 3. Establishing quality diagram on calibrated and computerized test stand

```
P R U E F P R O T O K O L L - E R G O M E T E R


K O N T R O L L K E N N D A T E N - E R G O M E T E R

TESTWERT:        SOLL:    100(+/-2)    IST:    101
DMS-NULLPUNKT:   SOLL:     10(+/-2)    IST:     11
PULSFREQUENZ:    SOLL:    60    120    180(+/-3)
    OHR-KANAL:   IST:     62    122    183
    EKG-KANAL:   IST:     62    122    183


L E E R L A U F - R E I B U N G S V E R L U S T E - E R G O M E T E R

DREHZAHL (U/MIN):     40     50     60     70     80     90

VERLUSTLEIST. (W):   5,0    6,4    8,9    9,4   12,0   16,5


L E I S T U N G S K E N N L I N I E N F E L D  -  E R G O M E T E R

DREHZAHL (U/MIN):        40     50     60     70     80     90

SOLL= 50WATT
ISTLEISTUNG   (W):     49,1   49,2   48,4   50,6   50,7   51,8
TOLERANZ      (%):     -1,8   -1,6   -3,2    1,2    1,4    3,6

SOLL= 100 WATT
ISTLEISTUNG   (W):    100,8   98,5   97,4   96,5   99,1   98,8
TOLERANZ      (%):      0,8   -1,5   -2,6   -3,5   -0,9   -1,2

SOLL= 150 WATT
ISTLEISTUNG   (W):    151,7  148,7  147,7  145,4  148,0  146,4
TOLERANZ      (%):      1,1   -0,8   -1,5   -3,0   -1,3   -2,4

SOLL= 200 WATT
ISTLEISTUNG   (W):    202,8  197,4  197,1  196,3  197,6  197,6
TOLERANZ      (%):      1,4   -1,3   -1,4   -1,8   -1,2   -1,2

SOLL= 300 WATT
ISTLEISTUNG   (W):    310,4  305,5  301,5  299,3  297,3  294,1
TOLERANZ      (%):      3,4    1,8    0,5   -0,2   -0,9   -1,9




S E R I E N N U M M E R :  835103
```

Fig. 4. Printed quality diagram. (Comparison of wattage produced and wattage indicated, and mean deviation in %, i.e. Toleranz)

In case these provisions do not prevent the breakdown of the unit, the construction should be modularized for making the repair service rapid and efficient. It is important that the modules are standardized too, so that an exchange should not jeopardize the output.

References

1. Hoffmann K, Kuhlmann E (1980) Metrological analysis of pedal ergometers. Biomed Tech 25: 59–62
2. Goffloo K (1982) How safe are ergometers? ZMT-Novum 11: 35–37
3. Schnellbächer F (1983) Lungenfunktionsprüfungen und ergometrische Untersuchungen in der arbeitsmedizinischen Praxis. Ergo-Med 7: 15–20
4. Sontopski W (1982) Calibration and constant quality for electronic ergometers. In: Mellerowicz H, Franz IW (eds) Ergometry-calibration-standardizing-methodology. Perimed, Erlangen p. 26–33
5. PTB-draft (15. 10. 1982) Construction standards for ergometers. [Distributed by manuscript; PTB = Physikalisch-Techn.-Bundesanstalt (Federal Institute for Physical and Technical Investigations)]

Performance Requirements for Bicycle Ergometers

E. Cramer

Introduction

Since 1979 the Physikalisch-Technische Bundesanstalt (PTB) has investigated the performances of ergometers which are used in the medical field. This is done in cooperation with the manufacturers of ergometers. For this work a special testing setup is used, which was described together with some measurement results at the 4th *Internationales Seminar für Ergometrie* in Berlin in 1981 [1]. On the basis of these results and extensive discussions with physicians and manufacturers of ergometers, a draft for performance requirements [2] has been formulated. With these requirements, the PTB wants to support the standardization proposals of the physicians from the technical point of view. In the following section, the main points of the draft will be reported. In the subsequent section, some measurements obtained in the PTB are summarized and discussed in the frame of the requirements of the draft.

Draft for Performance Requirements

In establishing the performance requirements, attention has been paid to the following three points:

1. The requirements on ergometers by the users
2. The state of the art in ergometer construction
3. Production costs of ergometers

Up to the present state of the draft, many suggestions have been made for altering and improving it, and many of them have been incorporated. In the following we will discuss the main parts of this draft.

The first part contains the definition of ergometers, general requirements, and the physical units to be used; this will not be considered here. The main part contains construction requirements for ergometers and accuracy requirements.

Construction Requirements

The ergometer must be stable enough; the saddle and grasp should be adjustable without tools.

The length of the cranks should be 16.5 cm ± 1 cm.

At least one crank should be removable. (This is required to get a stable connection between the test setup and the ergometer.)

The actual values of speed and power should both be displayed simultaneously. For speed-dependent ergometers, the power must be displayed for at least the two speeds of 50 rpm and 70 rpm.

The minimum scale division for the power display should be 5 W for speed-independent ergometers and 10 W for speed-dependent ergometers. The minimum scale division for the speed display should be 2 rmp.

The flywheel should have a moment of inertion of $J = i^2 \cdot 5.5 \, \text{kg}^2$. The variable i is the ratio of transformation between the cranks and the flywheel. The flywheel has the function of making a uniform treadle operation possible. (The above value is the same as required in the ICSPE proposals of 1965.)

There should be a possibility to test the electronic equipment, which handles the signal from the gauging instrument.

The starting point of the brake torque regulation should be different for several power sections: 30 rpm for a power section up to 250 W, 50 rpm for a power section from 200 W up to 400 W, and 70 rpm for a power section of more than 400 W. The time constant of the regulation should be between 2 s and 4 s.

Accuracy of Measurements

The relative error of the displayed power should be lower than ±5% for speed-independent ergometers and lower than ±10% for speed-dependent ergometers. For low-power operation under 60 W, the absolute error is allowed to be ±3 W for speed-independent ergometers and ±6 W for speed-dependent ergometers.

The margins of error should not be exceeded during continuous running for 2 h. The test conditions for this requirement are 50 W and 150 W with 50 rpm, 150 W with 60 rpm, and 250 W with 70 rpm. For power steps of more than 400 W there will be a separate interval test with 10-min load and 5-min cooling over 2 h.

Comparison of Requirements with Actual Performance of Ergometers

For illustration of the compatibility of the requirements with the actual performance of existing ergometers, we now present some data obtained in our laboratory. With the testing device of the PTB the brake power of an ergometer can be determined with an accuracy of better than 1% in a speed range from 0 rpm up to more than 100 rpm. The results of such brake power measurements are summarized in Fig. 1 for speed-dependent ergometers and in Fig. 2 for speed-independent ergometers.

In these figures, the margins of error are shown for three different power ranges. The ergometers B through P, and U were brand new; the ergometers A and Q through S were used ones. Ergometer A was measured twice. The first test (A) was done after using it for 2.5 years; the second test (A*) was made 2 years later after repairing and calibration by the manufacturer.

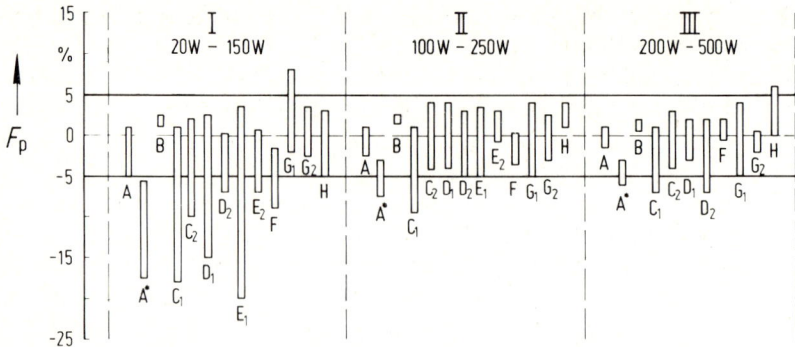

Fig. 1. Margins of error in power measurements with speed-independent ergometers. Different capital letters indicate ergometers from different manufacturers. Subscript 1 denotes the older versions; subscript 2 denotes the newer versions. Ergometers B through H are factory new; ergometer A is a used one, which was measured twice

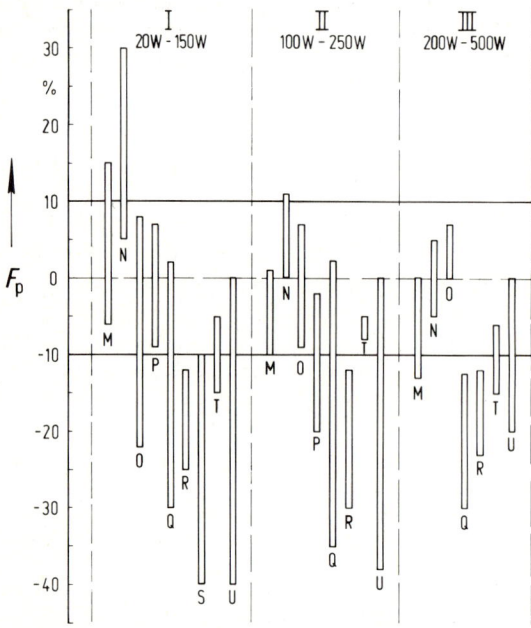

Fig. 2. Margins of error in power measurements with speed-dependent ergometers. Different capital letters indicate ergometers from different manufacturers. Subscript 1 denotes the older versions; subscript 2 denotes the newer versions. Ergometers M, N, O, P, T, and U are factory new; ergometers Q, R, and S are used ones

As Fig. 1 shows, many of the speed-independent ergometers from different manufacturers, indicated by different capital letters, will approximately come up to the requirements stated earlier. There are only problems for very low power sections. From certain manufacturers there are also two versions of ergometers. The older versions are indicated by a subscript 1 and the newer versions are indicated by a subscript 2. As a rule, the improved newer versions yield better results. The case of ergometer A shows the importance of careful calibration, especially after repairing the ergometer. Figure 2 shows that almost all speed-dependent ergometers will fail to meet the previously stated accuracy requirements. With a better calibration of these ergometers, it should become possible to fulfill the requirements for the accuracy of $\pm 10\%$.

Summary

A draft for performance requirements for ergometers was formulated on the basis of many discussions with physicians and manufacturers and investigations on accuracy of measurements made by the PTB during the last several years. The results from these measurements are compared with the reported requirements.

References

1. Hoffmann K, Kuhlmann E (1983) Meßtechnische Untersuchungen an Fahrradergometern. In: Mellerowicz H, Franz I-F (eds) Standardisierung, Kalibrierung und Methodik in der Ergometrie. Perimed, Erlangen, pp 14–22
2. Entwurf einer Bauartanforderung an Fußkurbelergometer (10.6. 1983) Physikalisch-Technische Bundesanstalt, Institut Berlin. (Prepared as manuscript)

Accuracy of Constant Load Electromagnetic Bicycle Ergometers: A Case Study*

F. Landry, C. Bouchard, D. Prud'homme, and D. Marceau

Introduction

Electromagnetic bicycle ergometers are said to produce constant load, independent of revolution. Their system is generally based on the principle of the hyperbolic relationship of braking power and revolution of a rotating disc in a magnetic field. Braking itself is effected by a Pasqualini eddy current brake. Braking energy is thus produced as one rides the bicycle. Mechanical transmission is by means of internally lubricated roller chains running on chain wheels with a flywheel to ease the start and finish. Researchers often use electromagnetic ergometers with only the manufacturer's original calibrations, as special equiqment required to check the hyperbolic relationship of braking power and revolution of a rotating copper disc in a magnetic field is not available in most exercise physiology laboratories. The matter is judged of importance with research-grade bicycle ergometers especially when data on physiological responses to set power outputs are wanted.

The purpose of this study was threefold: (a) to measure the true wattage produced by two constant load electromagnetic bicycle ergometers of the same type and commercial brand; (b) to measure the difference, as the case may be, between true wattage produced and wattage indicated on the scale meter of each ergometer; and (c) to create a system for the direct calibration of constant load bicycle ergometers.

Methodology

Two electromagnetic bicycle ergometers of the same brand used on a regular basis in research work were selected. The value of braking power on these ergometers is calibrated in watts on two scales ranging from 0–200 W and 0–400 W. With this brand of ergometer, the product of revolution and couple is said to remain constant, i.e., independent of RPM; hence the braking power, once adjusted, should not be altered by changing the speed.

The calibration equipment consisted basically of (Fig. 1):
1. A three-phase motor of 550 V
2. A set of sprokets and roller chains, speed ratio 7.84 total
3. A flywheel of 275 lb

---

* Research supported in part by Fitness Canada Grant 245-003-0

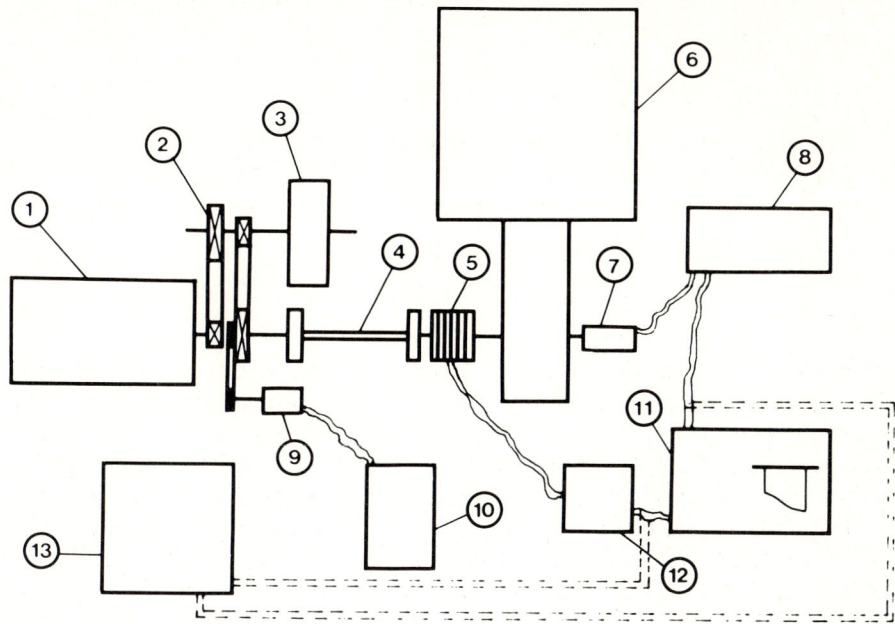

Fig. 1. Calibration system. *1* 3-phase motor 550 V; *2* sprokets and roller chains speed ratio 7.84 total; *3* flywheel (275 lb); *4* torsion bar 1/2″ with 4 strain gauges; *5* slip rings; *6* lanooy ergometer; *7* generator 7 V/1000 RPM; *8* digital voltmeter; *9* DC impulse generator; *10* electronic counter; *11* sanborn 2-channel recorder; *12* signal amplifier BAM 1; *13* Norland data computer and recorder

Table 1. System validity estimates on 2 different ergometers by test-retest at power settings ranging from 0 to 400 W

Conditions and variables	n	Test 1 $\bar{x} \pm SD$	Test 2 $\bar{x} \pm SD$	Difference between means	Validity estimate interclass (r)
Scale 200					
Power settings	11	124.55 ± 99.6	127.51 ± 102.2	NS	1.0[a]
Scale 400					
Power settings	11	206.94 ± 150.46	210.24 ± 147.42	NS	1.0[a]
Scale 200 and 400					
RPM 60	22	59.24 ± 1.83	56.66 ± 1.64	NS	0.6[a]
Power settings	22	165.74 ± 131.45	168.87 ± 130.84	NS	1.0[a]

[a] $p \leqslant 0.01$

4. A torsion bar of 0.5 in. with four strain gauges
5. Slip rings
6. The ergometer
7. A generator 7 V/1000 RPM
8. A digital voltmeter
9. A DC generator impulse
10. An electronic counter

11. A Sandborn two-channel recorder
12. A signal amplifier (BAM 1)
13. A Norland data computer and recorder

The validity of the calibration system was established by a series of 11 tests and retests on each of the 200- and 400-W scales of the two ergometers.

The calibration proper of each ergometer was checked by measuring the true wattage produced at specific settings from 25 to 500 W by 25- and 50-W increments, at given RPMs from 30 to 90.

Results

Validity Estimates of the System

The results are shown in Table 1. On the 200-W scale, the means of 11 different settings from 0 to 200 W (five on ergometer L_1 and six on ergometer L_2) were not statistically different in the test-retest situation. On the 400-W scale, no significant differences were noted. The interclass coefficients of correlation, in both cases, were of 1.0 ($p \leqslant 0.01$). The means of the measured RPMs in the test-retest situations were also not significantly different from each other, nor from the set level of 60 (59.2 ± 1.8 and 56.7 ± 1.6). The interclass coefficient of correlation was of 0.6 ($p \leqslant 0.01$).

Accuracy of Ergometer L_1

Figures 2a and 2b illustrate the results obtained on ergometer L_1 by 50-W increments in power settings on the 200-W scale and by 100-W increments on the 400-W scale at 60 RPM. It can be seen that at the four power settings from 50 W up, on the 200-W scale (Fig. 2a), there were no significant differences between the means of the power outputs measured and the powers set. The combined mean of the measures obtained corresponded exactly to that of the mean setting (105 ± 70 W). In contrast, on the 400-W scale, the values measured at five settings from 100 W up consistently tended to be lower than the corresponding power settings (Fig. 2b). The average difference was of the order of -10%. Considering the six power settings as a whole, the average difference between the power setting and the measured power output was of -30 W (260 ± 170 as compared with 230 ± 150, $p \leqslant 0.05$).

Accuracy of Ergometer L_2

Figures 3a and 3b illustrate the results obtained on ergometer L_2 in the same experimental conditions as was the case for ergometer L_1. It can be observed on Fig. 3a that on the 200-W scale, in six power settings ranging from 0 to 200 W, ergometer L_2 showed consistently and significantly higher outputs averaging $+64\%$ ($p \leqslant 0.05$). On the 400-W scale, the same tendency was observed, again consistent and significant, averaging this time $+39\%$ ($p \leqslant 0.05$).

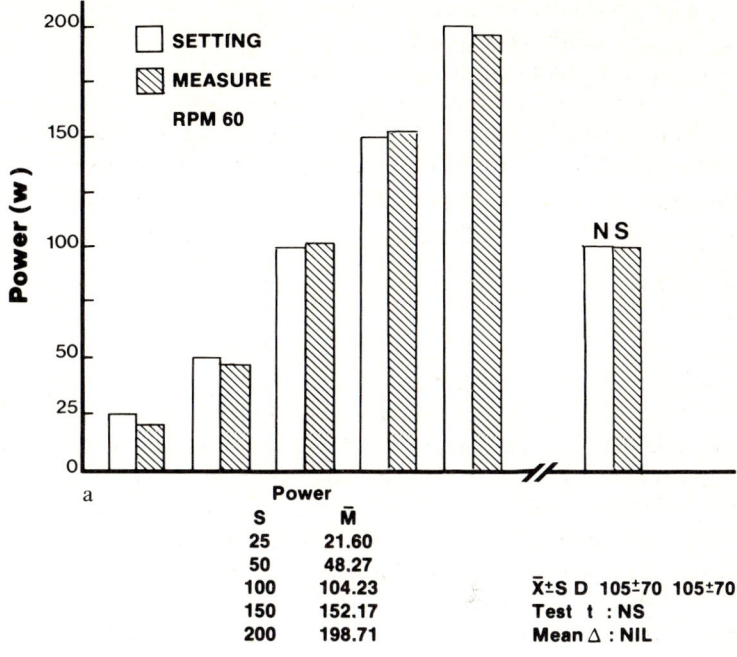

a

Power

S	$\bar{M}$
25	21.60
50	48.27
100	104.23
150	152.17
200	198.71

$\bar{X}\pm S\,D$ 105±70 105±70
Test t : NS
Mean $\triangle$: NIL

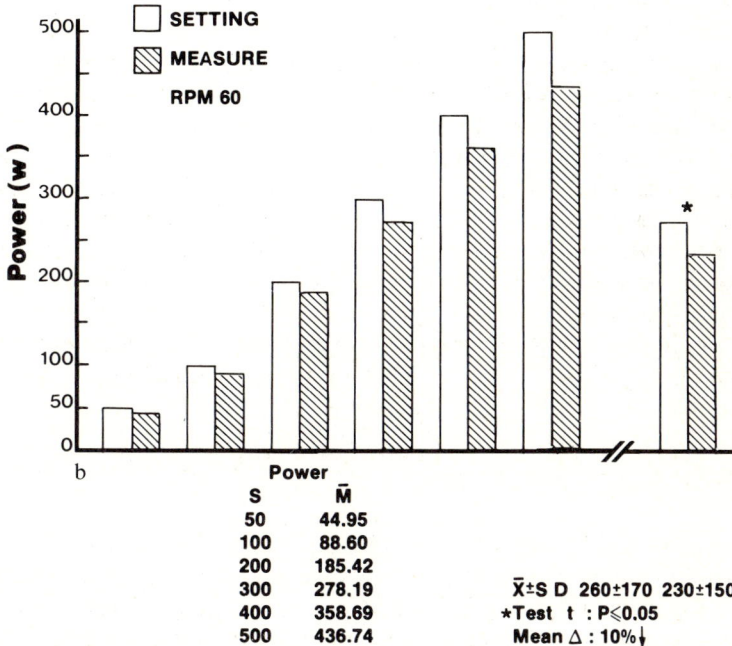

b

Power

S	$\bar{M}$
50	44.95
100	88.60
200	185.42
300	278.19
400	358.69
500	436.74

$\bar{X}\pm S\,D$ 260±170 230±150
★Test t : P⩽0.05
Mean $\triangle$: 10%↓

Fig. 2a, b. Sample data on ergometer L_1, five work-loads on 200-W scale *(a)* and six work loads on 400-W scale *(b)* (bicycle lanooy)

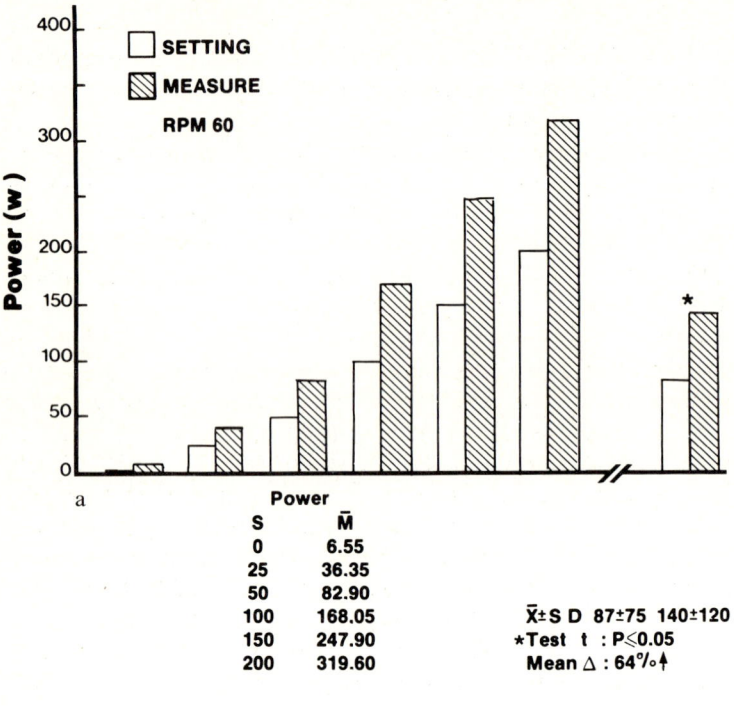

a

Power	
S	M̄
0	6.55
25	36.35
50	82.90
100	168.05
150	247.90
200	319.60

X̄±S D 87±75 140±120
*Test t : P≤0.05
Mean △ : 64%↑

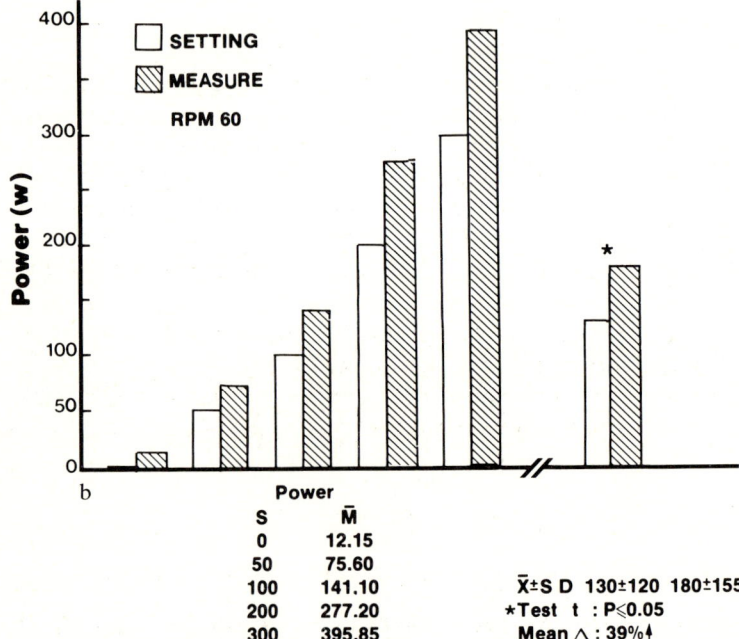

b

Power	
S	M̄
0	12.15
50	75.60
100	141.10
200	277.20
300	395.85

X̄±S D 130±120 180±155
*Test t : P≤0.05
Mean △ : 39%↑

Fig. 3 a, b. Sample data on ergometer L_2, six work-loads on 200-W scale *(a)* and five work-loads on 400-W scale *(b)* (bicycle lanooy)

Discussion

Increasingly, multiple research-grade bicycle ergometers are used concomitantly in research projects. Depending upon the situation, it may be just as important to measure physiological responses to given levels of exercise as their corollary, i.e., the work load tolerated in given physiological conditions. In both cases, one of the assumptions is that the ergometer used is "calibrated"; this may obviously not necessarily be the case. Potential interlaboratory variations between ergometers may be a matter of concern in cooperative studies [1].

In the experiment at hand, the intraergometer variance between power outputs measured on the two different scales of ergometer L_1 and the interergometer variance between ergometers L_1 and L_2 point to the importance of the issue. Relying on the power settings of ergometer L_1 on its 400-W scale resulted in a measured overestimation of the actual power output, which was of the order of 10%. For the same ergometer, this was not the case when the 200-W scale was used. With ergometer L_2, the converse was observed. Relying on the power settings on both the 200-W and 400-W scales yielded underestimations of the actual power output: more than 60% and about 40% for each scale, respectively. Blind faith in the "accuracy of measurement" of "constant load" ergometers may have serious implications. As an example, a subject's pre- and posttraining values for work tolerance, had he been tested on our ergometers L_1 and L_2, in that order, would have been "overestimated" at the beginning and "underestimated" at the end. Such a double bias is obviously capable of having disastrous random effects on individual and group trends in the data.

Conclusion

The calibration system described in this study proved itself to be a valid instrument in the determination of the power produced by constant load electromagnetic bicycle ergometers. It is postulated from the data that (a) the braking power of electromagnetic ergometers effected by Pasqualini eddy current brakes probably does not give constant performance with the passage of time and usage, and that (b) the braking power may vary significantly – according to the scale meter used for the power setting – between bicycle ergometers at the same scales as well as the same or equivalent power settings.

In research work or in training experiments where work load is to be measured and/or controlled as an important variable, special care should be given to the calibration status of the ergometers in use. So-called observed differences described as training effects (or lack of it) may at times be attributable to instrumentation discrepancies.

Reference

1. Wilmore JH, Constable SH, Stanforth PR, Buono MJ, Tsao YW, Roby FB, Lawdon BJ, Ratcliff RA (1982) Mechanical and physiological calibration of four cycle ergometers. Med Sci Sports Exerc 14: 322–325

Body Weight and the Evaluation of the Spiroergometric Test Criteria

I. Iliev

Both physical loads under natural conditions and spiroergometric tests reveal close correlations between the total body indices (body weight in particular), the energy consumption, and the working capacity of the vegetative systems supplying body energy. The obvious limiting effect of body size on the working capacity and performance of athletes has led to the introduction of weight classes in wrestling, boxing, weight lifting, and other sports and to specific morphological and functional selection in most sports and sports branches.

The functional diagnostics of athletes by spiroergometric tests anticipate:
1. An evaluation of the current state of the cardiorespiratory system with a view to the requirements of the coaching and training program
2. Prediction of the future state or evaluation of the potential capacity for development

In both cases the obviously accurate answer to the questions thus stated presupposes a basis for comparison, which would consider the morphological characteristics of the individual and the specific requirements of the particular sport or respective sport branch.

Body weight is usually accepted as the best single criterion for the morphological characteristics of the individual and it normally shows high correlation with the basic anthropometric parameters: height and diameters and circumferences of the trunk and limbs.

The body weight is obviously of dual importance in the spiroergometric evaluation. On one hand, it is a necessary criterion for the energy requirements and at the same time an interfering factor in the comparative evaluation of individuals of otherwise homogeneous populations, such as those framed by a particular sport or an individual sport branch. In order to neutralize the interfering influence of body weight when interpreting the spiroergometric test criteria, the absolute values of the used parameters, such as the maximal ergometric power (W_{max}), the maximal aerobic power ($\dot{V}O_{2\,max}$), the maximal oxygen pulse ($\dot{V}O_{2\,max}\,Fh^{-1}$), the physical working capacity at pulse rate 170 (PWC_{170}), and the like are usually substituted by the relative values of the respective parameters ($W_{max}\,kg^{-1}$, $\dot{V}O_{2\,max}\,kg^{-1}$, $\dot{V}O_{2\,max}\,Fh^{-1}\,kg^{-1}$, $PWC_{170}\,kg^{-1}$).

Such an approach, however, does not eliminate but only modifies the relationship between body weight and the respective spiroergometric test criteria, e.g., from $y = a\,w^b$ to $y = a\,w^{b-1}$ [1–4, 6], where w stands for body weight and a and b are constants. According to von Döbeln [1] there exist theoretical prerequisites to expect the coefficient b value in equations of the type $y = a\,w^b$ to be ⅔.

In practice, the empirically determined coefficients often differ substantially from the theoretically expected ones, as they carry latent information about the specific requirements of the type of sports, the preselection of the population investigated, the level and the methods of training of athletes, and other factors out of the investigator's control. One cannot neglect the importance of the protocol itself for the spiroergometric investigation, apparatus used, the methods of grouping, and the working out of data.

The specific requirements of the type of sport and the influence of training and body weight on the spiroergometric criteria, are with a few exceptions, still insufficiently studied and it is not yet possible to make theoretical generalizations and precise definitions.

This is probably the reason that in the functional diagnostics a preference is given to the simple linear models of the relationships we are interested in, and to the purely empirical classifications of sports into groups, defined according to some basic spiroergometric criteria [7].

Our approach towards solving problems which arise in practice for the evaluation and the interpretation of spiroergometric criteria is also a pragmatic one: we aim at isolating the body weight factor when evaluating the functional capacity of the cardiorespiratory system of the athlete within the framework of the given type of sport, not paying attention to the actual relationships between the spiroergometric criteria and the body weight.

Material and Methods

Subjects of investigation were elite male and female athletes from the Olympic sports teams of Bulgaria.

Data from thousands of investigations, done during the past years, were analyzed.

The maximal spiroergometric test criteria are determined directly at the progressively increasing work load till exhaustion on a bicycle ergometer or a treadmill according to a standard protocol with steps of 90 s each.

At work on a bicycle ergometer, the initial value is 60 W and every 90 s an additional 30 W are added (regardless of the type of sport or sex).

At treadmill running, the initial speed is 6 km h^{-1}, progressively increasing by 1.2 km h^{-1} without changing of the standard slope of 2.5%.

Gas exchange is recorded automatically without interruption by an open system for gas analysis and the heart rate is recorded by ECG.

The subject of the present paper is mainly the results from bicycle ergometer tests, but the method we describe here is equally valid for the results from treadmill tests.

The data are processed using conventional tests including variance, correlation, and regression analyses.

Results and Discussion

The results from the variance analysis of data for each sport showed specific characteristics of the different types of sports according to some or most standard spiroergometric criteria. This is the reason to give up whatever grouping of data from different types of sports for practical purposes and to interpret the results for each sport independently. We assess according to different normative tables not only the obviously specific branches of track-and-field athletics but also the freestyle and the Graeco-Roman wrestling. We even apply specific normative tables at the assessment of investigation data within the framework of a given type of sport at different periods of training.

Thus, the problems, as far as the suitability of the respective normative tables for a given sport are concerned, drop out beforehand.

An irremovable defect of these tables, at least for the present moment, is the purely statistical empirical approach to their juxtaposition.

The correlation analysis of multiple data from the different types of sports confirmed our conviction that the presentation of data in relative values does not solve the problem of eliminating the body weight when assessing and interpreting them. One can see in Table 1 that the absolute and relative test criteria correlate almost equally high with the body weight. The correlation is stronger with boxing and wrestling, due to the great dispersion of body weight, but is obvious enough with other sports too.

Table 1. Correlation coefficients between body weight, $\dot{V}O_{2\,max}$, and $\dot{V}O_{2\,max}\,kg^{-1}$ using straight line and power curve approximations

No	Kind of sport	Straight line		Power curve	
		$\dot{V}O_{2\,max}$	$\dot{V}O_{2\,max}\,kg^{-1}$	$\dot{V}O_{2\,max}$	$\dot{V}O_{2\,max}\,kg^{-1}$
1	Rowing (males)	0.505	−0.432	0.695	−0.662
2	Rowing (females)	0.370	−0.312	0.361	−0.309
3	Canoeing (males)	0.728	−0.030	0.729	−0.045
4	Road cycling	0.516	−0.377	0.517	−0.376
5	Basketball (females)	0.482	−0.579	0.476	−0.588
6	Swimming (males)	0.345	−0.701	0.374	−0.681
7	Boxing	0.833	−0.828	0.832	−0.840
8	Judo	0.747	−0.887	0.795	−0.866
9	Graeco-Roman wrestling	0.884	−0.891	0.918	−0.881
10	Freestyle wrestling	0.919	−0.877	0.924	−0.900

The plotting of investigation data of one and the same team during different training periods showed that the coefficients of the empiric equation, which describe the interrelation between $\dot{V}O_{2max}$ (and other test criteria) and the body weight, undergo considerable changes under the influence of factors not controlled by the investigator (selection, training methods, nutrition?). This is illustrated in Fig. 1 and 2.

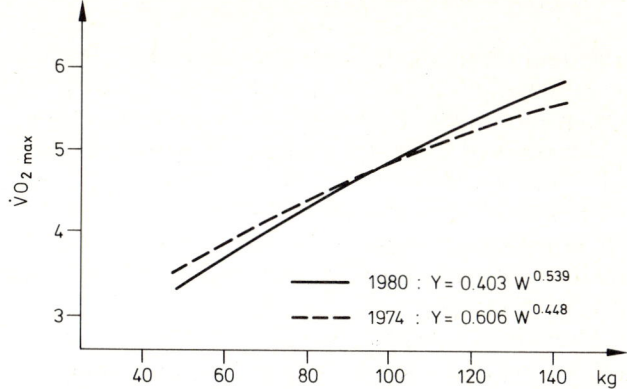

Fig. 1. $\dot{V}O_{2\,max}$ against body weight plot (data from free style wrestling team in 1974 and 1980)

1980 : Y = 0.403 W$^{0.539}$
1974 : Y = 0.606 W$^{0.448}$

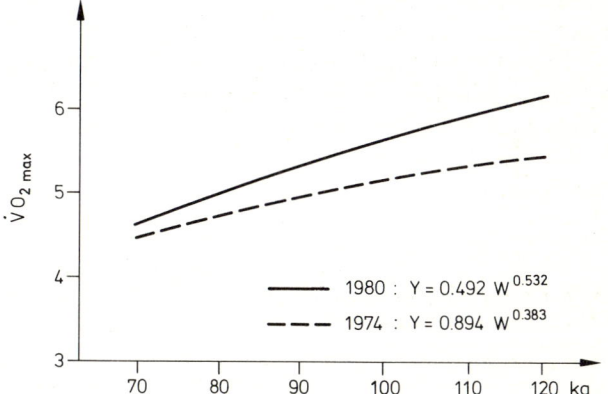

Fig. 2. $\dot{V}O_{2\,max}$ against body weight plot (data from men's rowing team in 1974 and 1980)

1980 : Y = 0.492 W$^{0.532}$
1974 : Y = 0.894 W$^{0.383}$

For that reason, the evaluation, which is based on analysis of regression equations (regardless of the type, linear or nonlinear), becomes gradually inadequate to the real relationships. There arises a necessity of current actualization of the normative base or of the corresponding regression equations.

The graphic presentation of $\dot{V}O_{2\,max}$, in absolute and relative values opposite to the body weight, gives us the possibility to see that in most cases we have curves with almost a mirror image (Fig. 3 and 4). This made us think that it is possible and convenient to eliminate the body weight factor at the evaluation of $\dot{V}O_{2\,max}$ (and the other test criteria), not knowing analytically the relationship $\dot{V}O_{2\,max}$ – body weight, by way of assessing separately the absolute and the relative values and taking the half-sum of the two evaluations. Obviously, if the individual values of $\dot{V}O_{2\,max}$ and of $\dot{V}O_{2\,max}$ kg^{-1} are a function of the body weight only, they will lie on the curves which represent the functions $\dot{V}O_2 = f\,(w)$, resp. $\dot{V}O_2$ kg$^{-1} = f^1\,(w)$, and the half-sum of the evaluations will tend towards the mean evaluation. If the absolute or the relative values (or both) are substantially different from the theoretically expected values $f\,(w)$, resp. $f^1\,(w)$, the half-sum of the evaluations, then, will be different from the

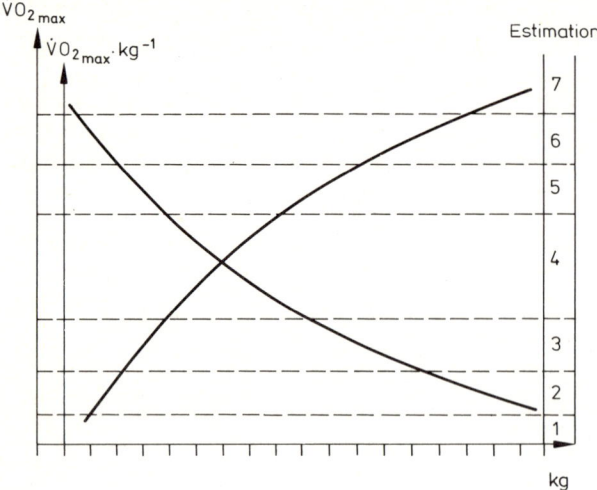

Fig. 3. $\dot{V}O_{2\,max}$ *(upward)* and $\dot{V}O_{2\,max}\,kg^{-1}$ *(downward)* against body weight plot and 7-graded estimating nomogram (data from Graeco-Roman wrestling team, 1974, schematically)

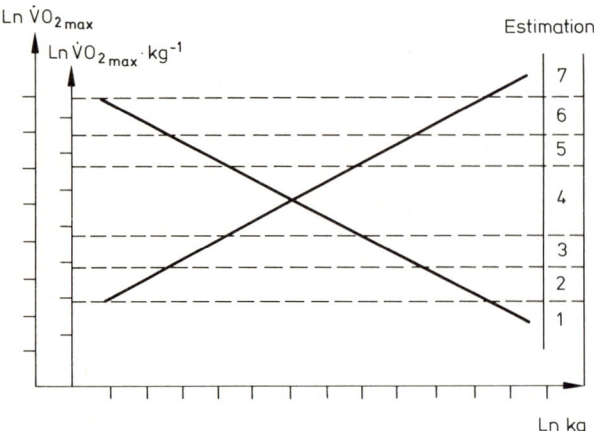

Fig. 4. The same as Fig. 3 on a logarithmic scale

mean value. And it is this value that will present the assessment of the cardiorespiratory function unrestricted by the disturbing influence of the body weight factor. In principle, it is not of great significance how $\dot{V}O_{2\,max}$ and $\dot{V}O_{2\,max}\,kg^{-1}$ will be assessed. An elementary rank evaluation is even possible for lack of better methods. We routinely apply a 7-graded evaluation scheme on the basis of the standard deviation of data. The absolute values of z evaluations, or the percentile evaluations, may be used also. In practice, the method of evaluation does not influence significantly the half-sum of the evaluations or the actual assessment of the cardiorespiratory system, unrestricted by the influence of the body weight, if it is properly conformed with the peculiarities of the statistical material.

A statistical evidence of the effective elimination of the body weight factor at the evaluation of the spiroergometric test criteria is the disappearance of the correlation between the estimation value and the body weight (Table 2). But if, nevertheless, it correlates, that means that the body weight has a positive (or negative) significance for the specific working capacity in the given type of sport and, because of that, stronger athletes with certain weight characteristics, who are in fact another population with its own special features, are selected in a natural way among the members of the team.

Table 2. Correlation between the body weight and the half-sum of $\dot{V}O_{2\,max}$ and $\dot{V}O_{2\,max}\,kg^{-1}$ estimation values

No	Kind of sport	Correlation coefficient
1	Rowing (males)	0.114
2	Rowing (females)	−0.078
3	Canoeing	0.304
4	Road cycling	0.052
5	Basketball (females)	−0.268
6	Swimming (males)	−0.190
7	Boxing	0.085
8	Judo	−0.069
9	Graeco-Roman wrestling	0.148
10	Freestyle wrestling	0.090

The problem of eliminating the influence of the body weight over the spiro-ergometric test criteria also arises with the use of submaximal laboratory (or clinical) investigations. We are old hands in using the test, proposed by Böhlau [5] for the determination of the so-called recovery coefficient (*Erholungsquotient* or EQ). Böhlau recommends $2\,W\,kg^{-1}$ as the standard 5-min work load. The application of the test with wrestlers from different weight classes showed that the work load is subjectively perceived easier by the wrestlers of the lighter weight classes. This was the cause for processing the data of EQ values and the body weight by a correlation analysis in the first place with wrestlers and boxers and then with other sports (for men and women). A considerable correlation of the order of 0.400 to 0.600 (with sports having weight classes it is even greater) between EQ and the body weight was established. When fitting the different curves to the data from different sports we got the best values of the determination coefficients (R^2) for the following logarithmic equations:

$$EQ = 6.3 - 0.573\,(\ln kg) \text{ for women, } R^2 = 0.536$$
$$EQ = 7.0 - 0.692\,(\ln kg) \text{ for men, } R^2 = 0.645$$

On the basis of these equation and our data for the mean working capacity of male and female athletes, we calculated out the following equations:

$$W\,kg^{-1} = 4.29 - 0.573\,\ln kg \text{ for women}$$
$$W\,kg^{-1} = 4.99 - 0.692\,\ln kg \text{ for men}$$

Table 3. Work load intensity of nonmaximal laboratory test on a bicycle ergometer for men and women

Body weight (kg)	Work load (W)	
	Men	Women
50	110	100
55	120	110
60	130	115
65	140	120
70	145	130
75	150	135
80	160	140
85	165	150
90	170	155
95	175	160
100	180	165
110	190	175
120	200	–
130	210	–

which made the basis for the standardization of the 5-min work loads for men and women (Table 3).

The practical application of the method showed that athletes, regardless of body weight, subjectively perceive the work loads, determined according to the above-mentioned table, as easy or moderate ones. Untrained persons perceive as rather intensive or heavy. Practically every healthy person, aged 15–45, is able to finish the test. The correlation between EQ and the body weight does not exist anymore.

Conclusions

1. The body weight substantially influences the spiroergometric test criteria at both maximal and submaximal work loads.
2. The relationship between body mass and the generally accepted spiroergometric test criteria, on principle, is a nonlinear one and may be of importance for the interpretation of test results.
3. The presentation of spiroergometric test criteria in relative values (per 1 kg body weight) does not eliminate but only modifies the relationship between the respective criteria and the body weight.
4. The use of the half-sum of the estimations of the absolute and relative values of the respective parameters is a practical and suitable method for neutralization of the body weight influence on spiroergometric test criteria.

References

1. Döbeln W v (1956) Maximal oxygen intake, body size and total hemoglobin in normal man. Acta Physiol Scand 38: 193
2. Michailow WW, Zaciorskij VM, Geselevic VA (1967) Untersuchung über die Abhängigkeit der funktionellen Möglichkeiten der Sportler von ihrem Körpergewicht. Med Sport 7: 169–172
3. Asmussen E (1969) The influence of size on physical performance. In: Karl J, Novotny V (eds) Physical fitness and its laboratory assessment. Charles University Press, Prague, pp 31–34
4. Koinzer K (1983) Zur Abhängigkeit des herzschlagfrequenzbezogenen Sauerstoffaufnahmevermögens von der Körpermasse bei 10 bis 14 jährigen Jungen und Mädchen. Med Sport 23: 240–244
5. Böhlau V (1955) Prüfung der körperlichen Leistungsfähigkeit. Thieme, Leipzig
6. Zatsiorsky BM (1966) Fizicheskie kachestva sportsmena. (Motor abilities of the sportsman). Fizkultura i Sport, Moscow
7. Szögy A, Lazaresku D (1977) Zur Beurteilung der maximalen O_2-Aufnahme bei Hochleistungssportlern, unter besonderer Berücksichtigung des Körpergewichtes. Sportarzt Sportmed 28: 163–165

Comparability of Absolute and Body-Related Performance Capacity in Ergometry

H.-V. Ulmer

Introduction

In the older literature concerning physiology, physiological performance indicators, such as $\dot{V}O_{2max}$, heart volume, or exercise performance are mostly published without relation to individual body parameters. In Germany it is still a convention to characterize the critical heart weight as "500 g" [6, 8, 11], without regard to the constitution of the athletes. In recent studies, however, there is an increasing tendency to uniformly calculate all physiological results in relation to body weight, in spite of the fact that, for example, Åstrand and Rodahl wrote in 1977: "In work and exercise where the body is lifted (as in walking or running), the oxygen uptake should be related to the body weight" [1].

Relation to Body Parameters

In general, results in the literature give three different reference values for the relation to individual body parameters:

– relation to body surface area
– relation to body weight
– relation to lean body mass

There is not only one answer to the question of whether or not ergometric results should be related to body parameters. The answer has to take into account the special problem to be solved. In some cases the relation to body parameters can improve comparability as well as validity of test results; in other cases this relation can impair it, especially when using body weight parameters.

A relation to body surface is common for values of basal metabolic rate or cardiac index. The body's surface is important as a thermal exchange area, and the values usually estimated include the body mass as the thermal productive factor. Concerning noninvasive ergometric results, however, it is unusual to relate them to the surface area.

Significance of Body Weight

The validity of physiological indicators for performance capacity can be optimized if the manner of work is taken into account. Three alternatives are to be considered:

1. Provided that the weight of the subject's own body is the only load, the performance parameters of different individuals can be compared best if the results are related to body weight.
2. If an additional weight is to be transported, the absolute performance capacity is also of importance. For example, when it is required to carry a heavy (30 kg) pack uphill, a small person with correspondingly light body weight will certainly perform less well than a tall, heavy person, even though their performance capacities per unit body weight are the same.
3. If the performance capacity of the skeletal muscle is to be evaluated, it is reasonable to express the performance test results with respect to the fat-free body mass (lean body mass) because this quantity is basically determined by the muscle mass.

Skinfold Measurement and Its Methodological Improvement

For estimation of body fat content (and by this, indirectly, the body mass) there is a very useful method. The method, measurement of skinfold thickness by a caliper, is well known in countries, such as Czechoslovakia, England, and the United States. However, it is not well known in Germany. Different types of calipers have been described in the literature. We surveyed 12 and found the differences of their results not very important. We use the Harpenden caliper (Fig. 1), which includes some advantages of standardization [4, 9, 13].

There is no doubt that the sum of the thicknesses of four representative skinfolds (Fig. 2) correlates with the total body fat content, determined by measurements of specific gravity. This agrees with the results of Durnin and Womersley [5] and others. Extensive experiments by Durnin and Womersley [5] are the basis for diagrams and tables for estimation of body fat content using skinfold measurements. These tables are separated for sex and age. The age, however, is only in steps of 10 years. A mathematical approximation by Berres et al. [2] avoids these steps. Their formula can be applied in computer programs (Table 1).

Fig. 1. Measurement with a Harpenden caliper [9]

Table 1. Formula for estimation of body fat content

Women

$$\% \text{ fat} = (\frac{4.95}{1.1572 - 0.0647 \cdot \log (B) - 0.00038 \cdot A} - 4.5) \cdot 100$$

Men

$$\% \text{ fat} = (\frac{4.95}{1.1739 - 0.06227 \cdot \log (B) - 0.000555 \cdot A} - 4.5) \cdot 100$$

Based on the sum of the skinfold thickness of the 4 areas shown in Fig. 2. The equations are fitted to the tables of Durnin and Womersley [5].
A, age in years; B, sum of the 4 skinfold thicknesses in mm [2].

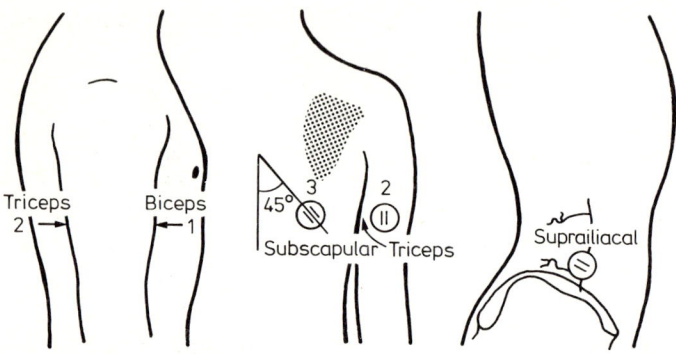

Fig. 2. Scheme of the 4 representative areas for measurements of skinfold thickness, according to Durnin and Womersley [5]

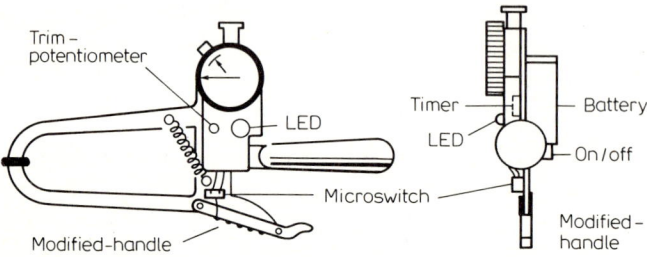

Fig. 3. Installation of an electrical timer on the Harpenden caliper [10]

A further methodological improvement was brought about in the reading of the micrometer [10]. In the past the time difference between loosening the handgrip and reading the micrometer was not standardized. In the literature we found values between 2 s [15] and 30 s [14], or characterizations such as reading until standstill of the indicator [7] or immediately after applying the caliper [3].

For the standardization of this time difference of the indicator reading, an electronic timer was connected to the handgrip (Fig. 3) in such a way that a light-emitting diode (LED) lighted up 2 s after loosening the handgrip, thus indicating the standardized moment for reading the micrometer scale. The 2-s interval was chosen in accordance with Weiner and Lourie [15] and with a systematic analysis by Zwerger [16].

Instead of the micrometer, a position pickup makes it possible to register the skinfold thickness by recorders, on-line computer systems, and so on [17]. Figure 4 shows a construction by Zwerger [16] and Fig. 5 gives an example of a graph, registered with a compensating recorder.

Conclusions

The decision of whether to relate physiological indicators of performance capacity to individual body parameters or not depends on the special problem to be solved.

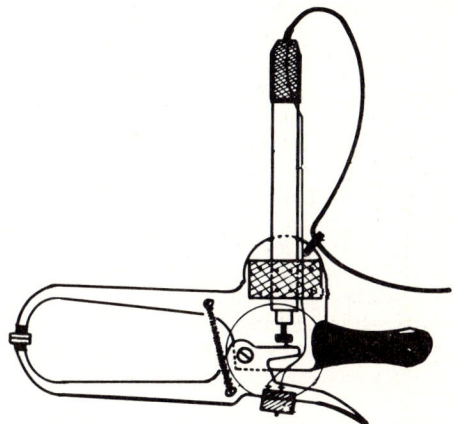

Fig. 4. Installation of an inductive position pickup (W 5 TK; Hottinger Baldwin, Darmstadt) on the Harpenden caliper [16]

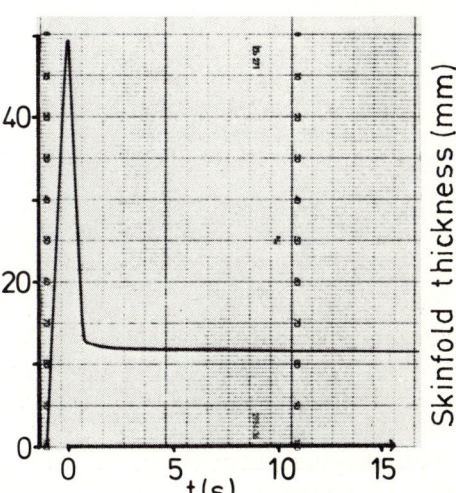

Fig. 5. Record of skinfold thickness with Zwerger's caliper system [16] (modified)

Depending on this problem, validity and comparability can be optimized if an individual factor is regarded. The choice of this factor depends, again, on this special problem. Concerning lean body mass, skinfold thickness measurements are a recommendable method. By new technical improvements the degree of objectivity, economy, and reproducibility might be improved.

Summary

Some years ago, it was unusual to refer ergometric results to body parameters. Today such results are often related uniformly to the body weight. The problem remains to be solved whether to relate or not to relate results to body parameters and if body weight or lean body mass should be preferred. Concerning skinfold measurements as an useful method to estimate body fat content and lean body mass should be preferred. new methodological improvements are presented. They enable the estimation of the body fat content by on-line systems, improving the quality of use and test criteria.

References

1. Åstrand P-O, Rodahl K (1977) Textbook of work physiology. Physiological bases of exercise. McGraw-Hill, New York
2. Berres F, Ulmer H-V, Lamberty M (1980) Calculation of total body fat from skinfold thickness by using an age-corrected formula. Pflügers Arch 384: Suppl 35
3. Bonatz H (1970) Über die Entwicklung des subkutanen Fettgewebes und seine Bedeutung als gruppenspezifizierende Körperbauvariable bei männlichen Jugendlichen. Math-naturwiss Dissertation, Universität Kiel
4. Brožek J (1956) Physique and nutritional status of adult men. Hum Biol 28: 124–140
5. Durnin JVGA, Womersley J (1974) Body fat assessed from total body density and its estimation from skinfold thickness measurements on 481 men and women aged from 16–72 years. Br J Nutr 32: 77–97
6. Findeisen DGR, Linke P-G, Pickenhain L (eds) (1976) Grundlagen der Sportmedizin. Barth, Leipzig
7. Fletscher RF (1962) The measurement of total body fat with skinfold calipers. Clin Sci 22: 333–346
8. Hollmann W, Hettinger T (1980) Sportmedizin – Arbeits- und Trainingsgrundlagen, 2nd edn. Schattauer, Stuttgart
9. Jürgens HW (1979) Hautfaltenmessungen als Indikator des Gesamtkörperfettes bei jungen Männern. Research Report BMVg InSan Nr 0476-V-073, Bonn
10. Krämer H-J, Ulmer H-V (1981) 2-second standardization of the Harpenden caliper. Eur J Appl Physiol 46: 103–104
11. Nöcker J (1980) Physiologie der Leibesübungen für Sportlehrer, Trainer, Sportstudenten,
12. Rehs HJ, Berndt I, Rutenfranz J, Burmeister W (1975) Untersuchungen zur Bestimmung der Hautfaltendicke mit verschiedenen Calipern. Z Kinderhk 120: 121–133
13. Tanner JB, Whitehouse RH (1955) The Harpenden skinfold caliper. Am J Phys Anthropol 13: 743–746
14. Walter GH, Zidek W (1980) Untersuchungen zur Beziehung zwischen Ergometerleistung (PWC$_{170}$) und fettfreier Körpermasse. Wehrmed Mschr 24: 336–341
15. Weiner JS, Lourie JA (1969) A guide to field methods. Oxford, Edinburgh
16. Zwerger M (1981) Zur automatischen Registrierung von Hautfaltendicken mit dem Harpenden-Caliper und zur Standardisierung des Ablesezeitpunktes. Diplomarbeit, University of Mainz
17. Zwerger M, Ulmer H-V (1980) Automatic registration with the Harpenden caliper. Pflugers Arch 384: Suppl 35

Physical, Physiological, and Body Compositional Differences of Male and Female Septuagenarians

L. P. Novak and H. West

Historically, aging has been a neglected area of scientific investigations in the United States. Physical activity or inactivity was one of many problems associated with the aging process. In recognition of the importance of exercise to overall well-being of people, the Senate Sub-Committee on Aging in 1975 provided money for the sole purpose of educating and enlisting participation of elderly people in physical activities.

If the changing pattern of aging is considered in the United States and if the birth rate sustains its decline, it is predicted that by the year 2000, nearly 50% of our citizens will be over 50 years old. Even though such an age cannot be considered as detrimental, it is of utmost importance to provide adequately objective appraisal of physical, physiological, and psychological capabilities of our elderly. Further, an equally important task lies ahead of all of us concerned with aging processes and that is to bring the focus of our attention to the improvement of physical and mental fitness, which in turn will, or should, improve the quality of life in aging people [14].

A brief summary of common physical manifestations of old age provides an insight into changes affecting muscular strenght, endurance, and agility [11]. According to Åstrand and Rodahl [4], strength seems to decline by 40%, and legs apparently suffer most from muscle cell loss. Muscular fatigue seems to be greater in the elderly [12] and reaction time suffers a significant decrease with advancing age [21].

Aging also involves redistribution of lean tissue and total body fat. In particular, body cell mass decreases with an age-concomitant increase in total body fat in both sexes, according to Novak [16]. Similar loss of so-called active protoplasmic mass was described by Shock et al. [18].

As far as the respiratory system is concerned, deVries [10] documented decrease of total lung capacity as well as vital capacity and maximal breathing capacity. This decrease might be due to weaker muscles engaged in facilitation of breathing or to reduction of the total number of alveoli and some decrease in flexibility and expandability of the lungs [17].

The major factor in reduced working capacity of the elderly is the decrease of cardiovascular function [20]. Practically every parameter of cardiovascular performance decreases as aging marches on [15]. Cardiac output is diminished due to a combination of a smaller stroke volume and to a decreased magnitude of achieving maximal heart rate. Maximal oxygen intake decreases by nearly 50% by the time 75 years of age is reached [10]. This parameter of cardiovascular capacity was studied by various authors [2, 7, 8, 9, 22] and others who were concerned primarily with the question of the highest achievable maximal oxygen intake in elderly males and females classified either as sedentary or active individuals. Despite these studies, very little is known about the maximal aerobic capacity of males in their 7th, 8th, and 9th decades. And, the total amount of data on the same subject for elderly females remains rather small even today.

The purpose of this study was not to document the highest maximal oxygen intake of elderly of both sexes. Rather, we were interested in finding out the response of cardiovascular-pulmonary functions to increasing work loads of the same magnitude for both sexes. Thus, the differences in responses between elderly male and female subjects could be obtained under examination eliciting conditions near to the vita maxima. The differences of body composition between male and female septuagenarians were obtained as a by-product of the study.

Subjects and Methods

The subjects for this study were informed volunteers, 15 males and 21 females, between the ages of 65 and 95 years who were recruited from two homes for the elderly where middle or upper socioeconomic status elderly live. All subjects had a complete physical examination including resting ECG prior to the beginning of the study. They were in relatively good health without any orthopedic problems or any debilitating diseases which would impair their neuromuscular and/or cardiovascular capacity. Signed release forms were obtained from all subjects prior to the beginning of testing procedure. The physical activity and smoking habits questionnaire indicated that the subjects walked about 30 to 60 min leisurely 3 times/week and the smoking habits of few male subjects were considered as very mild, i.e., less than one-half pack of cigarettes/day.

For ergometric examination, a Monarch bicycle ergometer was used. The starting work load was set at 150 kpm at 50 rpm for all subjects. This work load was increased every 3 min until 600 kpm were reached by all subjects. At that time, the heart rate achieved was used for calculations of oxygen intake, according to the method of Åstrand and Rhyming [3]. Correction factor for age was used as proposed by von Döbeln et al. [24].

Lead V_5 was used to monitor the heart rate on ECG as well as on an oscilloscope during the entire period of exercise and during 4 min of recovery. Blood pressure was determined using a sphygmomanometer every minute during exercise and during recovery. Vital capacity and 1 s forced vital capacity were determined with a Collin's spirometer with a 6-l capacity.

Anthropometrical measurements included height, weight, and lengths and circumferences of arms and legs. Skinfolds were measured with a Lange skinfold caliper with a constant pressure between the jaws of 10 g/mm². For assessment of lean tissues, corrected diameters of the upper arm and forearm were calculated according to the following formula:

$$\text{CD upper arm, cm} = \frac{C, \text{ cm}}{\pi} - \frac{(S_t + S_b)}{2}, \text{ mm}$$

where CD is the corrected diameter of upper arm, C is circumference of the upper arm, π equals 3.14145, S_t is skinfold measured on the triceps, and S_b is skinfold measured on the biceps.

$$\text{CD forearm, cm} = \frac{C, \text{cm}}{\pi} - \text{skinfold forearm, mm}$$

Arm muscle circumference, which also provides information about lean tissues, was calculated according to the following formula:

$$\text{AMC, cm} = C, \text{Ua, cm} - (0.314159 \times \text{skinfold triceps, mm})$$

As an estimate of static strength, grip strength was determined with a calibrated hand dynamometer. The maximal reading was used for recording the result after three trials were attempted. The preferred hand was used for testing static strength.

Body density of males was calculated according to the formula of Brozek et al. [5] and that for females according to the proposal of Young et al. [25]. Body fat was calculated as suggested by Brozek et al. [6].

Results

Table 1 presents means and standard deviations of age and physical measurements. The male subjects were significantly older and taller. Even though their weight was higher by 6.7 kg and their resting heart rate was lower by 12 beats per min, these dif-

Table 1. Age and physical measurements

Subjects	n	Age	Height	Weight	Heart rate	Blood pressure (mmHg)	
		(years)	(cm)	(kg)	(beats/min)	Systolic	Diastolic
Males	15	81.9[a]	169.5[a]	69.6	69	141	79
		±6.9	±7.0	±19.9	±17	±20	±6
Females	21	74.6	156.0	62.9	81	142	83
		±6.6	±10.0	±10.0	±10	±18	±7

$n \pm$ SD
[a] Significant at 0.01 level

Table 2. Lengths and circumferences of upper limb

Subjects	n	Length (cm)		Circumference (cm)	
		Upper arm	Forearm	Upper arm	Forearm
Males	15	34.2[a]	28.3[b]	28.4	25.1[b]
		±2.5	±1.4	±2.2	±1.9
Females	21	31.9	26.7	27.1	22.9
		±2.9	±1.2	±3.4	±1.7

$n \pm$ SD
[a] Significant at 0.05 level
[b] Significant at 0.01 level

ferences did not reach statistical significance. Resting blood pressure, both systolic and diastolic, was about the same in both sexes.

Lengths and circumferential measurements of upper arm and forearm are presented in Table 2. Males had significantly longer upper arms and forearms as well as greater circumference of forearm, compared with females.

Table 3 shows distribution of subcutaneous fat as measured by skinfolds. Males had significantly less subcutaneous fat deposited on the upper arm dorsal site (triceps) and on the iliac crest.

Table 4 presents the diameter of lean tissues of the upper limb and grip strength achieved by the subjects. Males had significantly larger lean diameter of the upper arm as well as forearm. Similar significant results from arm muscle circumference and grip strength in favor of males were obtained.

Table 5 shows calculated body density and leanness-fatness. Significantly higher body density and estimate of lean body mass of males were obtained in both abso-

Table 3. Skinfolds (mm)

Subjects	n	Upper arm		Forearm	Subscapular	Iliac crest
		Dorsal	Ventral			
Males	15	13.4[a]	6.9	7.1	19.7	23.0[a]
		3.3	±2.4	±2.4	±5.8	±7.3
Females	21	20.5	8.3	7.6	19.2	31.7
		±5.1	±3.5	±3.5	±7.7	±9.1

$n \pm SD$
[a] Significant at 0.01 level

Table 4. Lean tissues of upper limb and grip strength

Subjects	n	Corrected diameter upper arm (cm)	Corrected diameter forearm (cm)	Arm muscle circumference (cm)	Grip strength (kg)
Males	15	8.0[a]	7.3[a]	24.2[a]	34[a]
		±0.6	±0.5	±1.8	±8
Females	21	7.2	6.5	20.7	26
		±0.8	±0.4	±2.3	±5

$n \pm SD$
[a] Significant at 0.01 level
[b] Significant at 0.001 level

Table 5. Body density, fat, and leanmass

Subjects	n	Body density (g/m)	Total body fat		Lean body mass	
			kg	%	kg	%
Males	15	1.0541[a]	13.8[b]	19.5[a]	55.8[b]	80.5[a]
		±0.0161	±7.3	±6.7	±15.6	±6.7
Females	21	1.0327	18.2	28.4	44.6	71.6
		±0.0113	±5.7	±4.8	±4.9	±4.8

$n \pm SD$
[a] Significant at 0.001 level
[b] Significant at 0.05 level

Table 6. Exercise and postexercise heart rate (beats/min)

Subjects	n	Stage 1	Stage 2	Stage 3	Stage 4	Postexercise	
						2 min	4 min
Males	15	102[a]	108[b]	121[b]	132[a]	99[a]	90[a]
		±22	±17	±16	±9	±15	±13
Females	21	113	125	140	146	108	99
		±13	±14	±10	±10	±13	±13

$n \pm SD$
[a] Significant at 0.05 level
[b] Significant at 0.01 level

Table 7. Exercise and postexercise blood pressure (mmHg)

Subjects	n	Stage 1	Stage 2	Stage 3	Stage 4	Postexercise	
						2 min	4 min
Males	15	148/77	159/75	167/73	176/72	154/76	142/75
		±23/ 8	±25/ 7	±25/ 7	±21/ 7	±21/ 7	±20/ 5
Females	21	158/77	164/80	170/74	186/76	155/76	148/75
		±17/ 7	±19/ 7	±25/16	±19/ 7	±16/ 5	±21/ 9

$n \pm SD$

Table 8. Aerobic capacity and vital capacity

Subjects	n	Maximal oxygen intake		Vital capacity (ml)	1 s vital capacity (ml)
		l	ml kg-min^{-1}		
Males	15	1.86[a]	25.6	3052[a]	2364[b]
		±0.29	±4.8	± 700	±716
Females	21	1.61	24.8	2318	1629
		±0.39	±6.7	±1131	±511

$n \pm SD$
[a] Significant at 0.05 level
[b] Significant at 0.01 level

lute and relative terms. On the other hand, a significantly lower amount of body fat was found in males.

Table 6 presents changes of heart rate during exercise and in recovery. In all stages of exercise and in recovery, males achieved significantly lower heart rates under the same work load when compared with females.

Table 7 presents changes in blood pressure during exercise and in recovery. There were no significant differences found between the sexes, even though the blood pressure, primarily the systolic phase, was always slightly lower in males.

Table 8 shows aerobic capacity and vital capacity of subjects. Males achieved significantly higher maximal oxygen uptake in liters only. And, both the vital capacity and 1 s forced vital capacity were also found to be significantly higher in males.

Discussion

The results of aerobic capacity of elderly men and women in this study compared with Åstrand's [1] nomogram indicated that the cardiovascular-pulmonary fitness of both sexes was fair, rather than average. This achievement deserves clarification. The comparative design of the study in which the same work load was used for both sexes elicited cardiovascular-pulmonary responses near "vita maxima" rather than maximal effort to exhaustion. Our criteria was the achievement of predicted maximal heart rate based on the formula 220 beats/min minus age. Serial documentation of heart rates during stagewise increments of work load indicated that at stage 4 the males achieved heart rates near to 98% of the maximum with corresponding oxygen consumption of 1860 ml. On the other hand, the females in this study had already achieved predicted maximal heart rates at stage 3. An oxygen consumption of 1610 ml was achieved in stage 4 at a heart rate of 146 beats/min. Thus, the difference in oxygen consumption of the females compared with males of the similar age was about 14% lower when the same work load of 600 kpm was used, but their heart rates at stage 4 were about 11% higher.

When aerobic capacity of the elderly males in this investigation was compared with the results of a comprehensive study presented by Tlusty [23], it seems that the oxygen consumption of 1860 ml achieved by males in this study corresponds well to 1869 ml reached by elderly males aged 60 to 69 years in the latter investigation. Perhaps this comparison might indicate equally efficient cardiovascular capacity of the males who participated in our study and who were practically two decades older. If this fact is considered further, it points towards the necessity of describing previous participation in physical activities of subjects volunteering for a study. Our subjects were not only in reasonably good health but they also knew subjectively that they had reasonably good cardiovascular fitness. This point might be illustrated well if the study of Sidney and Shephard [19] is considered. Their elderly males – 63.7 years old, on an average – were as the authors quoted "sufficiently interested in physical activity." Their maximal oxygen intake was 2060 ml achieved by a "fair" effort. We know also that the subjects in our sample tried their best to finish all four stages of the bicycle test. Therefore, if we assume reasonably equal effort of both samples of subjects, it is possible to make a simple inference that our male subjects, despite their older age by about two decades, displayed similar aerobic capacity at the heart rate of 132 beats/min compared with 154 beats/min determined in the group of elderly males studied by Sidney and Shephard [19].

If we turn our attention to the aerobic capacity of females participating in this study, their aerobic capacity of 1610 ml was close to 1756 ml of oxygen consumption listed for females 50 to 59 years old. However, it was better than 1403 ml for females 60 to 69 years old reported by Tlusty [23]. It seems, as we mentioned earlier, that the cardiovascular capacity of our septuagenarian females was slightly better compared with their younger counterparts from Czechoslovakia. Sidney and Shephard [19] reported 1500 ml and 1640 ml of oxygen consumption achieved by elderly females 63.4 years old on an average with maximal heart rates of 155 and 161 beats/min, respectively. Female subjects in our sample a decade older were able to reach similar cardiovascular capacity at an average heart rate of 146 beats/min. Thus, it seems appropriate, at this point, to quote Åstrand [1] who proposed that with respect to studying the aging process of the cardiovascular system either by submaximal or maximal oxygen consumption tests, prior to a cardiovascular fitness test the degree of physical activity in which participants in the study were engaged must be taken into consideration. Such a statement was supported also by studies of Fischer et al. [13] who documented the favorable influence of habitual participation in sports activities, which manifested itself in elderly subjects even in the 6th and 7th decades of life. In terms of body composition, the elderly males and the females in this investigation displayed quite a favorable amount of lean body mass, i.e., 80.5% for males and 71.6% for females. If we consider the lean body mass of "reference man" and that of "reference woman," 25 years old [6], as "standards" with lean body mass of 84.0% and 75.0%, respectively, then, it is remarkable to conclude that the simple walking program of our subjects resulted in low body fat even at the 7th or 8th decade of life.

This important consideration of the beneficial role of physical activity (even at old age) on body fat stands out in view of the study presented by Novak [16] who demonstrated a rather high percentage of body fat – 36.2% – attained by sedentary males 65 to 85 years old and 44.8% found in sedentary females of the same age group.

Perhaps as more data from physiological and body compositional studies accumulate with pertinent information about the numerous benefits available to and attainable by the elderly through participation in simple physical activities such as walking, then we will be able to provide assistance in counseling the elderly based on the fact that relatively high cardiovascular fitness can be achieved in old age supplemented by appropriate leanness of the body. These attributes coupled with the psychological joy derivable from physical activities should contribute to the enhancement of the total well-being of the elderly.

Summary

Significantly higher values were found in males on measures of age, height, and weight. Even though resting heart rate was lower in males, it did not reach a significant level. Resting blood pressure was about the same in both sexes.

Significantly longer upper arm and forearm as well as greater circumference of forearm with lesser amount of subcutaneous fat on triceps and on iliac crest were found in males. Similarly, greater amounts of lean tissues on upper arm and forearm were noted in males, which resulted in greater grip strength in favor of males. Significantly higher lean body mass and lower body fat was demonstrated in males.

During the exercise test blood pressures at each stage did not differ significantly between sexes. However, males achieved significantly lower heart rates at all stages of exercise and in recovery. Oxygen intake in liters, vital capacity and 1 s forced vital capacity was found to be significantly higher in favor of males.

References

1. Åstrand I (1960) Aerobic work capacity in men and women, with special reference to age. Acta Physiol Scand [Suppl] 169: 1–92
2. Åstrand PO (1956) Physical fitness with special reference to sex and age. Physiol Rev 36: 307–335
3. Åstrand PO, Rhyming I (1954) A nomogram for calculation of aerobic capacity (physical fitness) from pulse rate during submaximal work. J Appl Physiol 7: 218–221
4. Åstrand PO, Rodahl K (1970) Textbook of work physiology. McGraw-Hill, New York, pp 32
5. Brozek J (1952) Changes in body composition in men during maturity and their nutritional implications. Fed Proc 11: 784–793
6. Brozek J, Grande F, Anderson JT, Keys A (1963) Densitometric analysis of body composition: revision of some qualitative assumptions. Ann NY Acad Sci 110: 113–140
7. Cotes JE, Hall AM, Johnson GR, Jones PRM, Knibbs AV (1974) Decline with age of cardiac frequency during submaximal exercise in healthy women. J Physiol (Lond) 238: 24–25
8. Cummings GR, Borysyk LM (1972) Criteria for maximum oxygen intake in men over 40 in a population survey. Med Sci Sports 4: 18–22
9. Davies CTM (1972) The oxygen transporting system in relation to age. Clin Sci 42: 1–13
10. deVries HA (1974) Education for physical fitness in the later years. In: Crabowski S, Mason DT (eds) Education for the aging. Educ Resources Div, Capital, Washington DC
11. deVries HA (1977) Physiology of physical conditioning for the elderly. In: Harris R, Frankel LJ (eds) Guide to fitness after fifty. Plenum, New York, pp 47–52
12. Evans SJ (1971) Electromyographic analysis of skeletal neuromuscular fatique with special reference to age. Ph.D. dissertation, University South California, Santa Barbara
13. Fischer A, Pařízková J, Roth Z (1965) The effect of systematic physical activity on maximal performance and functional capacity in senescent man. Int Z Angew Physiol 21: 269–304
14. Harris R, Frankel LJ (eds) (1977) Guide to fitness after fifty. Plenum, New York, p 3
15. Hodgson JL, Buskirk ER (1977) Physical fitness and age, with emphasis on cardiovascular function in the elderly. J Am Geriatr Soc 25: 385–392
16. Novak LP (1972) Aging, total body potassium, fat-free, and cell mass in males and females between ages 18 and 85 years. J Gerontol 27: 438–443
17. Rockstein M (1975) The biology of aging in humans. In: Goldman R, Rockstein M (eds) The physiology and pathology of human aging. Academic, New York, pp 1–7
18. Shock NW, Watkin DM, Yiengst MM, Norris AM, Gaffrey GW, Gregerman RI, Falzone JA (1963) Age differences in the water content of the body as related to basal oxygen consumption in males. J Gerontol 18: 1–8
19. Sidney KH, Shephard R (1977) Maximum and submaximum exercise tests in men and women in the seventh, eighth, and ninth decade of life. J Appl Physiol 43: 280–287
20. Simonson E (1977) Effect of age on work and fatigue. Cardiovascular aspects. In: Harris R, Frankel LJ (eds) Guide to fitness after fifty. Plenum, New York, pp 53–65
21. Spirduso WW (1975) Reaction and movement time as a function of age and physical activity level. J Gerontol 30: 435–440
22. Strandell T (1974) Circulating studies on healthy old men with special reference to the limitation of the maximal physical working capacity. Acta Med Scand [Suppl] 414: 4–44
23. Tlusty L (1969) Physical fitness in old age. I. Aerobic capacity and the other parameters of physical fitness followed by means of graded exercise in ergometric examination of elderly individuals. Respiration 26: 161–181
24. Von Döbeln W, Åstrand I, Bergstrom A (1967) An analysis of age and other factors related to maximal oxygen uptake. J Appl Physiol 22: 934–938
25. Young ChM, Blondin J, Tensuan R, Fryer JH (1963) Body composition studies of older women, thirty to seventy years of age. Ann NY Acad Sci 110: 589–607

Normal Values for Blood Pressure in Bicycle Ergometry

H. Heck, R. Rost, and W. Hollmann

Introduction

The measurement of blood pressure during exercise has received increasing interest recently as diagnostic and prognostic procedures for the development of hypertension, which will appear at a future time, as well as from the point of view of the necessity to treat the so-called exercise hypertension [6, 1].

The decision as to whether the blood pressure values during exercise are still normal or hypertensive requires that normal or standard values are known. Normal or standard values for blood pressures during exercise which are based on a large number of studies are difficult to find in the literature and are sometimes contradictory. We have conducted studies on blood pressure measurements during increasing work loads on the bicycle ergometer in our institute for the past 25 years. Using a large number of studies and tests on about 3000 subjects, we will attempt to give a standard range for blood pressures during exercise and with respect to sex, age, body weight, and type of sport.

The systolic blood pressure reaction was examined. The measuring of the diastolic blood pressure during exercise is most often done incorrectly when using a bicycle ergometer work load. This is shown by the comparison between the intraarterial measurements and the indirect method according to Riva Rocci [3, 5].

Methods

Experimental Period

The bicycle ergometer tests were conducted at the Institute for Circulation Research and Sports Medicine at the German Sports College in Cologne from 1968 to 1982.

Sample of Subjects

Persons with no evident blood pressure abnormalities and those in which the clinical examination found no indication of high blood pressure were tested. The sample of subjects can be divided into two groups: athletes and nonathletes. The group of nonathletes was composed of persons who did not participate in any sports, those who were recreational or leisure sport participants, and those who were school sports participants. The recreational or leisure sport participants were considered to be those persons who participated in sport or physical activity occasionally or regularly, but did not participate competitively. The group of athletes consisted predominantly of highly trained athletes from the national cadres. Furthermore, those persons who regularly trained and participated competitively were also included in this group. Table 1 shows the absolute frequency of the representation for the individual sport disciplines in the total group.

Table 1. Absolute frequency of tested persons in the individual sports

Type of sport	Male	Female
Nonathletes	112	179
Leisure athletes	628	164
School athletes	242	135
Combat sports	45	14
Long-distance running	54	9
Middle-distance running	18	48
Sprinting	10	33
Discus, javelin, shot put	7	11
Cycling	119	1
Players (Ball games)	469	105
Canoeing	109	101
Rowing	23	6
Swimming	57	129
Nordic skiing	9	4
Ice hockey	41	0
Diverse sports	46	44
Total	1989	983

Exercise Procedure

The ergometer tests were carried out with graded work loads while sitting on a bicycle ergometer. Three loading patterns were used. The most frequently used procedure (88.2%) was the standard test according to Hollmann and Venrath. The work loads were started at 30 W and increased by 40 W after every 3-min interval of exercise. In 6.4% of the cases the procedure outlined by the Federal Commission for Competitive Sports (BAL) was used for the testing. The work load began at 50 W with increases of 50 W after every 3-min exercise interval. The third loading procedure was used in 5.4% of the cases and began with a work load of 25 W and increased by 25 W after every 2-min interval. This method was based on the World Health Organization's scheme and will be abbreviated in the following as WHO.

Equipment

Ergometers of three different manufacturing companies were used, namely, E. A. Müller, Elema-Schönander, and Dynavit-Keiper. Blood pressure at rest and exercise was measured indirectly by a semiautomatic blood pressure measuring instrument (Elag Co., Schwarzhaupt Co., Metronik Co.) during the last 30 s of each level of loading, according to RIVA-ROCCI. A correction of the blood pressure values with respect to the upper arm circumference was not made. The cuff width was 12 cm.

Statistics

The data were analyzed with simple and multiple linear correlations and regressions. The level of significance of the probability of error *(p)* used in the following discussion and tables is: significant ≤ 0.05 (★); highly significant ≤ 0.01 (★★) [4].

Results and Discussion

Figure 1 depicts the linear regression between the systolic blood pressure and bicycle ergometer loading for the entire sample of subjects, which contained 2972 persons. The dotted lines have a distance of $\pm 2 \star S_{y \cdot x}$ (standard error of estimation)

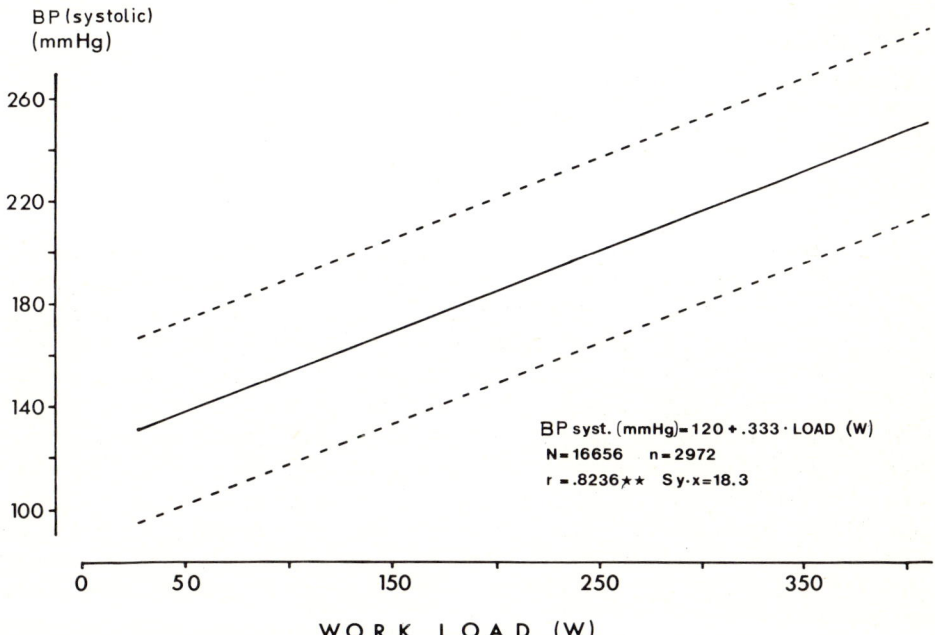

BP syst. (mmHg) = 120 + .333 · LOAD (W)
N = 16656   n = 2972
r = .8236★★   S y·x = 18.3

Fig. 1. Line of regression and the $\pm 2 \star S_{y \cdot x}$ areas for the total sample of subjects. *N*, number of value pairs; *n*, number of persons tested

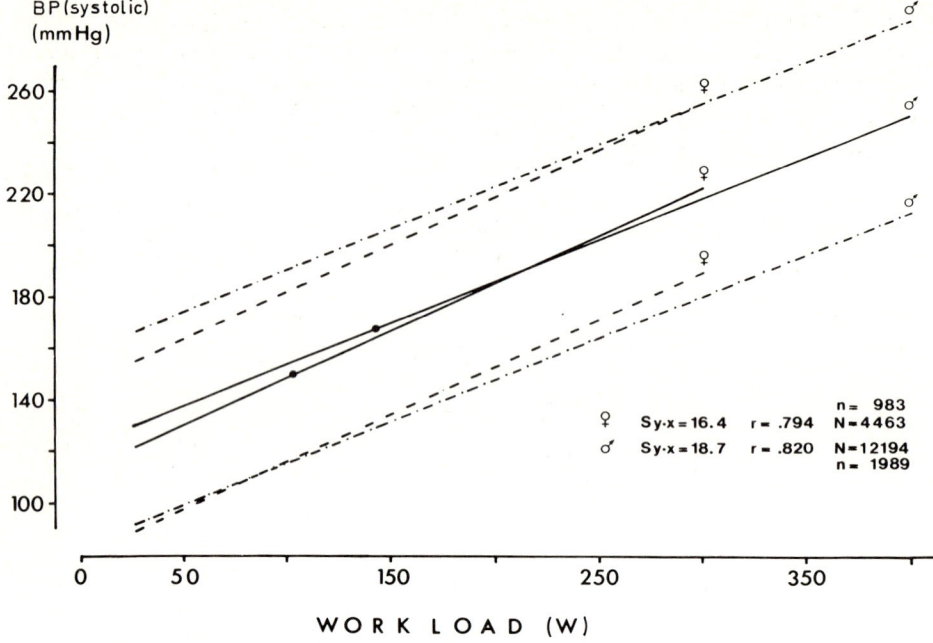

Fig. 2. Lines of regression with $\pm 2 \star S_{y \cdot x}$ areas for females and males

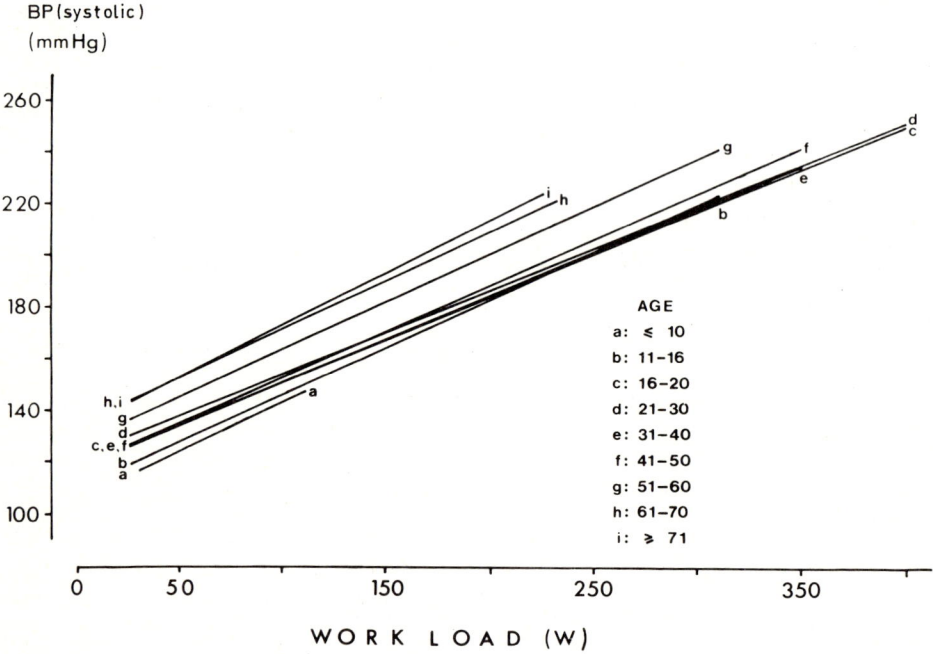

Fig. 3. Lines of regression for individual age groups

from the line of regression and the area between them encompasses 96% of the cases. Figure 2 shows the separation of values according to sex. Both lines of regression are significantly different from each other. At a work load of 50 W the females had an average blood pressure of about 7 mmHg lower than the males.

Table 2. Number and percentage of subjects by sex and age

Age	Male		Female	
	n	%	n	%
< = 10	57	2.9	17	1.7
11– 15	216	10.9	220	22.4
16– 20	395	19.9	336	34.2
21– 30	652	32.8	220	22.4
31– 40	285	14.3	104	10.6
41– 50	213	10.7	48	4.9
51– 60	112	5.6	24	2.4
61– 70	50	2.5	13	1.3
> =71	9	0.4	1	0.1
Total	1989	100.0	983	100.0

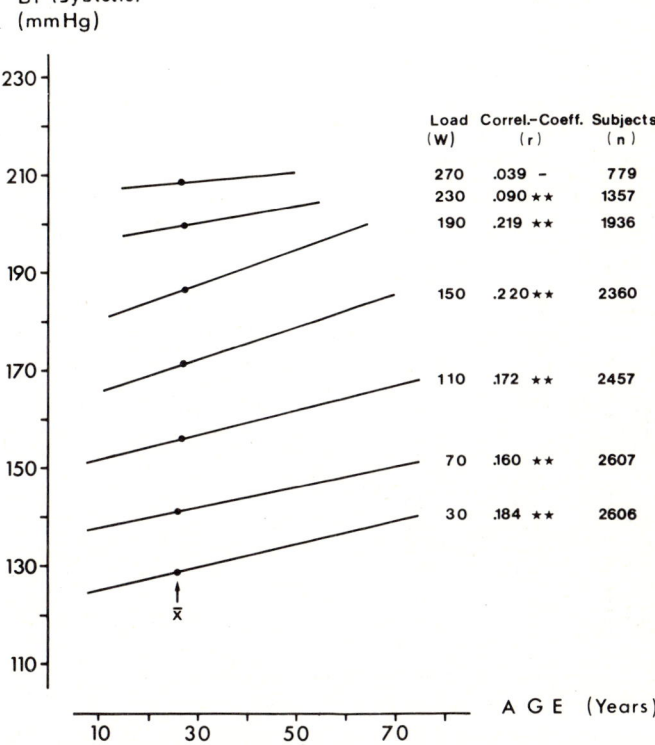

Fig. 4. Exercise blood pressure with respect to age for the individual work load levels

Figure 3 represents the effect of age on blood pressure. A line of regression is included in this diagram for each of the age groups listed in Table 2. Clear differences are noticeable for younger and older age groups. The age group 16–40 years shows only small differences in the position of the lines of regression with respect to each other. In Fig. 4, the regression between blood pressure and age for each of the individual levels of loading has been computed. The assessment was made only for the Hollmann scheme because both of the other procedures had a too small number of subjects included. Up to a work load level of 230 W, there is a highly significant correlation between age and blood pressure. The large decrease in the correlation coefficient at the loading intensities of 230 and 270 W can be explained by the smaller differentiation in age. These levels of work load were not reached by the young and older subjects.

In addition to the influence of age on blood pressure, the effect of body weight has to be discussed. This is partially due to the larger upper arm circumferences which are usually present in heavier persons. A positive correlation exists between upper arm circumference and resting blood pressure, which was measured indirectly, and with respect to the same intraarterial pressure. The correlation could further-

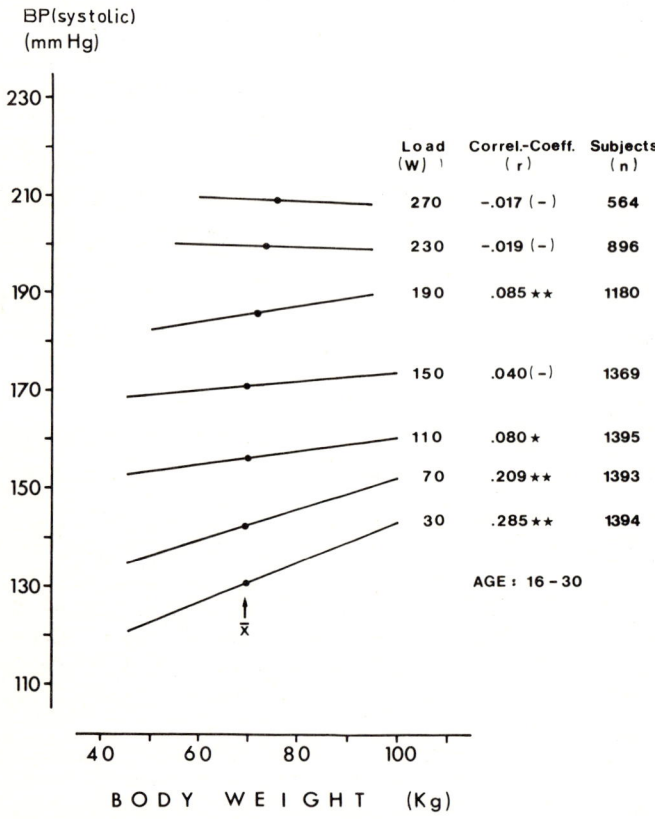

Fig. 5. Exercise blood pressures with respect to body weight for the individual work loads. Range of ages, 16–30 years

more be caused by a common dependency on age. This is to be especially expected when the group to be studied contains adults, children, and youth with lower body weights.

For purpose of eliminating the effect of age on the effect of body weight, we have only considered persons from 16 to 30 years of age in our calculations of the correlation between blood pressure and body weight. As shown in Fig. 3, there is no substantial age influence for this age group with respect to the group tested here. As a result, the lines of regression depicted in Fig. 5 only reflect the influence of body weight. A differentiation according to the effect of body weight and upper arm circumference cannot be made because no measurements of upper arm circumference are available. The closest correlation appears in the lower work load levels. As the work loads increase, the body weight range decreases, since persons with lower body weights are no longer capable of reaching high levels of loading. This is probably the reason for loss or absence of a significant correlation.

A further influencing factor on blood pressure is possibly found in the loading procedures used here. Figure 6 depicts the lines of regression for the three testing patterns applied in this study. The steepest increase with respect to work load is found in the WHO method, while the smallest is in the BAL procedure. The differences are probably due to the groups themselves which were tested. Mainly older or less performance capable persons were tested with the WHO procedure. The BAL procedure group was mostly composed of competitive athletes who were also younger. The Hollmann protocol was used for testing athletes and nonathletes. In order to get rid of a possible influence from age, the lines of regression were only

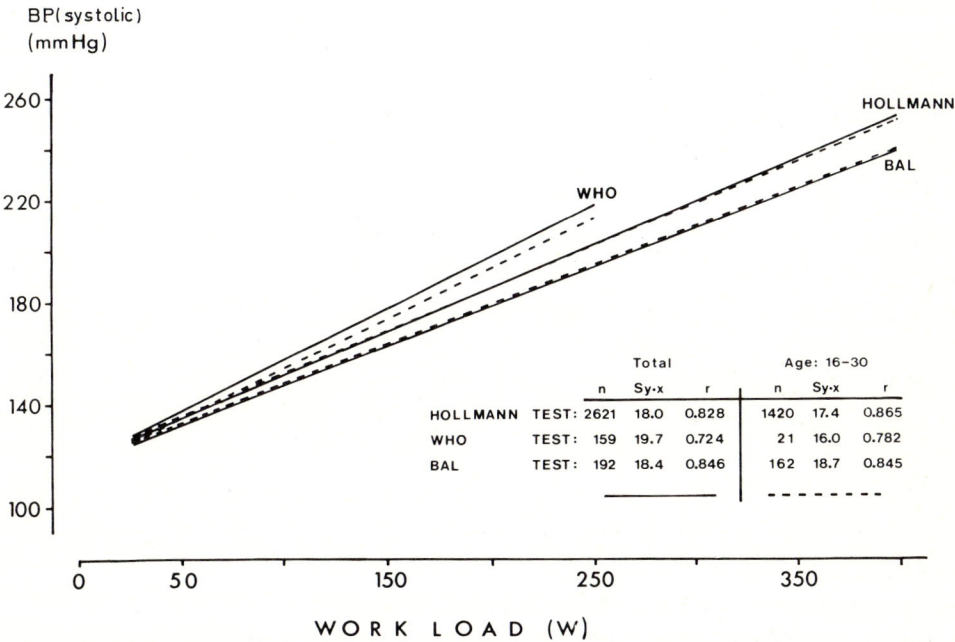

Fig. 6. Lines of regression for the three loading procedures (WHO, BAL, and Hollmann)

computed for the age group 16–30 years and are presented in Fig. 6 as dotted lines. A significant reduction in the slope of the lines is found in persons tested according to the WHO method. A substantial change is not noticeable in the other two testing procedures. Also after exclusion of the effect of age, there exists a significant difference between the three lines of regression. This is a surprising finding, which is not easy to explain. In order to clarify this, further statistical analyses were conducted.

It has been discussed that sport and physical activity, but especially endurance sports, lead to a decrease in resting and exercising blood pressures. In Fig. 7 the athletes as a whole and divided up according to sex are compared with the total sample of subjects and to the nonathletes. The lines of regression were computed only for the age group 16–30 in order to exclude the age influences. The females showed the same effect as the entire sample of females. Otherwise, the differences compared with the entire sample or with the nonathlete group were not significant. Figure 8 contrasts the lines of regression for selected sports or groups of sports represented by a large number of subjects with the entire group of subjects. Essential differences cannot be found. The maximum average difference between the cyclists and the total sample of subjects was 7 mmHg at a work load level of 250 W. This is approximately the same level as the maximum loading level of an untrained 20 year old. The ball game (football; handball; volleyball) players averaged approximately 4 mmHg more than the group as a whole. As a result, a clearcut blood-pressure-reducing effect by means of sport and physical activity in general or by specific kinds of sports cannot be proven by the data presented here.

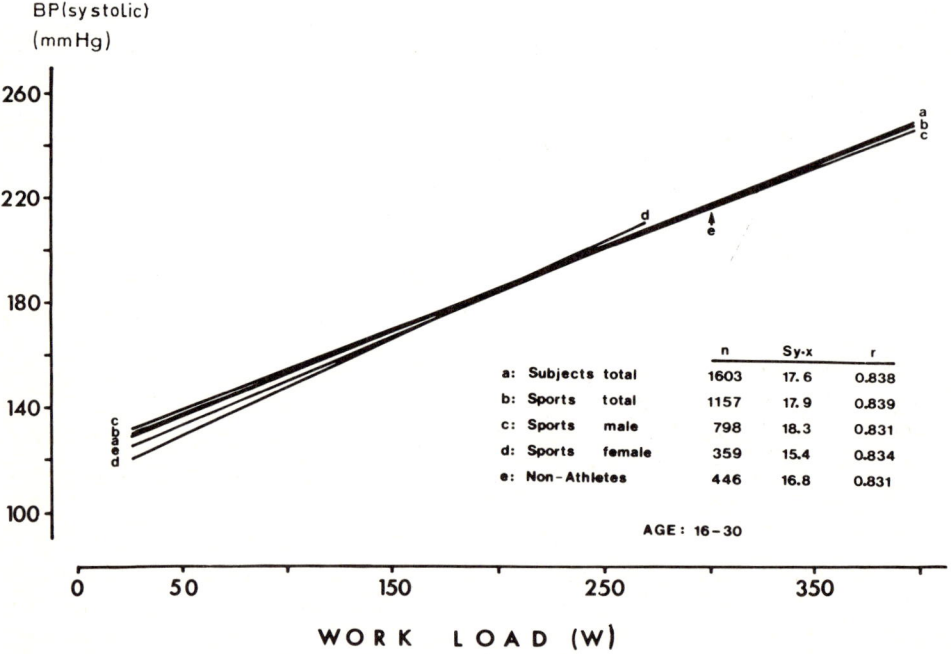

Fig. 7. Lines of regression for the entire group of subjects, all athletes, male and female athletes, and nonathletes

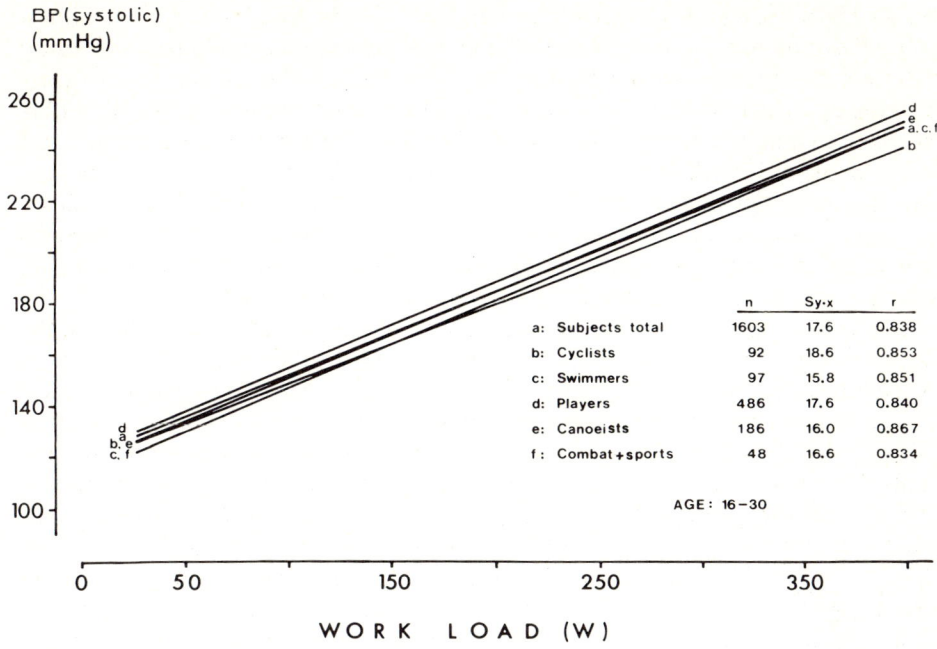

Fig. 8. Lines of regression for the entire group of subjects, cyclists, swimmers, ball or team sport players, canoeists, combat sports athletes

Table 3. Coefficients of multiple regression comparisons and standard error of estimation for the total sample and by sex

Multiple regression

Subjects	Intercept	Load coefficient	Age coefficient	Weight coefficient	Standard Error of estimate
Total	111.2	0.334	0.310	–	17.9
Female	104.9	0.373	0.322	–	16.0
Male	114.5	0.323	0.279	–	18.4
Total	105.2	0.324	–	0.228	18.1
Female	101.8	0.364	–	0.196	16.3
Male	109.4	0.316	–	0.192	18.6
Total	104.6	0.329	0.264	0.123	17.8
Female	101.1	0.370	0.306	0.077	16.0
Male	108.6	0.321	0.251	0.096	18.3

Table 3 shows the results of multiple regressions. For the entire group of subjects and separated according to males and females, the coefficients for the linear comparison of exercising blood pressure are given on the top with respect to work load and age, in the middle on work load and body weight, and in the bottom section on work load, age, and body weight. Although, as shown above, body weight has a significant influence on blood pressure behavior. This influence is still small if expressed in absolute values. The difference between subjects weighing 50 kg and 80 kg is only about 3–4 mmHg (comparison of regression, bottom of Table 3).

In the age distribution of the total sample, the older age groups are poorly represented. This was previously not taken into consideration by statistical analysis. In a secondary analysis the age coefficient may probably show a higher value.

The demarcation of a standard or normal range is specified by different statistical procedures. The one which is selected depends on the distribution of the measured values and on whether a wide or narrow area is desired. Considering a standard distribution, the 2-sigma range is frequently selected. This contains 96% of all cases. Figure 9 represents the upper 2-sigma areas, corresponding to the first comparison in Table 3 with respect to work load and age. Respective diagrams can be made as needed for the other comparisons.

The upper limit values for exercising blood pressures presented here are only valid for:

1. Persons who are represented by subjects in this sample
2. Persons who are sitting during bicycle ergometer testing. (The blood pressure values are clearly higher when lying down than when sitting. Kubicek and Gaul found a 16.1% higher value when in a recumbent position [2])
3. Identical loading patterns and procedures (???)

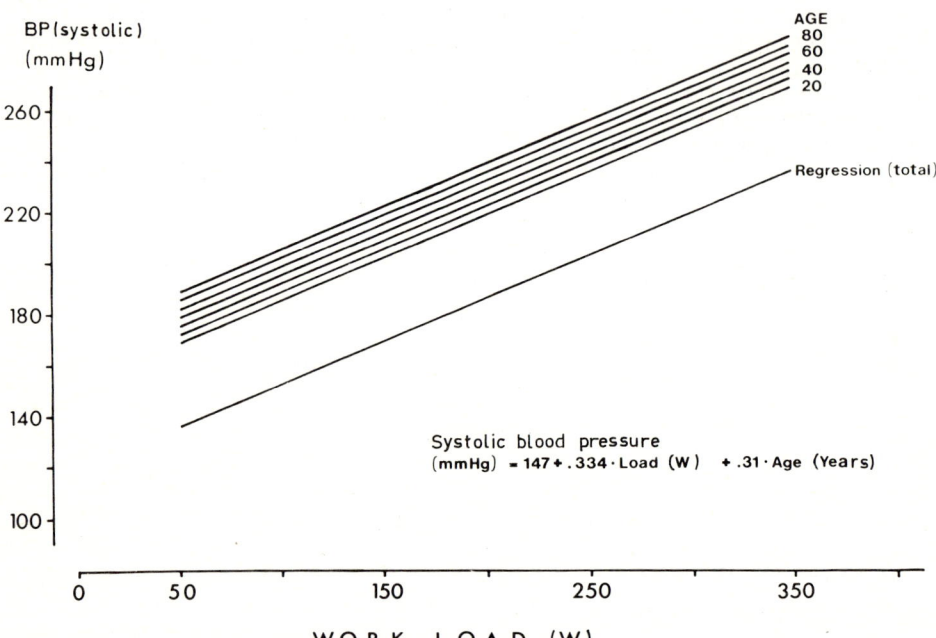

Fig. 9. Upper $2 \star S_{y \cdot x}$ area for exercise blood pressure with respect to age (first regression comparison in Table 3)

Summary

The evaluation of blood pressure during exercise plays an important role in diagnosis, assessment, and prognosis of hypertension. However, the prerequisite of an exact knowledge of normal behavior of exercise blood pressures is necessary. Blood pressure changes during bicycle ergometer work in the sitting position was tested and studied in 2972 athletes and nonathletes (1989 males, 983 females).

The most important findings are:

1. A significant difference exists between males and females, which decreases with increasing work load levels
2. There is a significant influence of age and body weight on blood pressure behavior and changes
3. The blood pressure during exercise is possibly dependent on the work load protocol
4. There are no substantial differences in blood pressure reactions between athletes and nonathletes

Normal or standard ranges are presented based on regression comparisons and the corresponding standard errors of estimation.

References

1. Franz IW (1982) Ergometrie bei Hochdruckkranken. Springer, Berlin Heidelberg New York
2. Kubicek F, Gaul G (1977) Comparison of supine and sitting body position during triangular exercise test. Eur J Appl Physiol 36: 275–283
3. Matthes D, Schütz P, Hüllemann KD (1978) Unterschiede zwischen indirekt und direkt ermittelten Blutdruckwerten. Med Klin 11: 371–376
4. Sachs L (1968) Statistische Auswertemethoden. Springer, Berlin Heidelberg New York
5. Schüller H (1976) Über das arterielle Blutdruckverhalten beim Menschen unter Belastungsbedingungen. Med Dissertation Köln
6. Wilson N, Meyer B, Albury J (1979) Early prediction of hypertension using exercise blood pressures. Med Sci Sports 11: 110 (abstract)

Prognostic Significance of an Overshooting Exercise Blood Pressure as an Indicator for Subsequent Manifestation of Hypertension

F. Amecke and R. Rost

Ergometry provides the possibility of performing a stress test of the arterial blood pressure under defined conditions. Hollmann [2] was the first to initiate the development of an automatic blood pressure device adapted to exercise blood pressure measurement as early as the 1950s. Particular interest in exercise blood pressure was raised by the investigations of Franz [1]. Besides his studies, only one follow-up study about the prognostic significance of the exercise blood pressure is available [4]. Therefore, the opinions about the diagnostic value of an overshooting exercise blood pressure in relation to a subsequent manifestation of hypertension are still controversial. Particularly, data on patients with exercise hypertension but normal pressure at rest are not available. The investigations of Franz [1] are based on borderline hypertensive subjects, who were divided into "exercise-positive" and "exercise-negative" ones.

For further elucidation of that question, we evaluated retrospectively the data of 4000 subjects who performed a stress test within the last decade in this institute. From this material we selected those persons who demonstrated an overshooting systolic pressure reaction without elevated blood pressure at rest and without any hint on hypertension in the past.

The consideration of an exercise blood pressure as overshooting was performed according to the data established in this institute [3]. The subsequent regression equation was developed:

Systolic pressure (mmHg) = 120 + 0.4 (W + age in years) (W = Watt)

This equation demonstrates the dependance of systolic blood pressure on the work load intensity and on the age. We calculated from the blood pressure value during exercise the individual regression line for each subject. A person was characterized as exercise-positive if this regression line crossed the pressure mark of 200 mmHg at a lower value than was given by the age-dependent general regression line, dislocated for $+ 2 s$. For the assessment of the pressure during resting conditions systolic as well as diastolic pressures were used. This was not true under exercise conditions because of the difficulties in evaluation of the diastolic pressure by indirect measurement. Here we only used the systolic value.

The investigation was restricted to male subjects. All persons with overshooting arterial pressure in a stress test which was performed earlier than a minimum of 2 years previously were invited to a control test. This oportunity was used by 57 patients, aged 16–72 years, the mean age being 39.3 ± 15 years. The mean time since the first investigation was 5.0 ± 1.8 years. The total sample was furthermore distin-

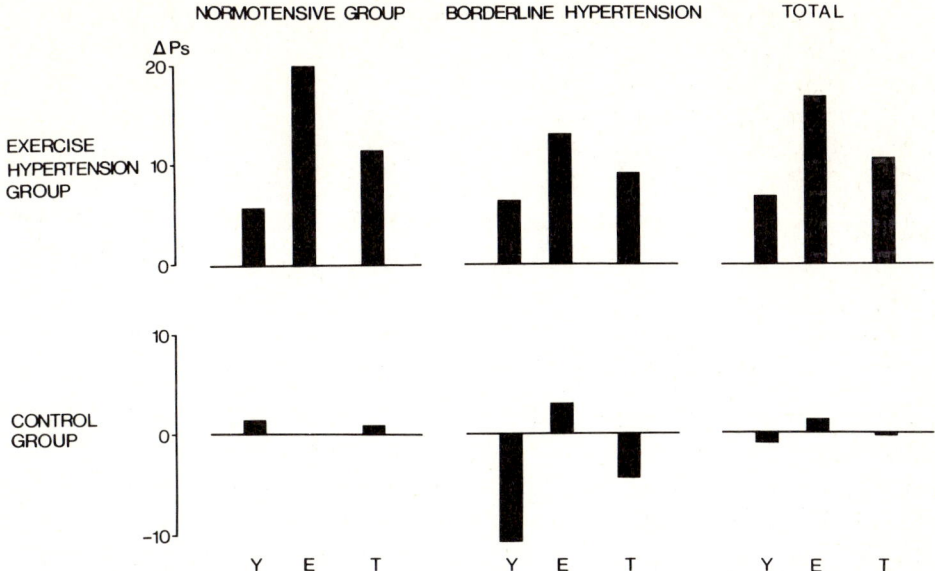

Fig. 1. Reproducibility of exercise-positive blood pressure in the second investigation. The *bars* symbolize the percentage of the total sample which was exercise-positive in the second test. The total sample *(right)* was subdivided in a normotensive group *(left)* and a borderline hypertensive group *(middle)*. These subsamples are further subdivided in a young group *(Y)* and an older group *(E)*. *T* means total. For further explanation see text

guished according to the resting pressure into 32 normotensives and 25 borderline hypertensives. Additionally we divided our group into 36 younger and 21 older subjects, the dividing line was 45 years. As control group, we selected an age-matched sample of 56 subjects, who were divided according to the same aspects.

The results deliver significant data about the reproducibility of measurement of blood pressure during exercise, which is demonstrated in the Fig. 1. The bar characterizes the percentage of exercise-positive subjects in the control investigations. From the exercise-positive sample in the first investigation, this was true in 65% in the second study. It had to be underlined that from 20 formerly exercise-positive subjects, who were exercise-negative in the second study, 18 belonged to the younger group. Of the young exercise-positive subjects, 50% were exercise-negative in the second investigation. In the control group, eight (or 14%) were exercise-positive in the second investigation. From these crossover subjects, the majority (5) belonged to the older group.

From this data the conclusion can be drawn that the exercise blood pressure is less reliable as a diagnostic criterium in younger subjects than in older ones. The same conclusion results in the assessment of the development of hypertension at rest. This is shown in the Fig. 2. The dark bar symbolizes the percentage of the patients in whom the control test showed a manifest hypertension. The percentage of the later borderline hypertensive subjects is characterized by the stippled bars. It should be pointed out that in the control group, the development of hypertension

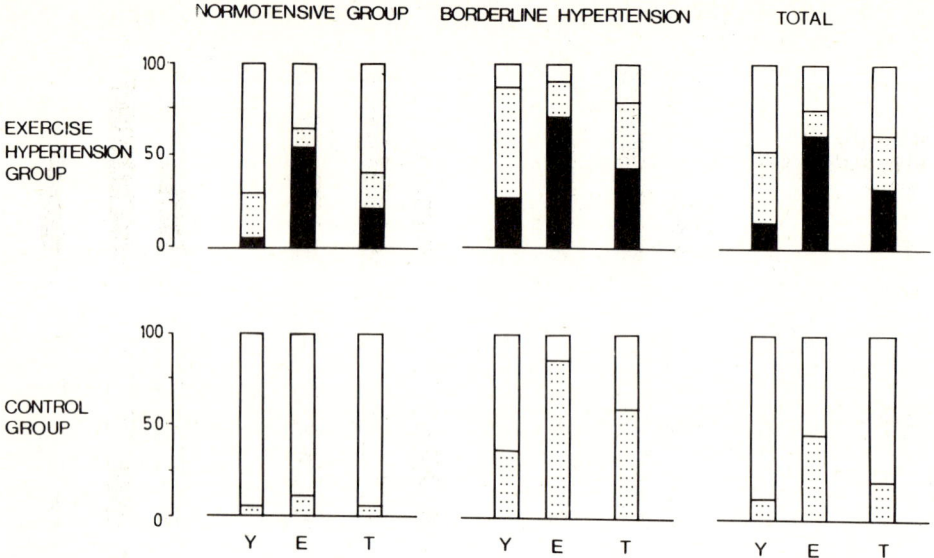

Fig. 2. Percentage of the samples (further described in Fig. 1 and in the text) demonstrating a manifest hypertension at rest *(dark bar)* or a borderline hypertension *(stippled bar)* in the second investigation. For further explanation see text

could not be observed. Conversely, 32% of the exercise-positive subjects demonstrated hypertension in the control test. This was true in 44% of the formerly exercise-positive borderline hypertensives, but only in 22% of the formerly normotensive subjects.

The results additionally demonstrate great differences in dependance on the age. In the older group, 62% developed hypertension, whereas this was the case in only 14% of the younger group. A very strong diagnostic hint by exercise blood pressure was delivered in the group of the older borderline hypertensive subjects, 70% of whom developed a manifest hypertension, whereas in the younger normotensive but exercise-positive group, this was only true in one patient (5%).

From these results it can be concluded that exercise blood pressure presents a useful prognostic index for the subsequent manifestation of hypertension in normotensive as well as in borderline hypertensive subjects. From a purely qualitative point of view, this concerns particularly the older subjects, over the limit of 45 years which was used here. The quantitative aspect shall be considered subsequently. For these reasons we can qualitatively confirm the results of Franz [1]; on the other hand it has to be stressed that we were unable to reach the prognostic reliability of his data, since in his sample, 96% of a group of exercise-positive borderline hypertensive subjects in an age range of 20–50 years developed manifest hypertension. It has to be discussed whether this difference may be explained by the higher limits of exercise hypertension which were used by the above-mentioned author. However, if this would be considered to be the reason, in our sample most of the exercise-positive subjects would have been shifted into the control group, leading to the consequence

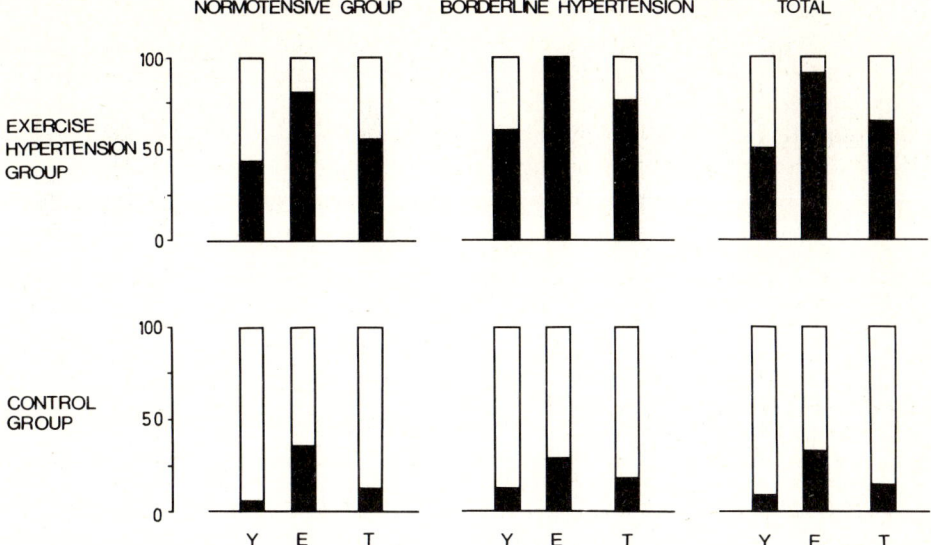

Fig. 3. Mean increase of systolic pressure at rest in the subsamples (described in Fig. 1 and text). It should be noted that in all subsamples of the exercise-positive groups, on the average an increase of the systolic pressure could be found, in contrast with the control samples

that in this sample some hypertensive subjects should have been expected in the second test.

In Fig. 3 the increase of the systolic resting pressure is demonstrated. This figure indicates the main advantage of measurement of arterial pressure during stress testing, which results, according to our opinion, from this trial. Up to now there has been no parameter which had a possibility of forecasting the increase of arterial blood pressure. This must also be confirmed for resting pressure. Its prognostic possibility related to the later development of arterial blood pressure is restricted to the indication that probably the age-dependent increase of peripheral resistance in follow-up investigations results in higher blood pressure. The borderline hypertensive subjects form a group which frequently demonstrate hypertension in the follow-up study, which only reflects the general tendency of the increase of arterial blood pressure. Therefore, the limit which is normally drawn as the borderline of hypertension will be surpassed more often.

However, the main parameter for assessment of the development of hypertension is not the fact that a voluntarily borderline will be surpassed but the real increase of pressure. Figure 3 demonstrates that no increase of arterial pressure could be observed in the total control group in contrast to the exercise-positive group. This can also be stressed for the younger sample. Even if only a small subsample of this group had developed during our observation time a manifest hypertension, the increase of the systolic pressure would have suggested a high risk of hypertension in these younger subjects as well. The results of a later control are expected.

Finally it should be mentioned that the significance of the exercise blood pressure as a prognostic indicator was further confirmed in this trial by statistical analysis. It could be shown by stepwise multiple regression calculation that the prognosis of the subsequent blood pressure development on the basis of the resting pressure can be significantly improved by simultaneous consideration of the stress pressure. Conversely, the prognostic reliability of the combination of resting and stress pressure cannot be further improved by additional variables, such as age, Broca index, and PWC_{170}.

References

1. Franz J (1979) Untersuchungen über das Blutdruckverhalten während und nach der der Ergometrie bei Grenzwerthypertonikern im Vergleich von Normalpersonen mit Patienten mit stabiler Hypertonie. Z Kardiol 68: 107
2. Hollmann W, Hettinger T (1980) Sportmedizin, Arbeits- und Trainingsgrundlagen. Schattauer, Stuttgart
3. Rost R, Hollmann W (1982) Belastungsuntersuchungen in der Praxis. Thieme, Stuttgart
4. Wilson N, Meyer B, Albury J (1979) Early prediction of hypertension using exercise blood pressure. Med Sci Sports 11: 110 (abstract)

Significance of Longitudinal Variance of Ergometric Measurements

H. Wollschläger, H. Löllgen, A. Zeiher, B. Wieland, and H. Just

Ergometric measurements are well-established, standard methods for assessment of cardiopulmonary parameters in healthy and diseased persons. Especially regarding long-term observations, ergometric data can give important information about functional changes due to training or disease. A very desirable goal would be to detect such functional changes in an individual as early as possible.

One major problem lies in establishing normal – or reference – values. Usually normal values are calculated from accumulated data from apparently healthy persons. A deviation from the normal is defined as exceeding the range of plus or minus a twofold standard deviation from the mean.

Because of the great variability of functional parameters in a large collective of so-called healthy persons the standard deviation is relatively high. Therefore marked changes of data are needed to recognize an abnormal behavior.

The aim of our study was to investigate whether the substitution of normal values by individual – or personal – reference values could minimize that limitation. Thus we studied the long-term variability of some ergometric data in individuals and calculated these individual reference values.

Methods

Eleven untrained men (mean age 26 ± 5 years) were investigated. Cardiopulmonary diseases were excluded by history, physical examination, ECG, and chest roentgenogram. Standardized exercise test was performed on a bicycle ergometer in the upright position to maximal work load starting at 50 W with increments of 50 W every 3 min.

Data collected were:

1. *Heart rate* from continuous ECG recordings every minute
2. *Systolic blood pressure* measured semiautomatically by standard cuff method every minute
3. *Respiratory minute volume* by an open-system pneumotachygraph at each work load

Exercise tests were repeated ten times during a period of 10 months. The minimal time between two consecutive tests was 3 weeks to exclude a possible training effect. Probands were not allowed to do unusual physical activities during the entire study.

EXERCISE   TESTS

The grid (Fig. 1):

	1	2	3	4	5	6	7	8	9	10			
1	D	D	D	D	D	D	D	D	D	D	$\overline{X}_1$	SD_1	$V\%_1$
2	D	D	D	D	D	D	D	D	D	D	$\overline{X}_2$	SD_2	$V\%_2$
3	D	.	.	.	.	.	.	.	.	D	$\overline{X}_3$	SD_3	$V\%_3$
4	D	.							.	D	"	"	"
5	D	.							.	D	"	"	"
6	D	.			$\overline{X}_G$, SD_G, $V\%_G$				.	D	"	"	"
7	D	.							.	D	"	"	"
8	D	.							.	D	"	"	"
9	D	.	.	.	.	.	.	.	.	D	"	"	"
10	D	D	D	D	D	D	D	D	D	D	$\overline{X}_{10}$	SD_{10}	$V\%_{10}$
11	D	D	D	D	D	D	D	D	D	D	$\overline{X}_{11}$	SD_{11}	$V\%_{11}$

(Left axis label: P R O B A N D S)

$$V\% = \text{COEFFICIENT OF VARIANCE} = \frac{SD}{\overline{X}} \times 100 \ [\%]$$

$$\mathbf{V\%_P}$$

Fig. 1. Calculation of reference values. For explanation see text. Abbreviations: $\bar{x}$, mean; *SD*, standard deviation; $\bar{V}\%$, coefficient of variability; *D*, data. Indices: *G*, global; *P*, personal; 1 to 11, no. of proband

Mean $(\bar{x})$, standard deviation (SD), and coefficient of variability (V%) for each parameter were calculated in the following manner (Fig. 1):

1. From the data (D) of all probands at every point of measurement, that is, from 110 measured values of each parameter at every work load. These statistics were defined as *global reference values* for the collective investigated.
2. From the data of each proband at every point of measurement, that is, from ten measured values of each parameter at every work load. These statistics ($\bar{x}_i$, SD_i, and $V\%_i$; for i = 1 to 11) were defined as *individual reference values* of each proband.

To demonstrate the variability of the individual reference values we calculated the mean of the coefficients of variability ($\bar{V}\%_P$) for each individual reference value.

Because of limitation of space, only the data at rest and at a 150 Watt work load for the three parameters mentioned above will be presented.

Table 1. Heart rate

	Variability of	
	global reference values (whole collective)	individual reference values (11 probands)
Rest:	$\bar{X}$: 87/min SD: 12.3 V%: 14.2	$\overline{V\%}_P$: 8.8 ± 1.9
150 W:	$\bar{X}$: 162/min SD: 18.2 V%: 11.2	$\overline{V\%}_P$: 3.8 ± 1.0

Results

Table 1 shows the data for the heart rate calculated as described in Fig. 1.

On the left, the so-called global reference values for the whole collective at rest and during exercise can be seen: mean, standard deviation, and coefficient of variability. These values correspond well to the data from cross-sectional studies [1] indicating that our probands can serve as a normal collective. Remarkable is the considerably great variability at rest and during exercise (14.2% at rest and 11.2% at a 150-W work load).

In contrast, the individual reference values from each of the 11 probands vary much less, as can be seen on the right of Table 1 (8.8 ± 1.9% at rest and 3.8 ± 1.0% at a 150-W work load). This effect is especially pronounced during exercise.

Table 2 shows a similar but less marked effect for the parameter systolic blood pressure: 8.2% compared with 6.1% at rest and 11.2% compared with 10.3% at exercise.

Regarding the parameter respiratory minute volume (Table 3), there is again a pronounced decrease of the coefficient of variability from global to individual reference values, particularly during exercise.

Discussion

The results of our studies suggest that calculating individual reference values decreases the variance of that reference value compared with the so-called global reference values, especially under exercise conditions. These results correspond well to the data presented by Löllgen [2] and von Nieding [3] for arterial blood gases and parameters of pulmonary function.

This probably offers an attractive concept for detecting minor functional changes in long-term observations, especially in clinical settings.

Figure 2 demonstrates hypothetically the long-term observations of a functional parameter – e.g., respiratory minute volume – of a diseased person at a distinct work load. The outer lines represent the range of twofold standard deviation using conventional reference values. A significant deterioration could only be detected at

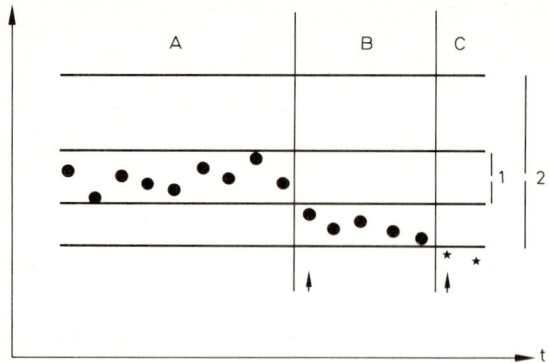

Fig. 2. Changes of a hypothetical functional parameter of an individual person. For explanation see text. *1* plus/minus twofold standard deviation of the individual reference value. *2* plus/minus twofold standard deviation of the global reference value. Adapted from [2] and [3]

Table 2. Systolic blood pressure

	Variability of	
	global reference values (whole collective)	individual reference values (11 probands)
Rest:	$\bar{X}$: 112 mmHg SD: 9.2 V%: 8.2	$\overline{V\%}_P$: 6.1 ± 1.7
150 W:	$\bar{X}$: 182 mmHg SD: 20.4 V%: 11.2	$\overline{V\%}_P$: 10.3 ± 4.1

Table 3. Respiratory minute volume

	Variability of	
	global reference values (whole collective)	individual reference values (11 probands)
Rest:	$\bar{X}$: 8 l/min SD: 1.9 V%: 24.6	$\overline{V\%}_P$: 20.6 ± 3.1
150 W:	$\bar{X}$: 69 l/min SD: 15.1 V%: 21.9	$\overline{V\%}_P$: 9.8 ± 3.2

point C when the measured values of repeated exercise tests leave the "normal range."

Using the individual reference values of that special person with their much lower variance (calculated from the first nine points) leads to an earlier recognition – at point B – of a functional change.

Therefore it might be useful, especially for persons with a detected disease of slow progression, e.g., cardiomyopathy or chronic lung disease, to establish such in-

Table 4. Variability of individual reference values ($\overline{V\%}_P$) after 3 examinations and after 10 examinations

		$\overline{V\%}_P$	
		After 3 examinations	After 10 examinations
Heart rate	rest	7.9	8.8
	150 W	3.1	3.8
Blood pressure	rest	6.3	6.1
	150 W	10.3	10.3
Respiratory minute volume	rest	23.0	20.6
	150 W	10.8	9.8

dividual reference values. Because of the gradual deterioration of functional parameters in those diseases a worsening could be detected much earlier.

But limited by socioeconomic implications it seems to be impossible to perform ten consecutive examinations over a long-lasting period only for establishing such reference values. Therefore, we compared the individual variability of the measured values of every parameter after three and after all ten examinations.

Table 4 demonstrates that there are only small differences between the mean of the coefficients of variability of each individual reference value for every parameter.

Accordingly, three baseline examinations will be sufficient to calculate individual reference values for exercise tests. Further studies, especially in pathological states, are needed to perfect this concept.

References

1. Löllgen H (1983) Kardiopulmonale Funktionsdiagnostik, Documenta Geigy, Wehr
2. Löllgen H, Haninger B, Just H (1980) Langzeitvariabilität ergometrischer Meßgrößen. In: Kindermann W, Hort W (eds) Sportmedizin für Breiten- und Leistungssport. Demeter, Gräfelfing, pp 273–278
3. Nieding G, Krekeler H, Löllgen H, Ripplinger E (1977) Individuelle Variabilität von Lungenfunktionsgrößen im Längsschnitt und ihre Bedeutung für arbeitsmedizinische Untersuchungen. Prax Pneumaol 31: 858–871

Low Work Load During Physical Stress Testing Is Mental Stress Testing

H. Rüddel, M. E. McKinney, J. C. Buell, R. S. Eliot, H. Otten, W. Schulte,
W. Langewitz, and A. W. von Eiff

Introduction

The typical physiologic response to exercise is a rapid increase in systolic blood pressure (SBP) and heart rate (HR). Diastolic blood pressure (DBP) does not increase significantly in healthy young persons. In the literature, the blood pressure response to physical exercise tests has yielded conflicting results regarding blood pressure (BP) reactions of hypertensives or normotensive patients. Most studies show no difference in the reaction pattern of hypertensives and normotensives [10, 11, 15] unless the patients are older or already have an impaired vascular system [17]. Pickering and co-workers [14] reported last year, for example, that 20 normotensive, 19 borderline hypertensive, and 15 hypertensive outpatients showed similar increases in BP (44/0, 52/ − 1, 48/ − 1 mmHg respectively).

However, reports of comparisons among normotensives, borderline hypertensives, and hypertensives (especially from Germany), have shown differences among these groups in response to exercise tests [8, 16, 17, 19]. In a prospective study by Franz [7], borderline hypertensives with a SBP of more than 200 mmHg during an ergometric exercise test in a supine position with a work load of 100 W were defined as hyperreactive. Of 26 hyperreactive patients with borderline hypertension, 25 manifested established hypertension 3.8 years later. Of the 19 normally reacting control patients with borderline hypertension, only six became hypertensive. Jandova et al. [9] and Wilson et al. [18] corroborated the hypothesis that exercise blood pressure could be of some importance in predicting hypertension.

Recently Dlin et al. [2] published their follow-up results of two groups of normotensive healthy young males. Group 1 had an exaggerated BP response to exercise, whereas the subjects in group 2 had normal BP reactions to any given work load on the bicycle. The patients were matched for age, BP, family history, and smoking history. After a follow-up of 5.8 years, eight of the 75 subjects in group 1 (10.6%) were hypertensive, whereas none of group 2 had elevated BP.

This leads to the following questions: is it the character of challenge, the duration of the resting prestress period, the work load, the duration on each work load, the vehicle employed in administering the exercise test (bicycle versus treadmill), or all or part of these factors that are important for discriminating between hypertensives and normotensives.

We investigated whether BP reactions to exercise are determined by age, casual BP level, or typical individual reactivity pattern to challenge by comparing the responses of healthy outpatients to a standard exercise test on the treadmill and to a primarily beta-sympathetic mental stress test.

Experiment 1: BP Reactions of Healthy Outpatients to Mental and Physical Stress

All male outpatients coming to the Department of Preventive and Stress Medicine between June 1981 and October 1982 ($n = 79$) who were free of chronic disease and unmedicated for at least 4 weeks prior to evaluation were included in this study. All had an intensive medical examination and only healthy normal persons and patients with elevated casual BP but without any target organ damages were included. Resting BP was taken in an upright position prior to the exercise tolerance test (ETT). Psychophysiological testing was performed in the morning before the ETT. During psychophysiological testing, BP was determined every 90 s by an automatic oscillometric device (Sentry automated sphygmomanometer). HR was continuously registered and averaged every minute from R-R intervals of standard ECG recordings.

After a baseline of at least 10 min, the following stress tests were done (all instructions presented via videotape):

(a) mental arithmetic, that is, serial subtraction of 7, starting at 777, for a 3-min period and (b) a video game (Atari Break Out) for 5 min. BP and HR were averaged over all stress tests for each patient. Later in the morning, all patients had a physical exercise test on the treadmill. A modified Balke/Naughton multistage maximal protocol was used. The protocol was:

A standard 12-lead ECG recording at baseline and every 2 min during ETT. BP and HR were taken at baseline and at the 90-s interval of each stage (Table 1).

BP was measured via auscultation (Baumanometer). The classification of the patients into groups 1 to 3 was based on the BP reading in the upright position prior to the treadmill test after attachment of all electrodes. Group 1: BP < 140/90, $n = 39$; group 2: BP = 140–159/90–94, $n = 18$; group 3: BP > 160/95, $n = 22$.

Table 1. Modification of the Balke/Naughton multistage protocol on the treadmill

METS	3	4	6	8	9	10	11	12	13	14
Speed (mph)	2	2	3	3	3	3.7	3.7	3.7	3.7	3.7
Grade (%)	3.5	7	7.5	12.5	15	12	14	16	18	20

Results of Experiment 1

The patients in this sample had an age range between 22 and 66 years with an average of 43 ± 10 (SD). Their weight was 183 ± 28 pounds and body surface area was 2.04 ± 0.18 m^2. Those with higher relative weight had higher resting BP and higher BP during physical and mental challenge ($r = 0.30$ and 0.48 respectively). Age was related to BP levels and BP increases during mental stress ($r = 0.35$).

When the white male outpatients were divided into three groups according to their resting BP, patients in group 1 showed a gradual, steady increase of SBP and HR with increasing work load. Groups 2 and 3 also had a gradually increasing SBP and HR at higher work load, but at the lower work load at the beginning of the ETT, especially during warm-up, they manifested a disproportionate rise in SBP, accompanied by an equally disproportionate rise in HR (Fig. 1).

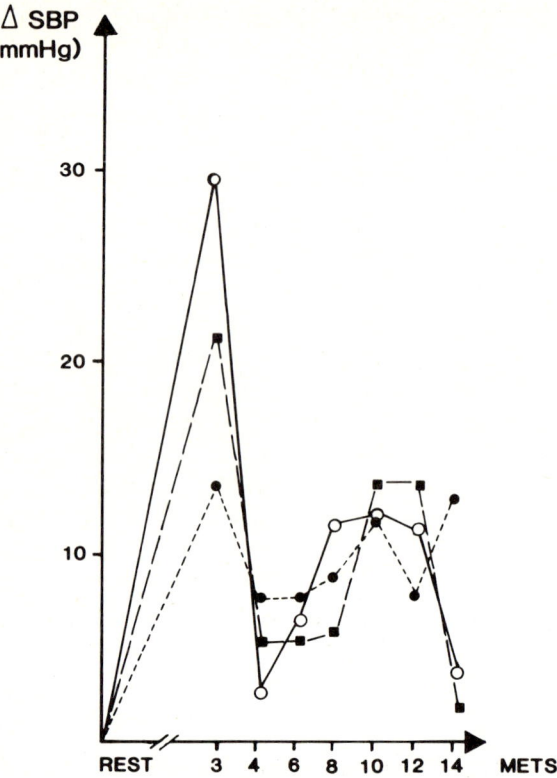

Fig. 1. Increase of systolic blood pressure *(SBP)* exercise testing for each stage on the treadmill for three groups. Group 1 *(dotted line)*, normotensives $(n = 39)$; group 3 *(dashed line)*, hypertensives $(n = 22)$; group 2 *(solid line)*, borderline hypertensives $(n = 18)$

This overreacting of SBP is even more obvious when the BP response to warm-up in those patients who increase SBP more than 1.5 times the standard deviation of the mean during mental challenge is compared with the average BP increase of all outpatients (Fig. 2).

Discussion of Experiment 1

It is well documented that during physical stress testing mean arterial pressure increases parallel to the work load. Also, the catecholamine response to exercise is fast, and only a hypothesis like "central command" can explain the very early rise in HR during exercise.

The fairly high correlations of BP responses to submaximal exercise or low work load stress tests on the bicycle [12] or on the treadmill with BP responses to mental arithmetics show that more than the "physiological" response is involved in the regulatory processes of BP increase to physical challenge. Of particular interest, therefore, are two findings concerning exercise testing in this study:

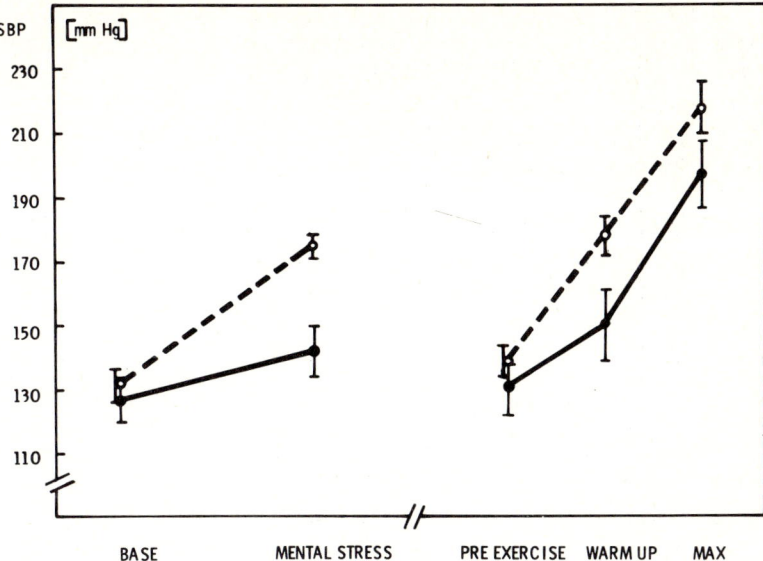

Fig. 2. Systolic blood pressure *(SBP)* before and during mental and physical challenge (see text for details) for all healthy outpatients (*n* = 77, *solid line*) and for the subgroup of patients with an increase in systolic blood pressure of more than 1.5 times the standard deviation of the mean (*n* = 7, *dashed line*)

1. During low-level exercise, patients with elevated resting BP increased their BP more than those with resting BP in the normotensive range
2. The differences in BP responses disappeared at high-level effort

These findings suggest that behavior influences BP responses during very mild physical activity or at the beginning of an ETT, when a patient is preparing for but not yet performing heavier activity.

This behavioral aspect, however, is only one clue for the very complex pattern of BP reactivity during mild physical challenge. One will miss it if the patient is not in a fairly good resting position prior to the ETT because hypertensives do not decrease BP to a resting level as do normotensives. It is also missed if the patient is allowed to speak, tell his medical history, or chat prior to the test – because this will increase his BP about 10/5 mmHg (unpublished observation) – and if the exercise testing protocol begins with large work loads.

Experiment 2: BP Reactions of Marathoners and Sedentary Executives to Mental and Physical Stress

We then investigated whether highly conditioned healthy men differed from sedentary outpatients in their BP response to mental and low work load physical challenge. If their BP reactions to stress would not be different we would regard this as a further clue for the hypothesis that low work load during physical stress testing is mental stress testing.

A sample of 12 highly conditioned white marathoners with an average "mileage" of 69 miles per week was compared with an age-matched group of untrained executives. The experimental protocol was similar to that used in experiment 1.

BP response during mental challenge and warm-up on the treadmill was not different in long-distance runners. During warm-up, BP increased an average of 25 mmHg in marathoners compared with an increase of 24 mmHg in the control group.

If physical conditioning does not attenuate BP response to low work load, can a typical cognitive pattern during ETT explain the hypothesized similarity between mental and low work load physical stress?

Experiment 3: Subjective Ratings About Low Work Load Physical Stress Testing

In a recent pilot study we interviewed 17 white male outpatients with mild hypertension in the age range of 25 to 50 years about their personal feelings during the first low work load of 40 W. They all had an ETT on the bicycle ergometer and were free of any medication for at least 4 weeks. This group had an average clinical casual BP of $146/94 \pm 13/8$ mmHg on the same day of the ETT. SBP during mental arithmetic was 170 ± 11 and increased from 140 ± 7 to 163 ± 12 mmHg during a 40-W work load. This average increase of 23 ± 8 mmHg during low work load correlated significantly with the personal significance patients ascribed to this particular stress test (Table 2).

Table 2. Correlations between the increase of systolic blood pressure during low work load physical stress testing (40 W) and statements about this kind of challenge ($n = 17$)

	r	p
I was determined to do it as well as possible.	0.43	< 0.05
I am fairly satisfied wieh my performance in this test.	0.41	< 0.05
I believe my physical condition is not good enough for even very slight exertion.	0.35	0.08
I was not particularly stressed at the low levels of exertion.	0.55	< 0.01
I can imagine that the slight exertion is pretty hard work for some people.	0.38	0.06

The patients tended to dissimulate their effort during the stress test while simultaneously regarding the procedure as very important and interesting and possibly difficult for others.

This might demonstrate that it is not only the character of challenge but the way subjects cope with this particular challenge that significantly influences the SBP response. We could also demonstrate a similar effect on SBP for coping with traffic noise [4] and with mental stress [13].

Conclusions

The conclusion that behavior substantially influences the response of SBP during mild physical challenge is based on these observations:

1. The correlation between the increase of SBP during mental challenge and sub-maximal exercise testing and warm-up on the treadmill, especially in patients who overreact to mental challenge
2. Hypertensives, especially borderline hypertensives, do have an increased BP reactivity to physical challenge when tested adequately. Data are accumulating that BP response to mental challenge is not only different for normotensives, borderline hypertensives, and hypertensives, but also for normotensives with or without a positive family history of hypertension [5]. Therefore, BP reactions during low work load could possibly be regarded as a risk factor for the development of hypertension (see also [6]) and/or cardiovascular diseases [3]. Recently, Criqui et al. [1] came to a similar conclusion when analyzing the data from the Lipid Research Clinics Program Prevalence Study.
3. Physical conditioning does not prevent overreacting to low work load physical stress
4. The coping strategies to specific challenge situations might be crucial for the physiological reactions during low work load

All these observations were made in healthy, middle-aged, white men. In adolescents and older patients, or in patients with altered arteries, a different pattern for BP reactivity to challenge is found.

Behavioral influences on BP reactions during physical stress testing can easily be missed by inappropriate exercise protocols, by talking during the testing procedure, or by taking BP readings only every 3 to 5 min, thereby missing the sharp initial rise in SBP.

References

1. Criqui MH, Haskell WL, Heiss G, Tyroler HA, Green P, Rubenstein CJ (1983) Predictors of systolic blood pressure response to treadmill exercise: the lipid research clinics program prevalence study. Circulation 68: 225–233
2. Dlin RAF, Hanne N, Silverberg DS, Bar-Or O (1983) Follow-up of normotensive men with exaggerated blood pressure response to exercise. Am Heart J 106: 316–320
3. Eliot RS, Buell JC, Dembroski TC (1982) Biobehavioral perspectives on coronary heart disease, hypertension and sudden cardiac death. Acta Med Scand [Suppl] 660: 203–213
4. von Eiff AW, Neus H, Münch K, Schulte W (1981) Verkehrslärm als Risikofaktor für Hypertonie. Verh. Dtsch. Ges. Inn. Med. 87: 549–551
5. Falkner B, Onesti G, Angelakos ET, Fernandes M, Langman C (1979) Cardiovascular response to mental stress in normal adolescents with hypertensive parents. Hypertension 1: 23–30
6. Falkner B, Kusher H, Onesti G, Angelakos ET (1981) Cardiovascular characteristics of adolescents who develop essential hypertension. Hypertension 3: 521–527
7. Franz IW (1980) Ergometrische Untersuchungen zur Beurteilung der Grenzwerthypertonie. Therapiewoche 30: 7857
8. Franz IW (ed) (1981) Belastungsblutdruck bei Hochdruckkranken Ausmaß, Bedeutung und Konsequenzen für die Praxis. Springer, Berlin Heidelberg New York
9. Jandova R, Widimsky J, Ressl J (1980) Hemodynamics in juvenile hypertension at rest and during exercise. Cor Vasa 22: 22–32

10. Julius S, Conway J (1968) Hemodynamic studies in patients with borderline blood pressure elevation. Circulation 38: 282–288
11. Lund-Johansen P (1967) Hemodynamics in early essential hypertension. Acta Med Scand [Suppl] 183: 1–105
12. Neus H, Schulte W, Friedrich G, Rüddel H, Schirmer G, von Eiff AW (1981) Relationship between blood pressure reactions on an ergometric and an emotional stress test. Klin Wochenschr 59: 47–48
13. Neus H, Otten H, Rüddel H, Schmieder R, Schulte W, von Eiff AW (to be published) Cardiovascular and emotional reactions to mental stress. In: Orlebeke IF, Mulder G, van Doornen LJP (eds) Cardiovascular psychophysiology. Therapy and methods. Plenum Press
14. Pickering TG, Harshfield GA, Kleinert HD, Blank S, Laragh JH (1982) Blood pressure during normal daily activities, sleep, and exercise. JAMA 247: 992–996
15. Sannerstedt R (1980) Hemodynamic responses to exercise in patients with arterial hypertension. Acta Med Scand [Suppl] 458: 1–83
16. Schulte W, Neus H, Rüddel H (1982) Ergometrie in der Diagnostik der juvenilen Grenzwerthypertonie. Med Welt 33: 942–944
17. Schulte W, Fehring C, Neus H (1983) Cardiovascular reactivity to ergometric exercise in mild hypertension. Cardiology 70: 50–56
18. Wilson NC, Meyer BM, Albury GW (1981) Early prediction of hypertension using exercise blood pressure. Prev Med 10: 62–68
19. Zerzawy R, Reis A, Bachmann K (1977) Belastungshypertonie bei Hochdruckkranken und Grenzwerthypertonikern. Verh. Dtsch. Ges. Kreislaufforsch. 43: 261 (abstract)

Reaction of Cardiopulmonary Parameters and Lactate During Submaximum and Maximum Work Loads Depending on Different Tests and Input Loads

A. Reinke, H. Heck, and R. Rost

Although bicycle ergometry has become a standard procedure for diagnosis in sport medicine, the standardization of test protocol remains a matter of discussion. Recently, new recommendations have been published [1, 4, 5]. In spite of all these efforts (and even in these new proposals), there are still a lot of different test protocols recommended, concerning the input load as well as the increase of performance during stress testing [6]. Therefore we asked if there is an influence on the reactions of cardiopulmonary and metabolic parameters, depending on the different test schedules and various input loads for submaximum and maximum work load.

Theoretically, too heavy or too light an input load, as well as too large or too small increments, could either increase or decrease the maximum work load regarding the effects of premature exhaustion or hyperlactatemia by nonsteady state reactions.

Method and Material

In order to answer these questions we performed the following tests. We investigated 18 subjects of different training level with two standardized stepwise tests. Groups of six untrained subjects and six students of physical education performed a modified WHO test with an input load of 25 W and a stepwise increase of 25 W every 2 min [3]. The same six students of physical education as well as six highly trained athletes were given the test which is recommended by the national committee for performance athletics (BAL) especially for the investigations of athletes (BAL test). This test starts with an input load of 50 W, and the load is increased by 50 W every 3 min [2].

Furthermore both test schedules had been varied by different input loads – the WHO test with input loads of 25, 50, 75, and 100 W, the BAL test with input loads of 50, 100, 150, and 200 W.

Besides the determination of heart rate and lactate, we measured oxygen uptake, ventilation, respiration equivalent, and oxygen pulse during the last minute of each step up to maximum.

We also measured PWC_{170} and aerobic-anaerobic threshold, e.g., work load at a lactate level of 4 mmol/l, and the above-mentioned parameters which were recorded at these work loads.

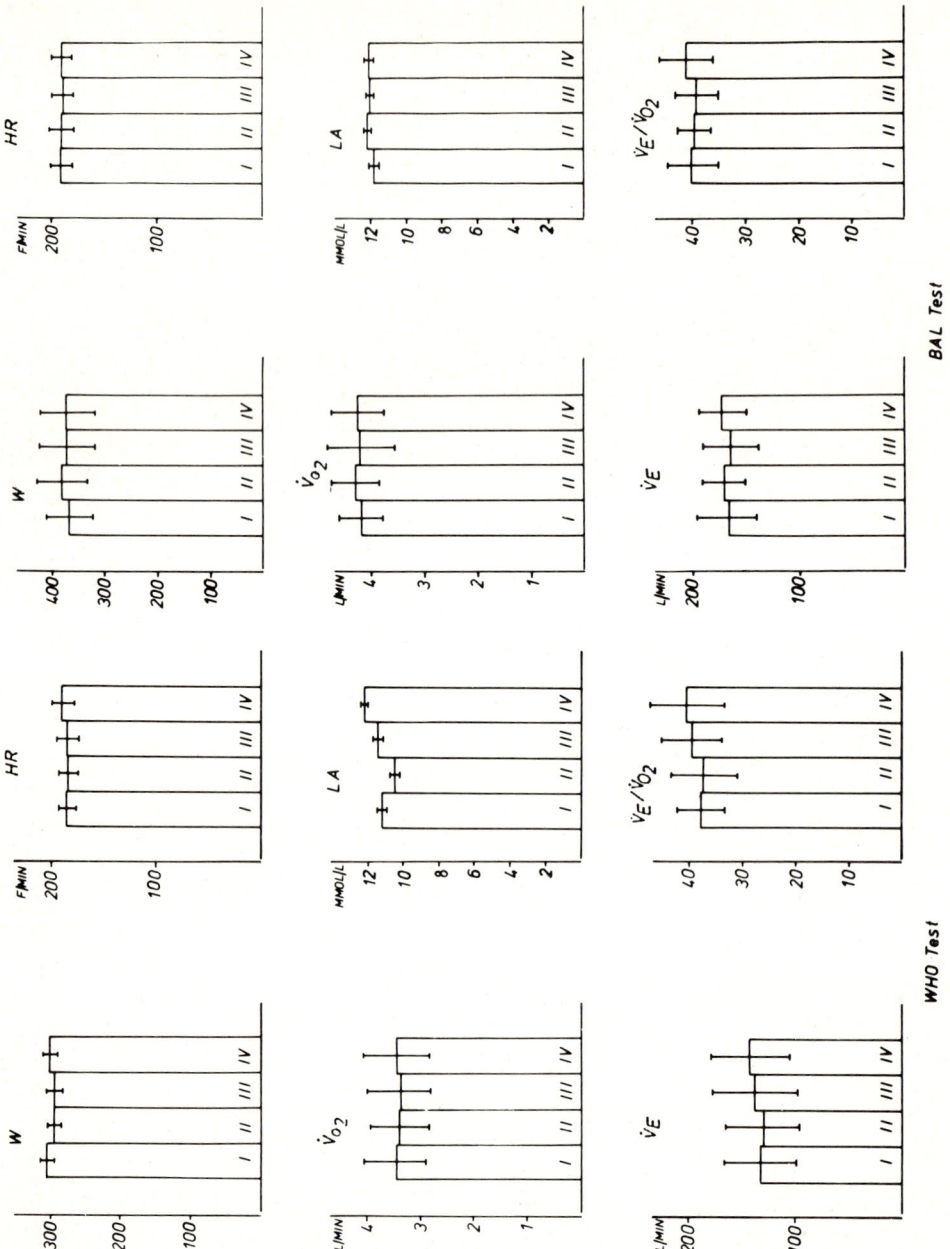

Fig. 1. Maximum values depending on the input load. WHO test: I, 25 W; II, 50 W; III, 75 W; IV, 100 W. BAL test: I, 50 W; II, 100 W; III, 150 W; IV, 200 W. $n = 12$ (six students of physical education and six untrained subjects) HR heart rate; $\dot{V}_{O_2}$, oxygen uptake; $\dot{V}_E$, minute ventilation; $\dot{V}_E/\dot{V}_{O_2}$, ventilatory equivalent; LA, lactic acid

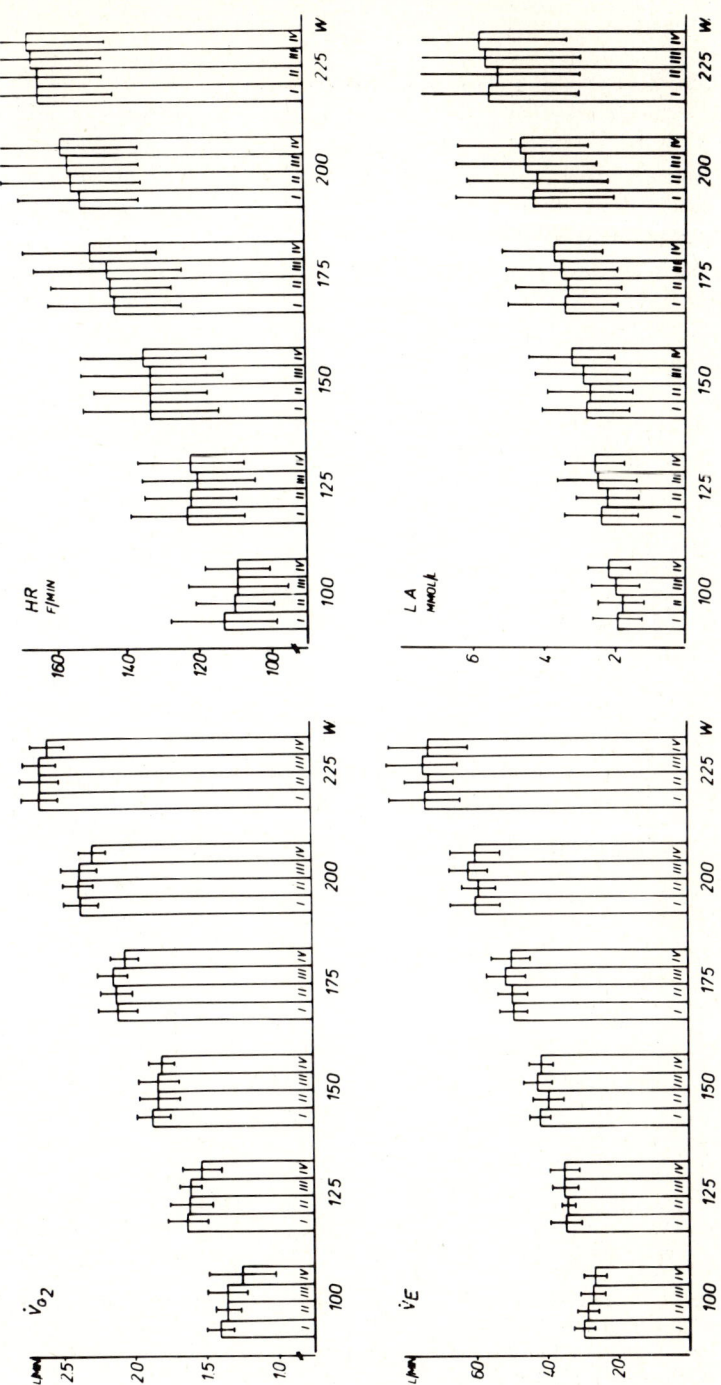

Fig. 2. Submaximum values depending on different input loads

Results

We were unable to find any influence of the different input loads on the maximum values during either the modified WHO test or the BAL test (Fig. 1).

In the WHO test on submaximum work load the lowest values for all spirometric parameters were found when we started with the highest input load (100 W). Conversely, we found the highest lactate values here (Fig. 2). In tests with higher input loads, heart rate was lower during the initial loads and higher during the upper steps. All these results could not be verified statistically; they just show tendencies only.

During the BAL test, we did not find an influence on the submaximum values by varying the input load – not even a tendency.

Corresponding to these results, there was no clearcut trend for all parameters either on the aerobic-anaerobic threshold (4 mmol/l lactate) or for the PWC_{170}

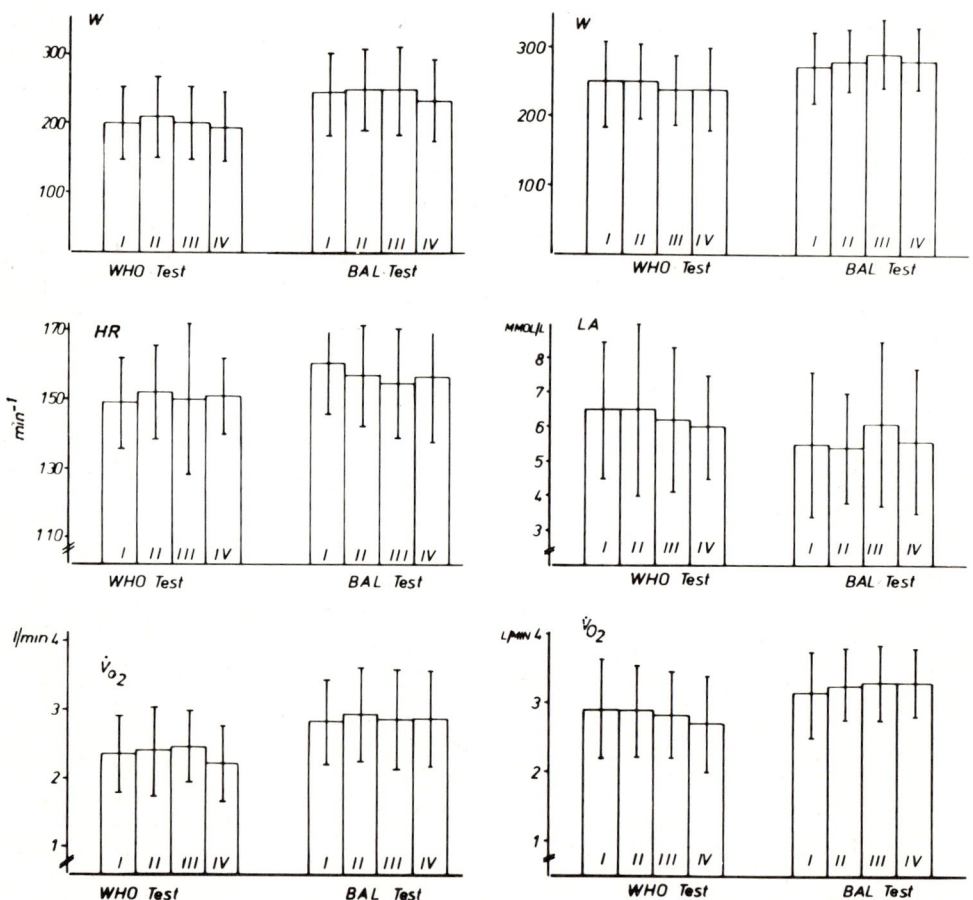

Fig. 3. Submaximum values of the aerobic-anaerobic threshold (4 mmol/l lactate) and for PWC_{170}

(Fig. 3). We also were unable to demonstrate find any influence of the increase and the duration per loading interval on cardiopulmonary and metabolic values.

The direct comparison of the maximum results of each test protocol revealed slightly higher lactate values for the BAL test compared with the WHO test (Fig. 4). On submaximum work load, lactate and heart rate show lower values in the WHO schedule than in the BAL schedule (Fig. 5). For some of the performance levels the results could be verified statistically. The respiratory parameters did not show any tendencies.

Corresponding to this result we found with the WHO test, e.g., the test with the slowly increasing rate, a higher performance at the aerobic-anaerobic threshold of about 10 to 20 W. This difference was not significant. We measured nearly the same differences for the values for the PWC_{170}.

Discussion

These findings demonstrate that there is almost no influence of the applied test schedules on the results of stress testing. Regarding the maximum values, no difference could be shown, although the input loads varied strongly between 25 and 100 W or 50 and 200 W. We also found no influence of the increasing rate of perfor-

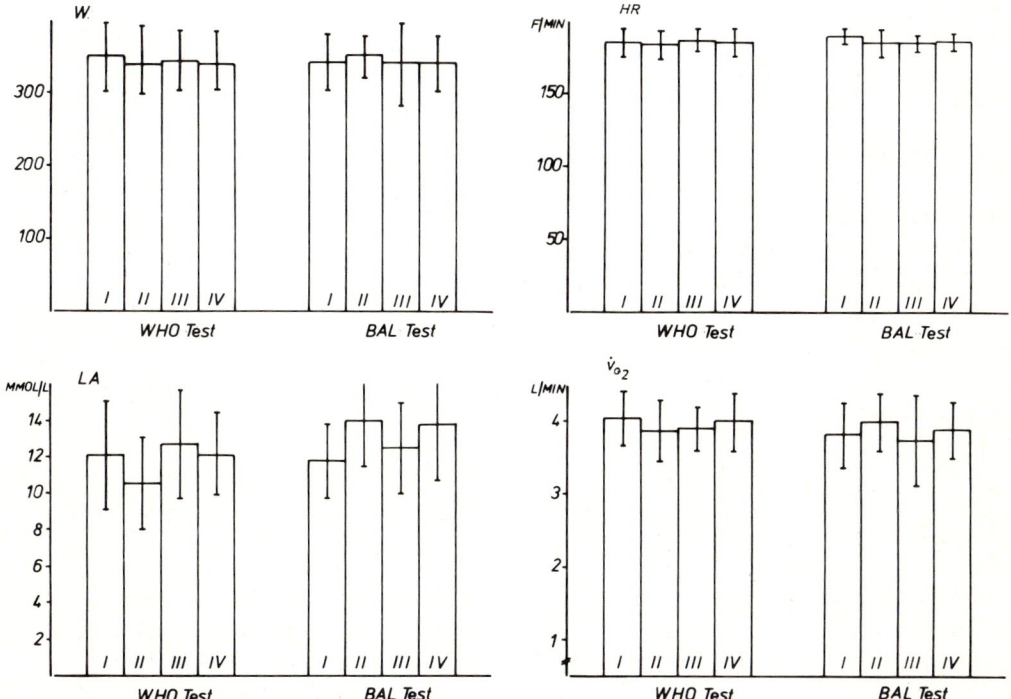

Fig. 4. Maximum values depending on different test schedules. $n = 6$ (students of physical education)

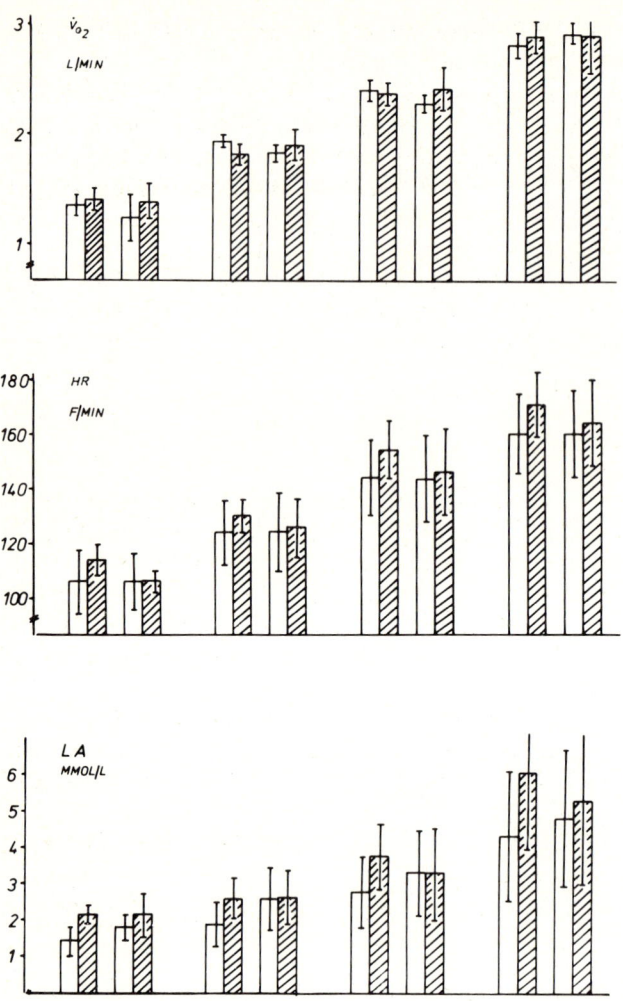

Fig. 5. Submaximum values depending on different test schedules. $n = 6$ (students of physical education) ☐ WHO test, ▨ AL test

mance on the maximum values. But it is worth-while to mention that the increasing rate varied little by the stepwise increase of either 25 W every 2 min (12.5 W/min) or 50 W every 3 min (17 Watt/min).

Only the submaximum values showed some differences, e. g. for the lactate level. It could be shown that the lactate concentration was lower during the WHO test than during the BAL test. We found corresponding tendencies also for the different input loads. It was provable for the WHO test that the lactate values were a lower by starting the test with a 25 or 50 W input load. This resulted in higher performance values at the aerobic-anaerobic threshold by a slow increase of the performance. This finding was expected considering the longer time which is available for the adaptation of the cardiac system. The differences were indeed so moderate that there would be no effect on the ergometric results in practice.

From our investigations we draw the conclusion for testing of young and healthy subjects that there is no influence on the maximum values depending on the test protocol and only little influence on the submaximum values within the limits of the test protocols investigated in this trial. On the other hand these results are nontransferable to ergometric testing with older, especially cardiac patients. Here muscular exhaustion may be the limiting factor. Additionally, in tests of cardiac patients it has to be taken into consideration that pathological reactions of the cardiac system might be observed even at very low levels of load. Therefore, in such cases a low input step should be recommended.

Summary

The discussion about the standardization of ergometric investigations questions the reactions of cardiopulmonary and metabolic parameters depending on different tests and various input loads by submaximum and maximum work load.

Groups of six highly trained athletes, six students of physical education, and six untrained subjects were tested. Besides the determination of heart rate and lactate concentration, we measured oxygen uptake, ventilation, respiration equivalent, and oxygen pulse on a closed system.

Each subject was tested up to maximum by two different standards, first beginning with an input load of 25 W and a stepwise increase of 25 W every 2 min (modified WHO test) and second with an input load of 50 W and a stepwise increase of 50 W every 3 min (BAL test).

Further on, both tests were modified by the initial loads (WHO test: 25, 50, 75, and 100 W, and the BAL test: 50, 100, 150, and 200 W).

According to our results, the obtained maximum depends on the test standard but not on the input load.

References

1. Böhm H et al (1978) Empfehlungen für eine standardisierte Ergometrie. Öst Ärztetag 33: 333
2. Böhmer D et al (1975) Das sportmedizinische Untersuchungssystem. In: DSB Bundesausschuß Leistungssport (ed) Beiheft zum Leistungssport. vol 4
3. Lange-Andersen KR et al (1971) Fundamentals of exercise testing. World Health Organization, Geneva
4. Mellerowicz H et al (1964) Vorschläge zur Standardisierung der ergometrischen Leistungsmessung, 2. Mitteilung. Z Kreislaufforsch 53: 856
5. Niederberger M et al (1978) Methodische Begründung für den ergometrischen Standardisierungsvorschlag. Öst Ärztetag 33: 345
6. Smodlaka L, Mellerowicz H, Horák J (1982) Revidierte Standardisierungsvorschläge für Ergometrie 81. Herz und Gefäße 2: 63

A Comparison of Bicycle, Paddling, and Treadmill Spiroergometry in Top Paddlers

J. Heller, V. Bunc, J. Novák and I. Kuta

Introduction

Functional predispositions of top sportsmen can be estimated by various types of spiroergometric tests. There is wide evidence that specific functional and metabolic adaptations could be verified only by means of adequate, specific performance tests [3, 11]. On the other hand some authors did not confirm this opinion [1]. The problem is discussed especially in paddlers, whose adaptation for arm work was tested by cranking [5, 12, 13] or on a specificly adjusted ergometer [3] or isokinetic pully-trainer [1]. Another problem pertains to the possibility of extrapolation of laboratory results to the racing performance where many other factors (e. g., technique, tactics, psychological factors) are involved.

The aim of the present study was to compare functional and metabolic parameters during a maximal work load on a bicycle ergometer, on a paddling ergometer, and on a treadmill with special attention to their relationship to the racing performance in a group of top paddlers.

Material and Methods

Twenty-four paddlers of the top performance level were examined in the laboratory at the end of the preparatory period. Fourteen subjects (group A, seven canoe and seven kayak paddlers) were tested on a bicycle and a paddling ergometer, ten subjects (group B, four canoe and six kayak paddlers) performed on the paddling ergometer and the treadmill. The recovery period between both tests exceeded 2 h.

The test on the bicycle ergometer (type Jaeger) started with two submaximal loads $2 \, W \cdot kg^{-1}$ and $3 \, W \cdot kg^{-1}$, each lasting 4 min. The subjects then pedaled to the maximal performance from the load W_{170} by an increment of 20 W every minute. The treadmill test started with two 5-min submaximal runs (at 0% level) of $11 \, km \cdot h^{-1}$, and $13 \, km \cdot h^{-1}$, respectively. The maximal test followed at a slope of 5%; the speed was increased by $1 \, km \cdot h^{-1}$ each minute, starting with $13 \, km \cdot h^{-1}$, till the maximum. The paddling ergometer of our own construction [6] was suitable for kayak as well as for canoe performance. Two submaximal loads, each of 4-min duration, attained 30% and 50% of the mean of a 1-min maximal force test. The incremental maximal test was initiated at 60% of the maximal force and was increased by 10 W each minute.

Expired air was analyzed in an open-circuit system on a Dataspir Jaeger apparatus. The gas analyzers were calibrated at the beginning and at the end of each test using cylinders of standard gases. Calculations of oxygen uptake and carbon dioxide output were based on $\dot{V}_E$ (STPD) and true O_2 and CO_2 concentrations. The highest $\dot{V}_{O_2}$ obtained during incremental exercise test was recorded as $\dot{V}_{O_2max}$. Heart rate was continuously monitored on an ECG recorder.

Capillary blood samples were obtained from a fingertip. After deproteinization, blood lactate was analyzed using an enzymatic method (Boehringer, Mannheim). The percentage of body fat was estimated by measuring ten skinfolds according to Pařízková [9]. Statistical evaluation was performed by paired and unpaired t tests and Spearman's correlation analysis of ranks.

Results

Comparing the weight, height, and percentage of body fat (Table 1) with normal values of the Czechoslovak population [10], the men's weight was higher and the percentage of body fat was lower than in the normal population. There were no significant differences between group A and group B in all the anthropometric parameters.

Table 1. Physical characteristics of paddlers

	Group A ($n = 14$)		Group B ($n = 10$)	
	$\bar{x}$	SD	$\bar{x}$	SD
Age (years)	20.5	2.4	21.4	3.0
Height (cm)	184.1	5.7	179.5	4.7
Weight (kg)	80.4	7.2	77.7	6.2
Body fat (%)	7.6	2.3	9.5	2.2

A comparison of maximal values reached by different types of spiroergometry (Table 2) showed that the results in group A and group B in the paddling performance were not significantly different, except for the higher respiratory frequency, respiratory quotient, and lactate, which were higher in group B. In the treadmill test, the highest values of $\dot{V}_{O_2}/kg$, $F_{IO_2}-F_{EO_2}$ and $\dot{V}_{E_{O_2}}$ were observed, values of $\dot{V}_{O_2}$, $\dot{V}$, V_T, and $\dot{V}_{O_2}/f_H$ were comparable with the bicycle ergometer results. Only f_H, R, and lactate from the treadmill test were comparable with the paddling ergometer results. The comparison of bicycle and paddling ergometer showed a higher $\dot{V}_{O_2}$, $\dot{V}_{O_2}/kg$, $\dot{V}$, $\dot{V}_{O_2}/f_H$, and maximal performance on the bicycle ergometer and, on the contrary, higher V_T and f_H in the paddling ergometer test.

Only few significant correlations were found between the maximal values of the parameters investigated in the different types of spiroergometry (Table 3). In group A a positive correlation for ventilation during bicycling and paddling was found; in group B (running and paddling) four significant correlations were found (positive for $\dot{V}_{O_2}$, $\dot{V}$, and $\dot{V}_{O_2}/f_H$ and negative for f_H).

When calculating the racing performance rank of paddlers, no significant correlations were found with the results of bicycle and treadmill tests (Table 4). In both

Table 2. Maximal values of functional parameters obtained on the bicycle and paddling ergometer (group A) and on the treadmill and paddling ergometer (group B)

	Group A (n = 14)				Group B (n = 10)				Significances[a]			
	Bicycle ergometer (1)		Paddling ergometer (2)		Treadmill (3)		Paddling ergometer (4)		1–2	3–4	1–3	2–4
	X̄	SD	X̄	SD	X̄	SD	X̄	SD				
$\dot{V}O_2$ (l · min⁻¹)	4.45	0.5	4.16	0.36	4.87	0.55	4.01	0.41	x	xx	NS	NS
$\dot{V}O_2/kg$ (ml · min⁻¹ · kg⁻¹)	55.6	4.4	51.9	3.4	62.5	3.7	51.6	4.5	x	xx	xx	NS
$\dot{V}E$ (l · min⁻¹)	147.2	18.2	136.8	15.8	146.3	22.3	136.7	16.6	xx	xx	NS	NS
$F_{IO_2}-F_{EO_2}$ (%)	3.68	0.36	3.7	0.3	4.04	0.35	3.57	0.3	NS	xx	x	x
f (min⁻¹)	51.0	8.0	56.4	4.9	56.7	4.5	61.6	3.9	NS	x	x	NS
V_T (l)	2.39	0.52	2.47	0.34	2.59	0.43	2.23	0.29	xx	xx	NS	NS
f_H (min⁻¹)	185.9	7.2	191.4	6.3	193.9	5.9	187.0	11.4	x	NS	xx	NS
$\dot{V}O_2/f_H$ (ml)	24.5	2.7	21.7	1.7	25.3	3.0	21.7	3.4	x	xx	NS	NS
$\dot{V}_{EO_2}$	33.2	3.2	32.7	3.0	30.0	2.7	34.4	1.9	NS	xx	x	xx
R	1.03	0.04	1.02	0.04	1.12	0.04	1.14	0.07	NS	NS	xx	xx
Lactate (mmol · l⁻¹)	8.6	1.6	8.8	1.2	13.2	1.6	12.5	3.0	NS	NS	xx	xx
Performance (W)	347.7	49.6	168.9	21.7	–	–	169.7	18.6	xx	–	–	xx
(km/h at a slope of 5%)	–	–	–	–	18.6	1.0	–	–	–	–	–	–

[a] Significances: x at level $p < 0.05$; xx at level $p < 0.01$; NS, not significant

Table 3. Correlations between maximal values of functional parameters obtained on the bicycle and paddling ergometer (group A) and on the treadmill and paddling ergometer (group B)

	Group A ($n = 14$) r	Group B ($n = 10$) r
$\dot{V}O_2$	0.46	0.75[b]
$\dot{V}O_2/kg$	0.08	0.41
$\dot{V}$	0.76[b]	0.72[a]
$F_{IO_2}-F_{EO_2}$	0.50	0.35
f	0.46	−0.48
V_T	0.52	−0.65[a]
f_H	0.45	0.56
$\dot{V}O_2/f_H$	0.02	0.84[b]
$\dot{V}_{EO_2}$	0.54	0.56
R	−0.14	0.35
LA	0.35	0.39
Performance	0.18	0.53

[a] Significant at $p < 0.05$
[b] Significant at $p < 0.01$

Table 4. Correlations between yearly competitive performance rank and maximal values of functional parameters

	Group A ($n = 14$)		Group B ($n = 10$)	
	Bicycling r	Paddling r	Running r	Paddling r
$\dot{V}O_2$	−0.04	0.30	0.48	0.31
$\dot{V}O_2/kg$	−0.41	0.11	0.55	0.24
$\dot{V}$	0.38	0.23	0.53	0.09
$F_{IO_2}-F_{EO_2}$	−0.41	0.01	−0.21	0.14
f	−0.30	−0.61[a]	0.48	−0.86[b]
V_T	0.33	0.41	0.46	0.47
f_H	0.21	−0.28	−0.52	−0.42
$\dot{V}O_2/f_H$	−0.20	0.44	0.60	0.41
$\dot{V}_{EO_2}$	0.40	0.02	0.49	0.18
R	−0.12	0.26	−0.31	−0.08
Lactate	0.09	0.20	−0.30	−0.20
Performance	−0.45	0.45	0.38	0.31

[a] Significant at $p < 0.05$
[b] Significant at $p < 0.01$

paddling tests (group A, group B), only a negative correlation between the respiratory frequency in the laboratory and racing performance was observed.

Discussion

Because of both the specific muscle mass and the oxidative capacity of working muscles are increased due to the specific training, it is expected that "sports-specific" ergometric loads give more information about the sports-specific abilities [11].

Comparing the work of legs and arms it can be shown [7] that $\dot{V}_{O_{2}max}$ in the untrained subjects during arm work reached 60%–62% of the values obtained on the bicycle ergometer. Israel and Brenke [5] reported that $\dot{V}_{O_{2}max}$ during cranking attained in cyclists and runners was 73%, whereas in paddlers it was 85%–89% of the maximal value reached on the bicycle ergometer [12, 13]. During work on the specific, adjusted paddling ergometer in our experiment, the maximal oxygen consumption averaged 95% of the value obtained on the bicycle ergometer, which corresponds to the data of Dal Monte and Leonardi [3], who found in canoeists, during paddling on a specific ergometer, 93% of the maximal value of $\dot{V}_{O_{2}}$ obtained on the bicycle ergometer.

The differences between running and paddling are greater in relation to the working muscle mass. In our study, paddlers during paddling attained 83% of the $\dot{V}_{O_{2}max}$ attained on the treadmill, which corresponds to the data of Tesch et al. [12] who found 85% $\dot{V}_{O_{2}max}$ reached on the treadmill in elite canoeists during paddling.

During paddling, ventilatory functions are also of considerable importance [2, 8]. Although maximal minute ventilation and tidal volume obtained in the bicycle ergometer and treadmill tests were not substantially different, the efficiency of respiration ($\dot{V}_{E_{O_{2}}}$) was apparently higher in running than in cycling or paddling. In paddling, lower minute ventilation was accompanied by a higher respiratory frequency. The circumstances of ventilatory work, especially the close relationship between respiratory frequency and frequency of paddling [2], could also affect the level of maximal "specific" aerobic capacity.

When discussing the values of maximal heart rate, the circulatory maximum was reached in the running test, but heart rate during cycling and paddling also attained a high level (96% of the maximum) in correspondance to the reported data of Dal Monte and Leonardi [3].

The values of the respiratory quotient and lactate could be considered to a certain extent as parameters of the anaerobic capacity. The differences were found only in groups A and B, and not between "specific" and "nonspecific" tests. This finding could possibly be connected with a different actual state of training abilities of the two groups.

The discrepancy between the small difference in oxygen consumption and the considerable difference in performance expressed in watts during cycling and paddling could be explained by a lower mechanical efficiency during the paddling performance [3].

A negative correlation between maximal oxygen consumption, pulse oxygen, and laboratory performance in cycling and yearly competitive performance rank indicates that testing on the bicycle ergometer seems to be inappropriate for well-trained paddlers. Contrary to this, testing on the treadmill or paddling ergometer could be relevant to this purpose, although the relationship to the racing performance did not reach a significant level. According to our experience the relationship between the racing performance results and laboratory spiroergometrical values could be affected by many factors (level of technique, tactics during races, psychological factors, especially a different level of motivation in racing and laboratory conditions, etc.). It is often difficult to explain the differences in racing performance rank only due to the spiroergometrical data in well-trained, relatively homogenous groups. Nevertheless, a high level of functional predisposition is a prerequisite for the top performance [4]. Laboratory spiroergometrical testing seems to be neccessary for estimating of maximal functional capacity of sportsmen, but more complete information, especially during the racing period, could be obtained by testing paddlers under field conditions, where the interference of other factors is likely to be smaller than in the laboratory.

Conclusions

A comparison of treadmill, bicycle ergometer, and paddling ergometer spiroergometrical tests showed that maximal values can be reached on a treadmill. Bicycle ergometer testing seems not to be adequate for well-trained paddlers. The differences between specific and nonspecific tests confirmed the high level of specific adaptation in paddlers investigated. All types of laboratory tests were not significantly related to the racing performance; nevertheless, the functional profile of the paddler makes an important part of the whole complex of racing performance predispositions. Testing under field conditions was suggested to obtain more information about specific functional predispositions of top canoe and kayak paddlers.

References

1. Apor P, Faludi J, Miháliřy P, Kozma Á (1982) Spiroergometry on treadmill vs. cranking and "kayak-specific" loads in kayakers. Hung Rev Sports Med 23: 259–266
2. Čermák J, Kuta I, Pařízková J (1975) Some predispositions for top performance in speed canoeing and their changes during the whole year training program. J Sports Med 15: 243–251
3. Dal Monte A, Leonardi LM (1975) Sulla specifità della valutazione funzionale negli atleti: esperienze sui canoisti. Med Sport (Torino) 28: 213–219
4. Heller J, Novák J, Bunc V, Kuta I, Šprynarová Š (1983) Functional profile of elite canoeists (in Czech). Trénér Meth [Suppl] 3: 1–5
5. Israel S, Brenke H (1967) Das Verhalten spiroergometrischer Messgrößen bei Läufern und Radsportlern sowie Kanuten bei Hand- und Fusskurbelarbeit. Med Sport (Berlin) 7: 104–108
6. Kuta I (1981) Testing of specific performance in speed canoeing (in Czech). In: Proceedings of scientific council of central committee of Czechoslovak union of physical culture, vol 11. Olympia, Prague, pp 95–106
7. Magel JR, McArdle WD, Toner M, Delio DJ (1978) Metabolic and cardiovascular adjustment to arm training. J Appl Physiol 45: 75–79
8. Mihályfi P, Apor P, Faludi J, Kozma Á (1981) Effect of respiratory frequency of oxygen uptake under steady state loading. Hung Rev Sports Med 22: 261–266
9. Pařízková J (1977) Body fat and physical fitness. Nijhoff, Hague, p 297
10. Seliger V, Bartůněk Z (1976) Mean values of various indices of physical fitness in the investigation of Czechoslovak population aged 12–55 years. Czechoslovak Union of Physical Culture, Prague, p 117
11. Stomme SB, Jugjer F, Meen HD (1977) Assessment of maximal aerobic power in specifically trained athletes. J Appl Physiol 42: 833–837
12. Tesch P, Piehl K, Wilson G, Karlsson J (1976) Physiological investigations of Swedish elite canoe competitors. Med Sci Sports 8: 214–218
13. Vrijens J, Hoekstra P, Bouckaert J, Van Uytvanck P (1975) Effects of training on maximal working capacity and haemodynamic response during arm and leg exercise in a group of paddlers. Eur J Appl Physiol 34: 113–119

Activity of Energy Metabolism Enzymes in the Vastus Lateralis of Young Men of Different Performance Levels

E. V. Macková, Š. Šprynarová, J. Melichna, A. Bass, and K. Vondra

Introduction

The level of activity of those enzymes involved in energy metabolism is considered as a sensitive indicator of the level of energy release in muscle tissue. This level represents the adaptive response to the intensity and duration of physical exercise [11, 5, 4, 6, 9]. Gollnick [3] showed that the properties of muscle tissue differ according to muscle fiber distribution and/or to different enzyme activities, which characterize athletes from each another.

In this paper some results of muscle enzyme activities are presented which are connected with energy-liberating processes under different conditions of training.

The following two questions were studied in particular:
1. The extent to which the observed enzyme activities increase and/or decrease under different exercise conditions
2. Whether a relationship exists between the indicators of muscle metabolic activity on the one hand and the functional and morphological indicators on the other

Experimental Groups and Methods

The study was performed on nationally ranked cross-country skiers who had an average age of 24.3 years ($n = 17$), downhill skiers of good performance who had an average age of 21.8 years ($n = 6$), middle-distance runners who had an average age of 17.7 years ($n = 13$), and a group of untrained men who had an average age of 22.0 years ($n = 6$). A group of nine cross-country skiers was examined twice during the annual training cycle, namely at the end of the preparatory period (October) and at the end of the winter season (March). During the summer, before the first stage of investigation, their training consisted of cross-country running, simulated skiing on roller skis, and cycling. During the winter months, the period between the first and second investigations, their training involved mainly ski-running, both for training purposes and in ski-running contests. During this period, five athletes trained mostly at a relatively high intensity, i.e., 140–180 heart beats/min (velocity training) and another four athletes trained at a relatively low intensity, i.e., performance not higher than 140 heart beats/min (endurance training) (see Fig. 2). For this purpose, the treadmill test was used to assess $\dot{V}O_{2max}$, and body composition was estimated by measurement of skinfold thickness. Enzyme activities were determined and histochemical analysis was performed on a muscle sample taken by biopsy from m vastus lateralis [6, 7].

Activities of the following enzymes were determined: hexokinase (HK) (EC 2.7.1.1), glycerol-3-phosphate dehydrogenase (GPDH) (EC 1.1.1.8), triosephosphate dehydrogenase (TPDH) (EC 1.2.1.12), lactate dehydrogenase (LDH) (EC 1.1.1.27), citrate synthase (CS) (EC 4.1.3.7), malate dehydrogenase (MDH) (EC 1.1.1.37), and hydroxyacyl-CoA dehydrogenase (HOADH) (EC 1.1.1.35). Three types of muscle fibers were determined: fast glycolytic fibers (FG), fast oxidative-glycolytic fibers (FOG), and slow oxidative fibers (SO).

Results and Discussion

Figure 1 shows the average activities of the observed enzymes in three groups of young subjects with different physical performance. The highest activities were found in the group of ski-runners, in whom the value of $\dot{V}O_{2max}$ was 67.7 ± 1.1 ml min^{-1} kg^{-1}. The lowest enzyme activities were found in the untrained subjects who also had lowest value of $\dot{V}O_{2max}$ (45.5 ± 4.1 ml min^{-1} kg^{-1}). In the group of downhill skiers, where the training is aimed more at technique and speed than at endurance, the values of the enzyme activities as well as the value of $\dot{V}O_{2max}$ (53.6 ± 1.8 ml min^{-1} kg^{-1}) corresponded to intermediate values. It is obvious that enzyme activities of practically all investigated enzymes, especially of the mitochondrial enzymes (CS, MDH, HOADH), were highest in athletes performing under endurance conditions.

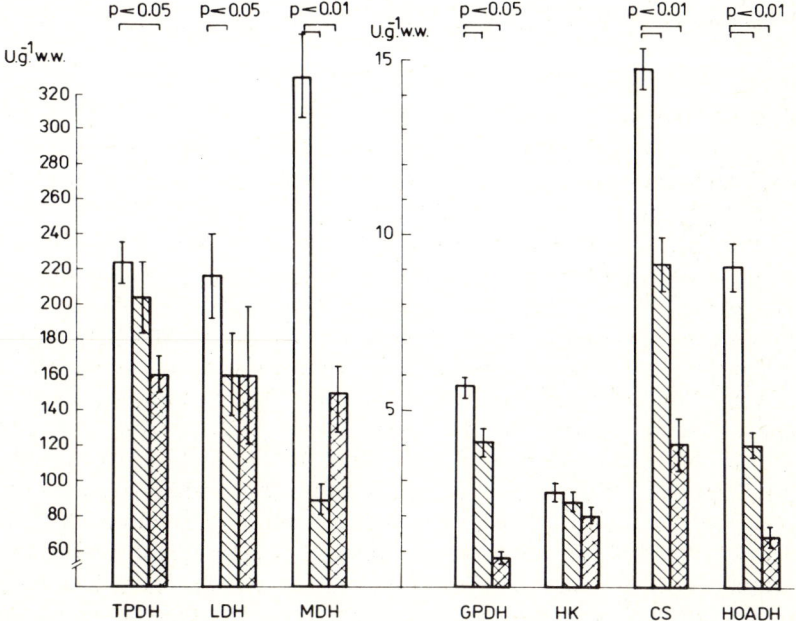

Fig. 1. Muscle enzyme activities in men of different performance levels (m. vastus lateralis). Values are means $\pm$ SEM; p, statistical significance; U·g.$^{-1}$ w.w., units per gram (wet weight). □ = cross-country skiers ($n=17$); ◨ = downhill skiers ($n=6$); ◪ = untrained m. ($n=6$)

Some authors selected the activity of other enzymes as the criterion of oxidative cellular capacity. Thus Gollnick, Moesch, Costill, and Rusko [3, 8, 2, 10] found higher activity of succinate dehydrogenase in endurance athletes. Relatively stable values were displayed by hexokinase [8], as was also confirmed by our results [1]. When comparing the glycolytic enzymes TPDH and LDH in untrained subjects with our group of ski-runners, it can be seen that the rise of glycolytic activity is markedly smaller than that of aerobic enzymes. The glycolytic enzyme GPDH, which determines interconversion between carbohydrates and lipids, shows a response similar to the aerobic enzymes CS and HOADH.

In the group of nine cross-country skiers of the national performance class we observed the influence of different training periods on muscle metabolic activity (Table 1). Although the $\dot{V}O_{2max}$ kg^{-1} practically did not change, the enzyme activities of aerobic as well as anaerobic metabolism were significantly reduced by 27%–59% at the end of the competition period as compared with values obtained at the end of the preparatory period. This finding was consistent with subjective feelings of fatigue and exhaustion of athletes because the biopsies were performed immediately after termination of the competition period. It is known that the mode of life of athletes during the preparatory and competing period differs markedly. The moderate and systematic training during the preparatory period (April–October) is often disturbed during the competing period (November–March) by fatigue caused by competitions or deteriorated health status (higher morbidity during winter) as well as by lower training activity.

However, in sportsmen who trained during the competing period with an intensity which did not induce an increase of the heart rate higher than 140 heart beats/min, the activity of HOADH increased by 34%. This finding underlines the importance of fatty acids as an energy substrate during endurance work and also the lower sensitivity of this enzyme to unfavorable conditions.

Table 1. Muscle enzyme activities in the group of cross-country skiers. U·g^{-1} w.w., units per gram (wet weight)

	Anaerobic		Aerobic		Aerobic fatty acids		$\dot{V}O_{2max}$/kg (ml)
			carbohydrate		HOADH		
	TPDH	LDH	MDH	CS			
Preparatory p. (October) n=9	211 ±11.5	186 ±25.1	341 ±31.7	14.2 ±0.74	8.8 ±0.70		67.7 ±1.1
	+ ↓	+ ↓	+ ↓	+ ↓	+ + V	E	
Competing p. (March) n=9	116 ±10.7	77 ±6.0	158 ±35.2	10.4 ±0.67	5.4 ±0.6	11.8 ±1.5	70.0 ±1.4
Changes %	45	59	54	27	39	34	3.4

V, velocity training; E, endurance training; means, ±SEM; *arrows* = significant changes; +p < 0.001; ++p < 0.05

Experimental data concerning the correlations between metabolic activity and functional and morphological indicators are demonstrated in Table 2 and 3. A significant relationship exists between activities of some enzymes on the one hand and cardiorespiratory indicators and muscle fiber types on the other hand. The activities of aerobic enzymes (MDH, CS, HOADH) are positively related to functional indicators ($\dot{V}O_{2max}$, O_2 pulse$_{max}$, and laboratory performance), which is not the case with enzymes of anaerobic glycolysis (TPDH, LDH, GPDH, HK). An inverse relationship between the activity of HOADH and the percentage of depot fat suggests that significantly more depot fat is utilized in athletes. It was also found that there is a significant relationship between muscle fiber types and enzyme activities. Figure 3 shows that predominantly glycolytic fast fibers (FG + FOG) are positively related to the glycolytic enzyme TPDH and that there is a negative correlation with oxidative enzymes CS and HOADH. On the other hand there is an inverse correlation between oxidative slow-twitch fibers (SO) and the glycolytic enzyme TPDH and a highly positive relation of these fibers to oxidative enzymes (CS and HOADH).

The results indicate that the activities of enzymes involved in energy metabolism, especially those of mitochondrial enzymes, react very sensitively to the intensity and duration of training and to various unfavorable conditions.

Table 2. Coefficients of rank correlations between muscle enzyme activities and between muscle enzyme activities and functional parameters ($n = 23$)

Enzyme	TPDH	LDH	GPDH	HK	MDH	CS	HOADH
TPDH	–	0.3108	0.4467^a	0.5571^b	0.2368	0.1216	−0.1990
LDH	–	–	0.0372	0.0340	0.2524	0.0398	−0.0193
GPDH	–	–	–	0.4426^a	0.3637	0.5613^b	0.1907
HK	–	–	–	–	0.3779	−0.1909	−0.2464
MDH	–	–	–	–	–	0.4746^a	$−0.4337^a$
CS	–	–	–	–	–	–	$−0.6528^b$
$\dot{V}O_{2max}$ (ml min^{-1}) kg^{-1}	0.0690	0.2527	0.3331	0.0349	0.6207^b	0.5606^b	0.7164^b
HR$_{max}$ (pulse min^{-1})	−0.0147	−0.2522	−0.2097	0.0784	$−0.7539^b$	$−0.5054^a$	$−0.5047^a$
O_2 pulse$_{max}$ (ml)	0.0882	0.2348	0.2949	0.0463	0.7469^b	0.5489^b	0.6525^b
Body fat (%)	0.1036	−0.3130	−0.2829	0.0901	$−0.5205^a$	−0.4029	$−0.6108^b$
Laboratory performance (J)	−0.2175	0.0096	0.4312^a	0.0132	0.4964^a	0.5006^a	0.4857^a

Significance by correlation coefficients: $^a p < 0.05$; $^b p < 0.01$

Table 3. The relationship between three muscle enzyme activities (glycolytic enzyme TPDH and aerobic enzymes CS and HOADH) and muscle fiber types (glycolytic types FG + FOG and aerobic types SO)

	TPDH	CS	HOADH
FG + FOG	$+0.61^a$	$−0.58^a$	−0.49
SO	$−0.69^a$	$+0.94^a$	$+0.92^a$

$^a p < 0.05$

References

1. Bass A, Vondra K, Rath R, Vítek V, Teisinger J, Macková E, Šprynarová Š, Malkovská M (1976) Enzyme activity patterns of energy-supplying metabolism in the quadriceps femoris (vastus lateralis). Pflügers Arch 361: 169–173
2. Costill DL, Fink WJ, Pollock ML (1976) Muscle fibre composition and enzyme activities of elite distance runners. Med Sci Sports 8: 96–100
3. Gollnick PD, Armstrong RB, Saubert CW, Piehl K, Saltin B (1972) Enzyme activity and fiber composition in skeletal muscle of untrained and trained men. J Appl Physiol 33: 312–319
4. Fox EL (1975) Differences in metabolic alteration with sprint versus endurance interval training programs. In: Howald H, Poortmans JR (eds) Metabolic adaptation to prolonged physical exercise. Birkhäuser, Basel
5. Keul J (1973) Muscle metabolism during long-lasting exercise. In: Howald H, Poortsmans JR (eds) Metabolic adaptation to prolonged physical exercise. Birkhäuser, Basel
6. Macková EV, Bass A, Šprynarová Š, Teisinger J, Vondra K, Bojanovský I (1982) Enzyme activity patterns of energy supplying metabolism in skiers of different performance levels (M. Quadriceps Femoris) Eur J Appl Physiol 48: 315–322
7. Melichna J, Havlíčková L, Vránová J, Bartůněk Z, Seliger V, Bartůňková S, Vodička P, Karas V, Otáhal S, Štichová J (1982) Muscle fibre composition and physical performance of sprinters and long-distance runners. Acta Univ Carol 18: 95–123
8. Moesch H, Howald H (1975) Hexokinase (HK), glyceraldehyde-3P-dehydrogenase (GAPDH), succinate-dehydrogenase (SDH), and 3-hydroxyacyl-CoA-dehydrogenase (HAD) in skeletal muscle of trained and untrained men. In: Howald H, Poortmans JR (eds) Metabolic adaptation to prolonged physical exercise. Birkhäuser, Basel
9. Oberholzer F, Classen H, Moesch H, Howald H (1978) Ultrastrukturelle, biochemische und energetische Analyse einer extremen Dauerleistung (100-km-Lauf). Schweiz Z Sportmed 24: 71–73
10. Rusko H, Havu M, Karvinen E (1978) Aerobic performance capacity in athletes. J Appl Physiol 38: 151–159
11. Saltin B (1971) Oxygen transport by the circulatory system during exercise in man. In: Keul J (ed) Limiting factors of physical performance. Thieme, Stuttgart

Comparative Measurements of $\dot{V}O_{2max}$ and PWC_{170} in Schoolchildren

I.-W. Franz, D. Wiewel, and H. Mellerowicz

Introduction

The determination of maximal oxygen uptake ($\dot{V}O_{2max}$) is generally accepted as a global index of aerobic work capacity and cardiorespiratory fitness. However, this procedure requires expensive apparatus.

For that reason, the Council of Europe organized four European Seminars on Testing Physical Fitness, the last one having been held in Athens in May 1982, in order to establish a simple ergometric test method which enables comparable and reproducible determination of cardiopulmonary capacity in schoolchildren. The investigation we will now present is the official contribution of the Federal Republic of Germany to the research program of the European Council.

Physical working capacity (PWC_{170}) is often used as a simple measure of cardiopulmonary function [1, 2, 3]. Studies in adults show [3, 6, 7] that the PWC_{170} is highly correlated with $\dot{V}O_{2max}$.

However, for testing schoolchildren the methods and procedure concerning the duration of work load and increment are still not standardized and the PWC_{170} is not sufficiently validated against $\dot{V}O_{2max}$.

In 100 untrained males, aged 18 to 39 years, we could prove a significant correlation ($r = 0.95$) between PWC_{170} (measured with increments of 25 W/2 min) and $\dot{V}O_{2max}$ [3]. Klissouras et al. [4] obtained from a small sample of pubertal boys a correlation coefficient of 0.9 between PWC_{170} and $\dot{V}O_{2max}$, using a bicycle ergometer.

The purpose of the present study was to investigate if schoolchildren also show a significant correlation between $\dot{V}O_{2max}$ and PWC_{170}, when ergometrically measured with increments of 25 W/2 min.

Material and Methods

Eighty-eight boys ($\bar{x}$ age + SD: 14.2 ± 0.9 years), randomly selected from a secondary school, participated in the study after giving their informed consent. The mean body height and weight were 168.2 ± 9.2 cm and 55.3 ± 11.2 kg, respectively.

The testing procedure had the following design: the subjects were familiarized with ergometric methods and the standardized working conditions for ergometry [2] (ICSPE 1967) one day before the investigation.

The boys exercised with a constant pedal rate of 50 rpm on an electromagnetically braked bicycle ergometer (Siemens Elema) in an upright sitting position. Heart rate was calculated every consecutive minute from the ECG, and oxygen uptake was continuously measured with use of an open system (Fa. Jaeger). The determination of PWC_{170} was carried out with increments of 25 W/2 min, starting with an initial load of more than 1 W/kg body weight. PWC_{170} was extrapolated from the heart rates of the first three or four work loads (heart rates should exceed 150 beats/min at the last work load).

Increments of 25 W/min were then used for measurement of $\dot{V}O_{2max}$. Attainment of two of the following criteria were necessary for determination of $\dot{V}O_{2max}$:

1. A heart rate of more than 200 beats/min
2. A respiratory exchange ratio of more than 1
3. A quotient of maximal ventilation and oxygen uptake of more than 30 ml

Results

Table 1 shows the results of $\dot{V}O_{2max}$, of maximal work load, maximal heart rate, maximal respiratory exchange ratio, maxima $\dot{V}E\ \dot{V}O_2^{-1}$, and $PWC_{170} \cdot$ kg body weight^{-1}.

Table 1. Mean $(\bar{x})$ and standard deviation (SD) of maximal oxygen uptake ($\dot{V}O_{2max}$), maximal work load, maximal heart rate (HR_{max}), respiratory exchange ratio (R_{max}), quotient of maximal ventilation and oxygen uptake (Max $V_E \cdot \dot{V}O_2^{-1}$), and physical working capacity$_{170}$ (PWC_{170})/kg body weight

$n = 85$	$\bar{x}$	$\pm$	SD
$\dot{V}O_{2max}$	3.131		$0.35\, l\,min^{-1}$
$\dot{V}O_{2max}$	56.60		$7.40\, ml \cdot kg\,BW\,min^{-1}$
Maximal work load	254.90		40.1 W
Maximal work load	4.61		$0.71 \cdot W\,kg\,BW^{-1}$
HR_{max}	202		$6.10\, beats \cdot min^{-1}$
R_{max}	1.09		0.06
Max $\dot{V}_E/\dot{V}_{O_2}^{-1}$	31.9		3.80 ml
$PWC_{170} \cdot kg\,BW^{-1}$	2.97		0.49 W

From the data we can conclude that the criteria for attainment of $\dot{V}O_{2max}$ were achieved. Three boys had to be excluded from the statistical analysis because they did not fulfill the above-mentioned criteria.

Figure 1 shows the relation between directly measured $\dot{V}O_{2max}$ and determination of PWC_{170} with work steps of 25 W/2 min. There was a significant ($p < 0.001$) correlation between $\dot{V}O_{2max}$ and PWC_{170}. This was true for low and high values of PWC_{170}, revealing a correlation coefficient of 0.87.

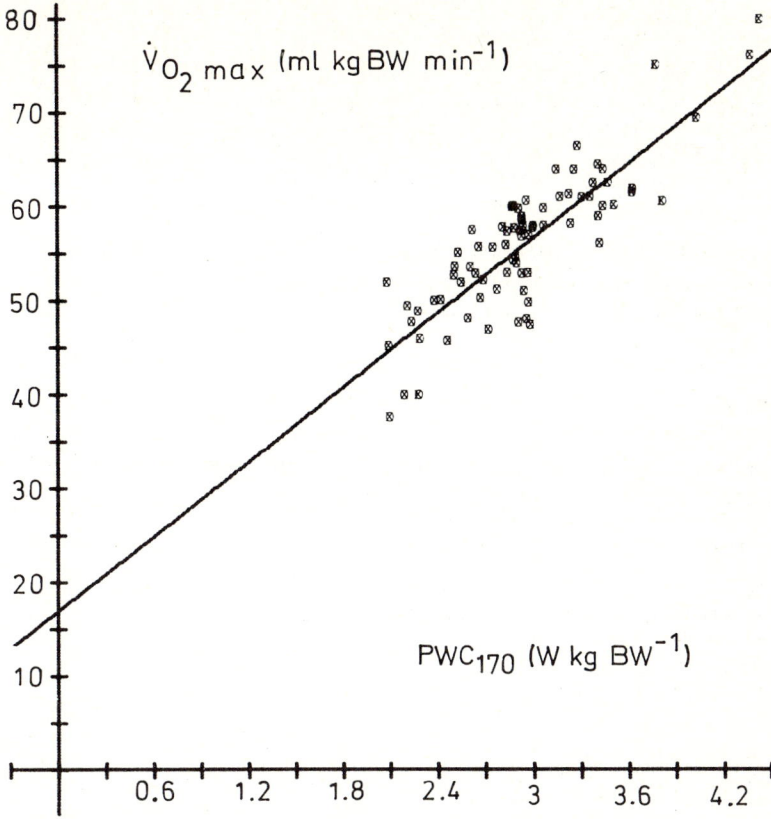

Fig. 1. Correlation between directly measured $\dot{V}O_{2max}$ and the PWC_{170}, estimated with increments of 25 W/2 min. *BW*, body weight. Regression equation: $y = 13.18\,x + 17.38$. Coefficient of correlation: $r = 0.85$ ($p < 0.001$)

Discussion

The close correlation between $\dot{V}O_{2max}$ and PWC_{170}, measured with work steps of 25 W/2 min, implies that increments of 25 W/2 min allow for comparable and reproducible results also in schoolchildren. This is in agreement with the results of Klissouras et al. [4]. In conclusion, cardiopulmonary capacity can be estimated reasonably well by means of a PWC_{170} test in a comparable group.

For practical application the time saving increments of 25 W/2 min should be recommended. Work steps of 1 W/kg body weight/6 min result in a significantly lower PWC_{170}. In a comparative study [2] with increments of 25 W/2 min we could show that at the second load of 2 W/kg body weight/6 min there was a considerable and significant ($p < 0.01$) increase in heart rate of 10.2 beats/min from the 1st to the 3rd min, but also of 7.7 beats/min from the 3rd to the 6th min. That means that the increase in work load with increments of 1 W/kg body weight is too excessive for schoolchildren and takes too long to reach a steady state.

At the 4th European Research Seminar on Testing Physical Fitness the delegates of the different countries therefore recommended using increments of 25 W or 0.5 W/ kg body weight with a duration of increments of 2 and 3 min, respectively.

The PWC_{170} should be extrapolated from the heart rates measured on three or four work loads (heart rates should exceed 150 beats/min at the last work load). The first load should be greater than 1 W/kg body weight in order to reach a steady state of heart rate [5].

A prediction of $\dot{V}O_{2max}$ is possible by multiplying the value of PWC_{170} with the calculated factor of $F = 19.1$. However, predicted values of $\dot{V}O_{2max}$ cannot be used as a substitute for the actual measurement of precise values when scientific exactness is required.

References

1. Franz IW, Mellerowicz H (1977) Comparative measurements of PWC_{170} with work steps of different load increase and duration. Z Kardiol 66: 670
2. Franz IW, Wiewel D, Agrawal B (1982) Comparative measurements of PWC_{170} with work steps of different load increase and duration in Schoolchildren. In: Council of Europe (eds) 4th European research seminar on testing physical fitness. Cardio-respiratory aspects. Strasbourg, p 104
3. Franz IW, Meyer-Rosorius R, Mellerowicz H (1982) Zur Methodik der Bestimmung der PWC_{170}. In: Mellerowicz H, Franz IW (eds) Kalibrierung – Standardisierung – Methodik in der Ergometrie. Perimed, Erlangen, p 145
4. Klissouras V, Tsarouchas E, Hadjiconstantinou S (1981) The physical work capacity of Greek schoolchildren. In: Council of Europe (eds) 2nd European seminar on testing physical fitness. Strasbourg, p 24
5. Mocellin R, Rutenfranz J (1970) Methodische Untersuchung zur Bestimmung der körperlichen Leistungsfähigkeit (W_{170}) im Kindesalter. Z Kinderheilk 108: 61
6. Szögy A, Rosca E, Artanescu J, Minescu T (1970) Die Aussagekraft einiger bei submaximaler Belastung ausgeführter ergometrischer Tests für die Beurteilung der Leistungsfähigkeit bei Spitzensportlern. Med Sport 10: 17
7. Teräslinna P, Ismail AH, MacLeod DF (1966) Nomogram by Astrand and Rhyming as a predictor of maximum oxygen intake. J Appl Physiol 21: 513

Response to Maximal Ergometric Load of Different Types and Relation of Cardiorespiratory Parameters to Specific Performance in Young Swimmers

J. Novák, T. Jurimae, V. Bunc, E. V. Macková, M. Čermák, and T. Paul

Introduction

In competitive swimming, 77% of disciplines last about 2 min at the longest. According to Åstrand's design [5], the relative contribution in percent of total energy yield is covered from 65%–70% by anaerobic processes and from 30%–35% aerobically in 100 m races, while in 200 m races both processes are approximately equally important. Other authors [6, 25] presume that anaerobic sources participate to a higher extent. Therefore, it is important for competitive swimmers to develop both aerobic and anaerobic power to reach maximal performance in 100 m and 200 m races. It seems to be expedient to distinguish functional predispositions for these two components for appropriate orientation of the training process. Moreover, in freestylers it is possible to choose the race length according to these characteristics. The aims of this study were (a) to distinguish differences in cardiorespiratory and metabolic response to two kinds of ergometric load: *first*, load corresponding to the duration of swimming races and *second*, a stepwise increased load to vita maxima conditions, and (b) to compare obtained data with specific performance level of the swimmers.

Methods

Subjects

The subjects were 24 young swimmers. The whole group was divided into three subgroups: subgroup 1 (men) and subgroup 2 (women) were swimmers participating in the selection procedure before admission to top sport swimming center, while subgroup 3 were male swimmers only and members of the top swimming center for 1–6 years. Mean characteristics of all three subgroups are presented in Table 1. All participants were well trained in swimming but not accustomed at all to prolonged cycling exercise.

Procedure

Two different bicycle ergometer test protocols were used. Test A was designed to imitate the character of the swimming race or track. It consisted of maximum length of time until exhaustion at high work load intensity, empirically estimated at

Table 1. Characteristics of young swimmers

Subgroup		Age (years)	H (cm)	W (kg)	BF (%)	LBM (kg)	W_{170} (W)	$W_{170} kg^{-1}$ (W)	PL (points)
1	$\bar{x}$	13.71	168.6	55.98	9.12	50.62	179.7	3.2	518
	SD	0.51	5.9	9.71	3.44	7.53	44.1	0.6	27
2	$\bar{x}$	13.00	162.2	49.52	14.10	42.20	153.2	3.2	505
	SD	1.15	6.5	8.16	4.09	5.18	55.3	1.4	115
3	$\bar{x}$	16.47	180.1	69.91	10.27	62.66	243.1	3.5	673
	SD	1.22	4.1	6.18	3.11	5.18	39.9	0.4	67

H, height; *W*, weight; *BF*, body fat; *LBM*, lean body mass; *PL*, performance level. Subgroups 1 and 3 were males, while subgroup 2 were females

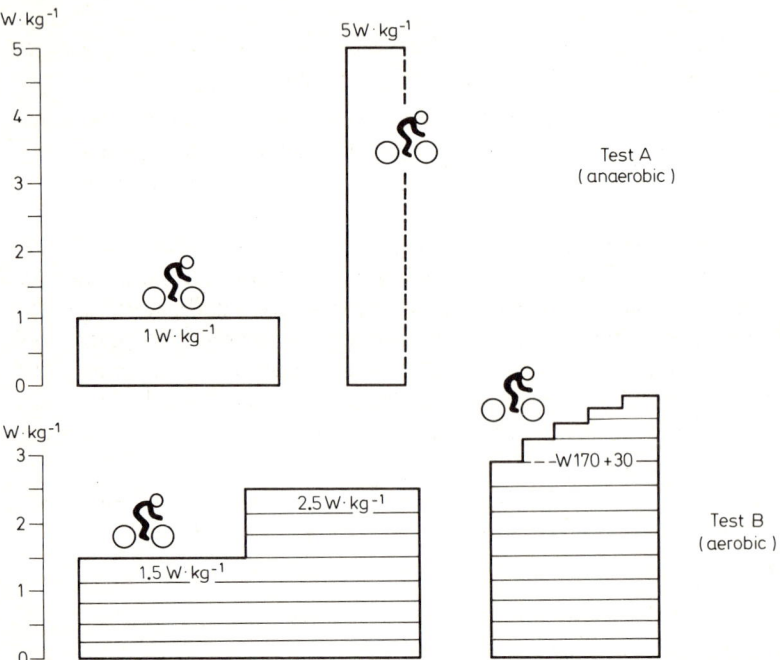

Fig. 1. Test procedures using bicycle ergometry used in the study. For explanation see the text

5 Wkg^{-1} body weight in subgroup C and 4 Wkg^{-1} body weight (BW) in subgroups A and B. A warming-up period of 6 min at 1 Wkg BW preceded the test in all subgroups. According to Fox [11] we presumed that this protocol would stress more pronounced anaerobic processes compared with aerobic ones. Test B consisted of stepwise increases in work load of 30 W/min, having started at the W_{170+30}W level. The load was increased until the subject was unable to continue. Two-step work load at 1.5 Wkg^{-1} BW and at 2.5 Wkg^{-1} BW in subgroup C and 1.0 Wkg^{-1} BW and 2.0 Wkg^{-1} BW in subgroups A and B without a rest interval preceded as a

warming-up and W_{170} was then calculated according to Wahlund [28]. There was at least a 3-h interval between tests A and B (Fig. 1). Only a light meal was allowed early in the morning and between both tests, but not later than 1 h prior to test B.

An open-air circuit system (Dataspir) was used for expired air analysis. An electrically braked bicycle ergometer with automatically regulated pedaling frequency braking product was used. The equipment was supplied by Jaeger (West Germany). In test A, basic cardiorespiratory parameters were registered every 10 s. In test B the intervals registered were prolonged to 30 s. Three minutes prior to the load and 7 min after the completion of each test resting values and/or recovery values of the followed cardiorespiratory parameters were also registered.

As criteria for maximal effort before exhaustion reached by the subject in test B, the following conditions were considered:

1. f_H (heart rate) reached a value higher then 190 beats min^{-1}
2. RQ reached a value higher then 1.1
3. $\dot{V}_{O_2}$ value in the last 30 s of the work load was at least 150 ml min^{-1} higher then the $\dot{V}_{O_2}$ value in the preceding 30 s, and so-called leveling-off appeared.

There were 62.5% of the subjects who fulfilled the first criterion, 75.0% of the subjects who fulfilled the second, and 68.8% of the subjects who fulfilled the third one. There was one subject who was not able to fulfill any of the crtieria mentioned above.

The time in seconds needed for 50% recovery of the $\dot{V}_E$, $\dot{V}_{O_2}$ (STPD), and f_H to resting levels was considered as recovery half-time (RHT). As recommended for evaluation of anaerobic lactic capacity [1, 20], arterialized capillary blood was drawn from the fingertip at the 1st, 3rd, 5th, and 7th min of the recovery period and Boehringer sets were used to determine lactate concentration. In the 3rd min of the recovery period, heparinized glass capillaries were used to collect blood samples for pH, base excess (BE), P_{CO_2}, and actual HCO_3^- determination with the help of a Radiometer (Denmark) pH meter.

Specific swimming performance was calculated in point score according to the formula used for score tables by the Czechoslovak Swimming Union:

$$b = \frac{10 \cdot {}^t 1000^3}{t}$$

where t is time achieved by the swimmer in his best evaluated discipline in s; t_{1000} is the average of the ten best times achieved in the world in the respective track, evaluated as 1000 points; and b is the resulting point score.

The peak value of $\dot{V}_{O_2}$ in both tests and also the peak values in other parameters were compared, the recovery half-times in $\dot{V}$, $\dot{V}_{O_2}$, and f_H were also compared. Finally, the data of blood lactate in the 1st, 3rd, and 5th min of recovery and those of acid-base balance were compared using the statistical evaluation by means of the Student's t test for paired and unpaired comparisons. A 95% level of confidence was accepted for all comparisons.

To further explore the significance of cardiorespiratory and acid-base balance parameters, correlation coefficients with the help of Spearman's analysis of ranks were calculated between selected predictor and criterion variables. The best point score for the swimming track was used as a criterion.

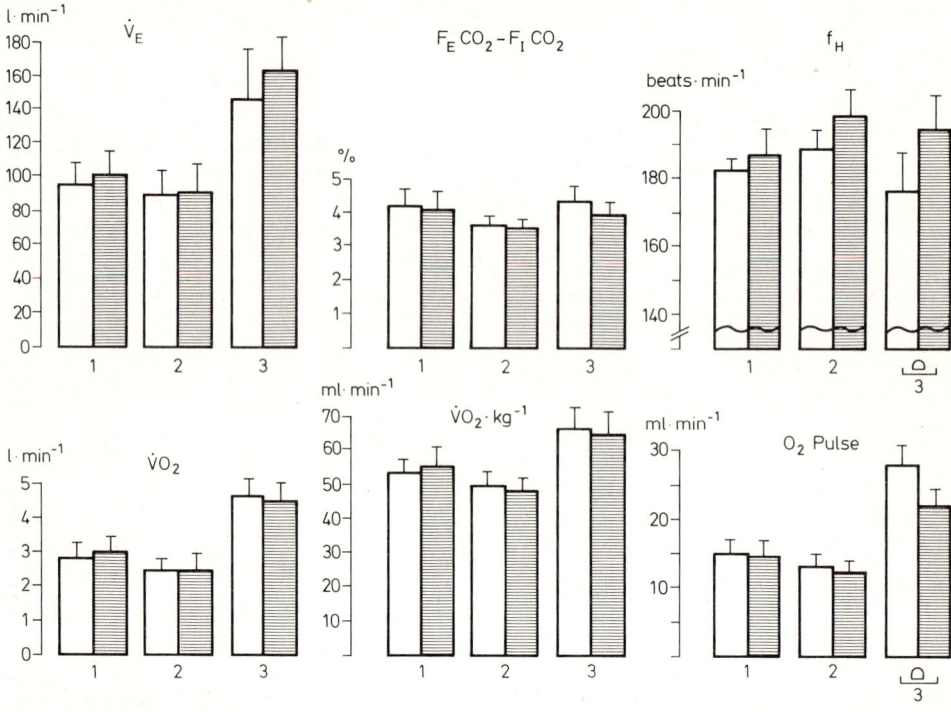

Fig. 2. Comparison of functional variables reached in test A *(white columns)* and in test B *(striped columns)* in subgroups 1 to 3. Significance signed by abscissa below the columns

Results

First of all, the data obtained in both types of bicycle ergometer protocols in each subgroup of swimmers were compared. Means for $\dot{V}_E$, F_ECO_2–F_ICO_{2max}, f_{Hmax}, and maximal aerobic power parameters are provided in Fig. 2. All parameters correspond to the highest values of $\dot{V}O_2$ measured in a 10-s interval in test A, and in a 30-s interval in test B, except f_H. The highest f_H value reached during each maximal load was considered without regard to $\dot{V}O_{2max}$ values. Duration of test A was on the average 96.7 ± 22.1 s in subgroup 1, 91.1 ± 23.3 s in subgroup 2, and 82.5 ± 20.9 s in subgroup 3. No significant differences in any of the mentioned parameters were found, except of f_{Hmax} and O_2 pulse $_{max}$ in subgroup 3. In that case, lower f_H and higher O_2 pulse in test A were found.

Figure 3 illustrates the means for acid-base balance parameters, lactate concentration, and hematocrit ratio. The data present relative changes of each parameter, being calculated as $x = x_{post\ load} - x_{rest}$. There were no statistically significant differences in any of these parameters, except La concentrations in subgroup 1 and 2 in the course of the recovery period. In all measurements Δ La was significantly higher in test B in these subgroups, while very slight, nonsignificant differences were found in subgroup 3. Hematocrit reached significantly lower values in test A compared with test B in subgroup 2, as Δ Htc ratio illustrates.

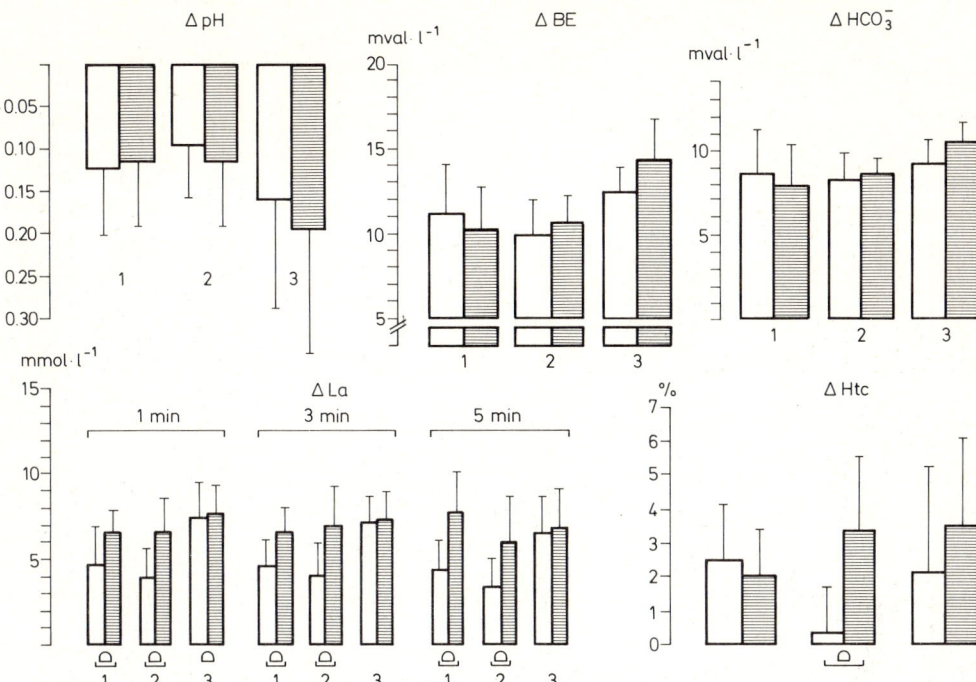

Fig. 3. Comparison of values of pH, base excess *(BE)*, actual bicarbonate *(HCO₃⁻)*, lactate *(La)*, and hematocrit *(Htc)* reached in test A *(white columns)* and in test B *(striped columns)* in subgroups 1 to 3. Significance signed by abscissa below the columns

Recovery half-times for $\dot{V}_E$, $\dot{V}_{O_2}$, and f_H are presented in Fig. 4. Significantly slower recovery in all parameters in group 3 was observed, while significantly slower recovery in $\dot{V}_{O_2}$ in all groups was found.

Spearman's analysis of ranks showed significant positive correlation between all parameters and performance level in both tests in subgroup 1 only (Table 2). On the contrary, there was no significant correlation found in subgroup 2, while a significant negative correlation was found between $\dot{V}_{O_2max}$ kg^{-1}BW in test A and $\dot{V}_{O_2max}$ in test B and performance level. No significant correlation was found in either group regarding the Δ La, Δ BE, or $\dot{V}_{O_2}$-RHT both in test A and in test B.

If all maximal aerobic power data ($\dot{V}_{O_2max}$ and $\dot{V}_{O_2max}$ kg^{-1}BW) in all male swimmers included in the study (subgroup A plus subgroup C) were subjected to correlation analysis, significantly positive correlation with performance level was found in test procedure B.

Discussion

Laboratory testing of swimmers' physiological predispositions to top specific swimming performance seems to be an often discussed problem. While Åstrand et al. [5] found higher values of $\dot{V}_{O_2max}$ during cycling on an ergometer, others [15] recently

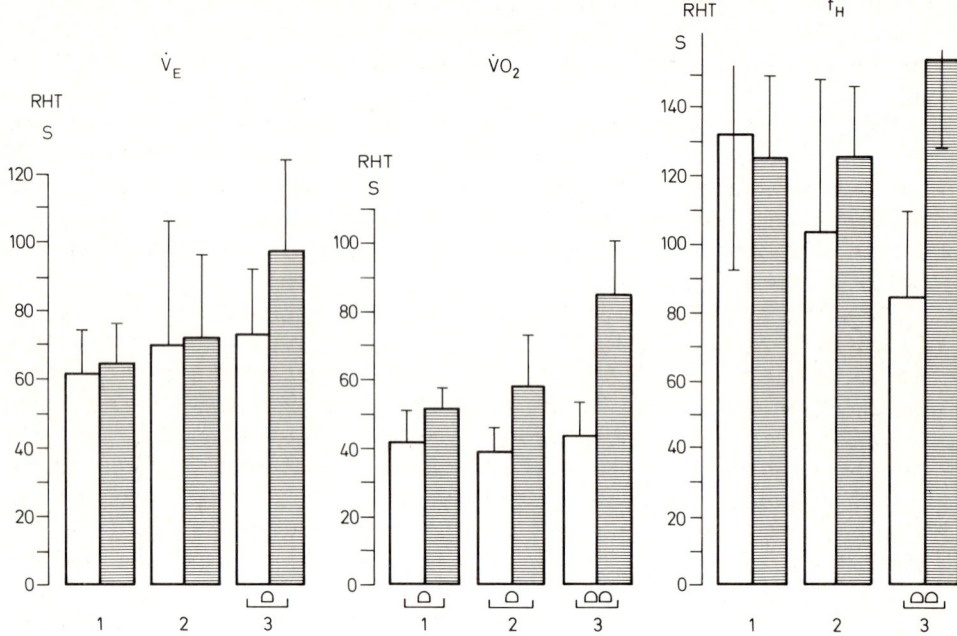

Fig. 4. Recovery half-times *(RHT)* for $\dot{V}$, $\dot{V}_{O_2}$, and f_H after exhaustion in test A *(white columns)* and in test B *(striped columns)* in subgroups 1 to 3. Significance signed by abscissa below the columns

Table 2. Correlation matrix between performance level and selected physiological data in young swimmers

Test A	$\dot{V}_{max}$	$F_E CO_2$ $-F_I FO_2$	f_{Hmax}	$\dot{V}_{O_{2max}}$	$\dfrac{\dot{V}_{O_{2max}}}{kg}$	O_2 pulse $\dot{V}_{O_2}$ RHT	La	BE
Subgroup 1 $n=6$	0.97[b]	0.95[a]	0.85[a]	0.99[b]	0.97[b]	0.99[b]	0.07	−0.20
Subgroup 2 $n=8$	0.72	−0.45	0.25	0.55	−0.30	0.46	0.23	0.23
Subgroup 3 $n=12$	−0.42	0.61[a]	−0.03	0.02	−0.67[a]	−0.15	−0.01	0.07	0.12
Male swimmers $n=18$				0.824[a]	0.61[a]			
Test B								
Subgroup 1 $n=6$	0.98[b]	0.88[a]	0.86[a]	1.00[b]	0.98[b]	0.98[b]	−0.04	0.10	0.60
Subgroup 2 $n=8$	0.50	−0.18	0.24	0.55	0.12	0.58	0.43	0.52	0.28
Subgroup C $n=12$	−0.72	−0.60	−0.31	−0.75	−0.51	−0.36	0.26	−0.25	−0.08
Male swimmers $n=18$				0.61[a]	0.73[b]			

La, lactate; *BE*, base excess
[a] $p < 0.05$
[b] $p < 0.01$

presented the opposite experience. In any case, top swimming performance on 100 m and 200 m tracks is highly dependent not only on high maximal oxygen uptake but also on the ability to support anaerobic processes as well [6, 12, 13, 16, 25, 26].

The level of maximal aerobic power in our group of swimmers largely exceeded the average Czechoslovak population data [19, 22] and comparable age groups of less trained swimmers [14, 17, 24] as well, but agree with the data of well-trained young swimmers in other studies [2, 3, 8, 9, 17, 19, 22, 23, 24, 27]. Unfortunately, very little attention was paid to the correlation between the functional data and specific performance level in those studies.

Regarding the data of anaerobic power reachable in swimmers both in training [21, 26] and in specific competitive performance [10], the results of our subgroups seem to be very poor in laboratory tests used. The reason for this phenomenon is presumably a low share of anaerobic power training in the total training amount in young swimmers. This problem is common in Czechoslovak swimming as a whole, and our results documented the consequences well: the stress on the lactic mechanism in training was apparently not sufficient to bring about significant alterations. Moreover, in subgroups 1 and 2 we observed higher Δ Lactate values in test B compared with test A, so that there was more pronounced mobilization of anaerobic processes in test B.

In subgroup 3 noticeable difference in the rise of f_H was found. It takes a longer time to reach the maximal value of f_H in older swimmers in test A. Therefore, maximal O_2 pulse was significantly higher in test A compared with test B, in spite of $\dot{V}_{O_2max}$ values being almost identical.

Regarding RHT values, the only significant difference appeared in all three groups: longer RHT for $\dot{V}_{O_2}$ in test B. This delayed repayment of the oxygen debt can be caused by elevated metabolism due to an increased tissue temperature, aerobic removal of anaerobic metabolites, and increased oxygen demand of activated respiratory muscles and the heart [5]. For the latter two points, there are well-documented arguments.

The presence of a significant correlation between performance point score and several physiological parameters has a logical background in subgroup 1 but not in the case of subgroup 3. It seems that in the small group of lower level swimmers with a rather wide range of performance level, physiological predispositions play a more important role than in the higher level sportsmen. It was also shown that the reaching of maximal exhaustion of functional systems in highly trained swimmers with the help of bicycle ergometry seems to be questionable in a relatively large number of subjects. This is perhaps the reason why in small groups of highly trained swimmers we can get false correlations, while having the larger group with wider range of performance the correlations correspond to commonly accepted assumptions. To explain the discrepancies in laboratory parameters and specific performance level in young swimmers, we must consider the following circumstances:

1. Specificity of the test regarding large muscle groups involved
2. Age group and performance level of the swimmers
3. Training structure and training phase during which laboratory examination was done

4. Relation to the date when the competitive result was reached
5. Motivation of the swimmer to work hard during laboratory tests
6. Level of exhaustion by means of objective parameters

Conclusion

In the training of three subgroups of swimmers the aerobic and possibly alactic anaerobic pathways were presumably the major energy sources utilized. That was the reason why there were no significant differences found in answer to two different test protocols (test A, maximal load of 1–2 min duration; test B, stepwise increased load to vita maxima) in $\dot{V}_E$, F_ECO_2–$FiCO_2$, $\dot{V}O_2$, $\dot{V}O_2 kg^{-1}$, and f_H in younger age groups. Despite the significant positive correlation between performance level and $\dot{V}O_{2max}$ and/or $\dot{V}O_{2max} kg^{-1}BW$ respectively in the subgroup A, or in the male swimmers as a whole, we presume that more valid results could be obtained with the help of more specific load. As it is untenable to use a swimming flume [4, 7] in swimmers' testing at the moment, the experience given by Lavoie et al. [15] using specific swimming performance in water might be an alternative possibility.

References

1. Apor P, Szabo-Wahlstab S, Miklos M (1973) Zusammenhänge zwischen einigen aeroben und anaeroben Parametern bei Spitzensportlern. In: Hansen G, Mellerowicz H (eds) 3. Internationales Seminar für Ergometrie 1970. Ergon, Berlin, pp 17–23
2. Andrew GM, Becklade MR, Guleria JS, Dates DV (1972) Heart and lung functions in swimmers and nonathlets during growth. J Appl Physiol 32: 245–251
3. Armstrong N, Davies B (1981) An ergometric analysis of age group swimmers. Br J Sports Med 15: 20–26
4. Åstrand P-O, Engström L, Eriksson DO, Karlberg P, Nylander I, Saltin B, Thoren C (1963) Girl swimmers. Acta Paed Scand [Suppl] 147: 1–74
5. Åstrand P-O, Rodahl K (1970) Textbook of work physiology. McGraw-Hill, New York
6. Bonner HN (1980) Energy systems used during swimming. Swimming Technique 17/3: 10–11
7. Burke ER, Falssatti VL, Kennedy JF, Harris OC, Potton GS (1982) The swimming flume. Application for research and training. Swimming World 23: 34–35
8. Cunningham DI, Eynon RB (1973) The working capacity of young competitive swimmers. Med Sci Sports 5: 227–231
9. Dietrich L (1967) Ergebnisse von spiroergometrischen Längsschnittuntersuchungen an trainierten Kindern und Jugendlichen im Schwimmen. In: Mellerowicz H, Hansen G (eds) 2. Internationales Seminar für Ergometrie. Inst. f. Leistungsmedizin, Berlin, pp 221–226
10. Di Prampero PE, Pendergast DE, Wilson DW, Rennie DN (1978) Blood lactic acid concentrations in high velocity swimmers. In: Eriksson B, Furberg D (eds) Swimming medicine IV. University Park Press, Baltimore, pp 249–261
11. Fox EL (1975) Differences in metabolic alteration with sprint versus endurance interval training program. In: Howald H, Poortmans JR (eds) Metabolic adaptation to prolonged physical exercise. Birkhäuser, Basel, pp 119–126
12. Foulkner JA (1966) Physiology of swimming. Res Q 37: 41–54
13. Holmér I (1972) Oxygen uptake during swimming in man. J Appl Physiol 33: 502–509
14. Jirka Z (1978) The development of basic functional parameters in boy and girl swimmers at age 12–15 years (in Czech). Acta Univ Palacki Olom 60: 125–150
15. Lavoie JM, Léger LA, Montpetit RR, Chabot S (1982) Backward extrapolation of $\dot{V}O_2$ from the O_2 recovery curve after a voluntary maximal 400 m swim. In: Biomechanics and medicine in swimming, Amsterdam, p 27 (abstract)

16. Magel JR, Foglia GP, McArdle ND, Gutin B, Pechar GS, Katch FI (1974) Specificity of swim training on maximum oxygen uptake. J Appl Physiol 38: 151–155
17. Nomura T (1979) Maximal oxygen uptake of age group swimmers. Swimming Technique 15: 105–109
18. Novák J (1982) Longitudinal development of the state of health in young swimmers. In: Semi-ginovský B, Tuček S (eds) Metabolic and functional changes during exercise. Charles University Press, pp 305–311
19. Novák J, Rouš J (1979) Health problems of swimming (in Czech). In: Health problems in water sports. Sportpropag, Prague, pp 61–78
20. Schönholzer G, Bieler G, Howald G (1973) Ergometrische Methoden zur Messung der aeroben und der anaeroben Kapazität. In: Hansen G, Mellerowicz H (eds) 3. Internationales Seminar für Ergometrie 1970. Ergon, Berlin, pp 17–23
21. Švarc V, Novák J (1975) The changes in acid balance during interval swimming training in trained and untrained men. In: Howald H, Poortmans JR (eds) Metabolic adaptation to prolonged physical exercise. Birkhäuser, Basel, pp 73–77
22. Seliger V, Bartůněk Z (1976) Mean values of various indices of physical fitness in the investigation of Czechoslovak population aged 12–55 years. ČSTV, Prague
23. Šprynarová Š, Juřina K (1967) Aerobic capacity and performance level of Czechoslovak internationals in swimming (in Czech). Teor Praxe Těl Vých 15: 412–419
24. Šprynarová Š, Pařízková J, Juřinová I (1978) Development of the functional capacity and body composition of boy and girl swimmers aged 12–15 years. Med Sport 11: 32–38
25. Stejskal P, Jirka Z (1983) Metabolic response to the training and competition load in swimmers (in Czech). Teor Praxe Těl Vých 31: 276–281
26. Trefene RJ, Craven C, Hobbs K (1976) Differences in sprint and endurance swimming plasma lactate and heart rate response in controlled swims. Swimming Technique 16/2: 34–36
27. Vaccaro P, Clarke DH, Morris AC (1980) Physiological characteristics of young well trained swimmers. Eur J Appl Physiol 44: 61–66
28. Wahlund H (1948) Determination of physical working capacity. Acta Med Scand 132 [Suppl 215]: 1–228

Determination of the Anaerobic Threshold in Various Ergometric Tests

V. Bunc, J. Heller, I. Bojanovský, Š. Šprynarová, and J. Novák

Introduction

Since the evaluation of the effects of training and rehabilitation is primarily based on submaximal energy expenditure, the corresponding work load in ergometry is in the range of the anaerobic threshold (AT). AT in this context means the transition from aerobic to anaerobic metabolism and is defined as the maximum work load with a constancy of lactate production and utilization [9, 14]. In addition to the approach using the AT as a measure of submaximal power output, relative values of maximal oxygen uptake ($\dot{V}O_{2max}$) are often used to characterize submaximal work intensity. For untrained subjects, AT is in the range of 50%–70% $\dot{V}O_{2max}$ [8, 10, 11, 18, 22], whereas in trained persons the AT lies in the range of 80%–90% $\dot{V}O_{2max}$ [11–13, 15, 18, 22]. $\dot{V}O_{2max}$ depends on the methodological approach to exercise testing and on the specificity of the exercise testing device [26, 27].

This paper presents data on some parameters at the AT examined in a group of highly trained athletes using treadmill and bicycle ergometers and some special testing devices.

Material and Methods

The study was performed in young and adult female canoeists, in young and adult middle distance and long-distance runners, and in football players. In addition, pentathlonists, marathon runners, and skiers were studied.

Measurements were performed on a treadmill (Fa. Jaeger; inclination 5%; initial speed 12 km/h–15 km/h, increased by 1 km/h every minute to exhaustion; the increase was 0.5 km/h every minute for young athletes). The bicycle test (Fa. Jaeger) was started from an initial load of W_{170} plus 20 W with an increase of 20 W every minute till exhaustion. Further, a paddling ergometer (own construction) was used, starting the tests with about 60% of $\dot{V}O_{2max}$ and increasing work load by about 10 W (adults) or 5 W (young athletes) every minute.

Respiratory gas exchange was analyzed in an open system (Fa. Jaeger) in a standardized way. The following parameters were assessed: minute ventilation, oxygen uptake, heart rate (cardiotachometer, Hellige). Results were printed every 30 s, the analysis was performed using a computer-assisted device (Fa. Jaeger).

The AT was derived from measurements of minute ventilation ($\dot{V}_E$) and $\dot{V}O_2$ [2, 4, 16, 17, 22–24] (Fig. 1). We used a discontinuous linear model with two parts [4] assuming that the initial load is about 40%–60% $\dot{V}O_{2max}$ [4, 8] and that anaerobic metabolism will be reached [1, 25].

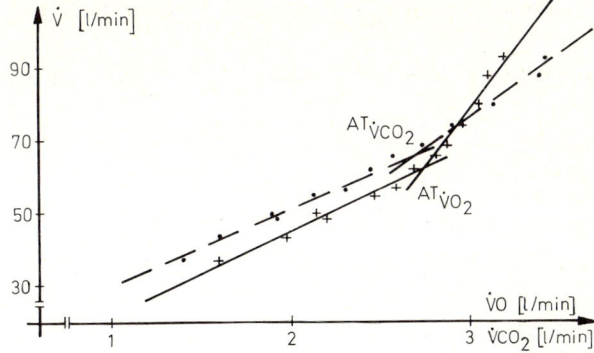

Fig. 1. Scheme of noninvasive determination of the AT from the relationship of $\dot{V}_E$ on $\dot{V}O_2$ and $\dot{V}CO_2$ by means of two-part discontinuous linear model

Table 1. Physical characteristics of subjects (treadmill at 5% grade)

	Age (years)	Height (cm)	Weight (kg)	$\dot{V}O_{2max}$ (l/min)	$\dot{V}_{Emax}$ (l/min)	HR_{max} (beats/ min)	v_{max} (km/h)	n
Marathon runners[a]	25.60	176.22	64.76	4.68	132.95	192.00	19.69	13
	2.94	6.10	3.10	0.39	17.77	6.56	0.61	
5–10 km runners[a]	24.10	176.00	62.02	4.64	143.15	189.86	21.29	7
	3.26	2.71	3.43	0.40	13.62	8.3	0.45	
Long-distance runners (young)[a]	17.00	178.03	63.80	4.25	118.63	195.06	17.85	17
	0.66	4.52	4.91	0.43	19.06	6.33	0.89	
Middle-distance runners[a]	23.80	180.29	66.40	4.71	149.76	196.25	21.00	8
	3.12	2.96	3.96	0.33	6.99	6.40	1.15	
Middle-distance runners[b]	23.10	166.20	53.24	3.65	113.55	190.33	17.00	6
	2.90	5.30	4.90	0.47	15.04	7.99	0.58	
Middle-distance runners (young)[b]	16.10	163.40	51.00	2.78	91.12	201.35	16.13	12
	0.60	4.60	5.30	0.34	11.43	7.11	0.89	
Cross-country skiers (young)[a]	17.30	175.20	65.47	4.27	131.50	199.00	19.57	7
	0.94	4.76	5.15	0.32	7.80	3.81	0.50	
Modern pentathlon[a]	25.70	183.40	79.30	5.16	163.05	197.89	17.90	10
	4.26	4.60	5.20	0.74	20.11	10.33	1.22	
Canoeists[a]	22.80	180.18	79.15	4.93	148.50	192.25	18.46	28
	3.30	5.60	5.46	0.64	27.72	8.31	0.98	
Canoeists (young)[a]	17.40	178.73	74.53	4.64	137.42	198.00	17.80	10
	0.80	7.30	5.11	0.43	20.06	7.09	0.75	
Canoeists[b]	21.05	169.40	67.32	3.13	111.04	194.67	15.17	6
	2.11	4.26	6.20	0.25	8.44	9.93	0.69	
Football players[a]	24.89	182.67	78.73	4.80	146.07	186.17	16.67	15
	3.41	5.50	6.22	0.41	15.10	8.92	1.07	

Means and standard deviations
[a] Male
[b] Female

Table 2. Physical characteristics of subjects

	Age (years)	Height (cm)	Weight (kg)	$\dot{V}_{O_{2}max}$ (l/min)	$\dot{V}_{Emax}$ (l/min)	HR_{max} (beats/ min)	Perfor- mance (W)	n
Table tennis[a]	25.20	174.30	76.20	4.33	150.33	180.50	328.33	6
	5.64	4.32	4.96	0.51	23.08	4.79	44.25	
Water slalom[a]	24.90	182.40	82.30	4.62	186.80	166.77	348.00	10
	4.12	5.24	5.24	4.36	7.87	21.11	29.16	
Canoeists (young)[b]	16.00	168.69	64.89	2.87	97.41	192.19	230.00	8
	0.87	4.25	5.70	0.24	12.81	10.12	38.15	
Canoeists (young)[b]	16.00	168.69	64.89	2.54	86.76	186.85	89.01	8
	0.87	4.25	5.70	0.18	7.06	7.92	7.12	
Rowers[a]	25.70	189.90	88.00	5.25	196.90	192.10	420.90	15
	3.54	4.93	7.49	0.41	17.80	9.81	25.61	
Rowers[a]	25.70	189.90	88.00	5.14	179.20	174.30	3.07	15
	3.54	4.93	7.49	0.16	11.03	7.12	0.12	
Ice hockey[a]	26.80	182.93	80.67	4.43	151.85	188.39	375.44	23
	3.60	6.24	6.94	0.41	20.97	10.88	28.20	

Means and standard deviations. Bicycle ergometer; paddling ergometer; rowing ergometer
[a] Male
[b] Female

Results

Data obtained by the treadmill test are shown in Table 1, those analyzed by bicycle and paddling ergometer are presented in Table 2. $\dot{V}_{O_{2}max}$ is highest in rowers and in modern pentathlonists due to their physical stature and condition. Relating oxygen uptake to body weight, high values for $\dot{V}_{O_{2}max}$ occur in long-distance runners. Similar results can be observed for minute ventilation (Table 1). Heart rate is highest in young athletes, and there is a decrease in maximal heart rate with age. Highest values for the speed of running occur in middle-distance runners (Table 1).

In addition, results of AT analysis are shown in Table 3 (treadmill) and Table 4 (bicycle and paddling ergometer). Maximum oxygen uptake is highest in those athletes with the highest body mass. There is a strong correlation between $\dot{V}_{O_{2}max}$ and $\dot{V}_{O_{2}}$ at AT level.

The values of $\dot{V}_{O_{2}}$ at AT are highest in athletes with long performance duration at competitive activity, that means in long-distance runners and in cross-country skiers. Similar observations can be made for $\dot{V}_{E}$. There is a good and close (positive) correlation between $\dot{V}_{Emax}$ and $\dot{V}_{E}$ at AT level. Again, highest values have been reached by long-distance runners or cross-country skiers. In young athletes, heart rate at AT and maximal heart rate demonstrate a close, positive correlation.

In all groups studied here, there was a negative correlation between parameters at maximal performance and relative results (percentage of maximal values) at the AT level. This also holds true for analysis of maximal running speed on the treadmill and for the maximal power output obtained on the bicycle ergometer (Tables 1–4). Results for oxygen uptake, ventilation, heart rate, and watts given in percentage of maximal values are higher on the paddling and rowing ergometer than on the bicycle ergometer.

Table 3. Cardiorespiratory variables at the anaerobic threshold (treadmill at 5% grade)

	$\dot{V}_{O_2}$ (l/min)	% $\dot{V}_{O_{2max}}$	$\dot{V}_E$ (l/min)	%$\dot{V}_{Emax}$	HR (beats/ min)	% HR_{max}	v (km/h)	%v_{max}
Marathon runners[a]	4.15	86.67	98.15	73.82	180.31	93.91	16.09	86.29
	0.37	1.75	13.04	2.46	7.02	4.15	0.42	3.26
5–10 km runners[a]	3.96	85.29	95.36	66.62	172.14	90.67	18.38	86.33
	0.27	5.13	15.56	4.16	6.73	3.26	0.42	2.12
Long-distance runners (young)[a]	3.62	85.16	87.31	72.69	181.82	93.41	15.31	85.77
	0.39	2.90	11.39	7.22	6.80	1.44	0.70	2.17
Middle-distance runners[a]	3.90	82.80	92.13	61.52	178.13	90.77	17.23	82.05
	0.20	3.96	7.57	5.12	6.03	4.26	0.99	3.12
Middle-distance runners[b]	3.02	82.73	79.19	69.74	178.00	93.52	14.12	83.06
	0.35	3.65	10.46	2.86	5.89	4.56	0.50	0.58
Middle-distance runners (young)[b]	2.30	82.73	59.42	65.00	186.00	92.38	12.93	80.16
	0.30	2.56	7.01	4.15	7.53	5.12	0.64	4.16
Cross-country skiers (young)[a]	3.65	85.39	100.50	76.36	182.71	91.81	15.53	79.36
	0.47	4.76	10.00	5.99	7.07	2.92	0.68	3.52
Modern pentathlon[a]	4.26	82.62	104.90	64.33	182.36	92.15	13.93	77.82
	0.67	3.08	15.64	6.12	9.86	4.96	1.09	3.12
Canoeists[a]	4.01	81.34	99.14	66.76	176.86	91.99	14.66	79.40
	0.46	6.26	16.56	4.94	7.70	5.36	0.79	5.12
Canoeists (young)[a]	3.69	78.91	92.40	66.79	181.80	91.82	14.10	79.28
	0.44	3.07	11.02	3.61	7.74	2.57	0.49	2.26
Canoeists[b]	2.59	82.74	70.67	63.67	181.67	93.33	12.35	81.41
	0.19	4.12	6.87	3.68	5.73	6.15	0.48	3.12
Football players[a]	3.87	80.53	96.73	66.22	173.20	93.03	13.33	79.94
	0.27	2.53	7.29	3.26	8.54	4.26	0.78	4.26

Means and standard deviations
[a] Male
[b] Female

Table 4. Cardiorespiratory variables at the anaerobic threshold

	$\dot{V}_{O_2}$ (l/min)	% $\dot{V}_{O_{2max}}$	$\dot{V}_E$ (l/min)	%$\dot{V}_{Emax}$	HR (beats/ min)	% HR_{max}	Perfor- mance (W)	%P_{max}
Table tennis[a]	3.55	81.92	89.33	59.42	161.73	89.60	232.90	70.96
	0.33	2.73	12.88	4.56	3.86	2.46	26.15	3.81
Water slalom[a]	3.67	79.52	93.50	56.00	169.73	90.86	238.45	68.50
	0.37	2.36	13.55	3.12	6.24	4.16	17.20	6.26
Canoeists (young)[b]	2.13	74.20	58.94	60.51	172.78	89.90	168.00	73.43
	0.19	5.60	6.26	8.54	7.36	3.20	34.21	8.16
Canoeists (young)[b]	2.15	84.80	69.63	80.26	178.27	95.41	79.38	89.18
	0.14	4.72	5.12	5.28	4.26	2.14	5.81	6.14
Rowers[a]	3.92	74.60	113.22	57.50	171.55	89.30	293.79	69.80
	0.40	6.23	12.60	3.28	7.42	4.26	19.80	4.78
Rowers[a]	4.37	85.00	140.50	78.80	174.30	96.20	2.61	85.00
	0.16	1.40	19.47	4.29	7.12	3.04	0.16	4.05
Ice hockey[a]	3.19	72.02	80.57	56.06	168.22	89.29	266.74	71.08
	0.38	5.50	15.03	4.20	10.77	4.96	29.59	5.64

Means and standard deviations. Bicycle ergometer; paddling ergometer; rowing ergometer
[a] Male
[b] Female

Discussion

Maximal oxygen uptake is generally accepted as a reliable criterion of cardiopulmonary performance capacity. Recently, the anaerobic threshold became of great interest as a measure of physical fitness. Anaerobic threshold deserves a submaximal stress test because only in this way can the necessity of a maximal exhausting stress test be avoided. In a previous study from this laboratory, the noninvasive approach to AT determination on the basis of $\dot{V}_E/\dot{V}_{O_2}$ ratio has been validated and this procedure has proved to be reliable [4].

In earlier studies, a fixed level of lactate accumulation (4 mmol/l) was suggested as the index of AT [14]. However, as was shown in some earlier reports [2–7], AT determined by analysis of the individual metabolic response is lower than the value of 4 mmol/l [2–7]. The range of AT in top athletes is about 2–3 mmol/l and decreases with age [4–7, 19].

The results of this investigation present another approach to measure physical fitness using a submaximal exercise test and the AT concept. As was shown, the relative values, i.e., percentages of values at peak stress test, correspond well to AT. High relative values (e.g, percentage of $\dot{V}_{O_{2max}}$) can be found in highly trained endurance athletes.

One more diagnostic finding of this study is that AT depends on the type of testing and the kind of sport which actually is performed by the athlete [19–21]. For example, AT or relative $\dot{V}_{O_2}$ is near normal in canoeists and rowing athletes when testing on a bicycle ergometer. However, the results are far better if the same athletes are tested with a specific sport-related test device such as a paddling or rowing ergometer [26]. This is confirmed by the findings in long-distance runners: Treadmill testing is more suitable and more linked to the type of testing and yields better results than testing on a bicycle.

The results of this study can be summarized as follows:

AT given as a percentage of $\dot{V}_{O_{2max}}$ is higher the longer the duration of the competition.

Relative values of parameters such as $\dot{V}_E$ or $\dot{V}_{O_2}$ can be determined and related to AT. These results are given as a percentage of a value at AT. The higher the relative values of AT, the better the physical fitness. However, these relative values depend on the type of testing and are strongly related to the specific kind of sports activity performed by the athlete.

Relative values at AT do not differ between the young and adults or between men and women. This might be assumed to be an advantage for daily routine analysis of athletes' physical fitness.

References

1. Bachl N, Reiterer W, Prokop L, Czitober H (1978) Bestimmungsmethoden der anaeroben Schwelle. Österr J Sportmed 8: 9–12
2. Bunc V, Leso J, Heller J, Novák J, Strejčková B, Novotný V (1981) Anaerobic threshold by specific and nonspecific load. Lékař a TV 3: 35–37 (in Czech)
3. Bunc V, Böswart J (1982) The use of anaerobic threshold at stress diagnostics. Čas Lék Česk 121: 1225–1229 (in Czech)

4. Bunc V, Heller J, Novák J, Šprynarová Š, Leso J (1983) Comparison of invasive and non-invasive determination of anaerobic threshold. Teorie a Praxe TV 31: 114–118 (in Czech)
5. Bunc V, Heller J, Leso J, Novák J (1982) Determination of the individual anaerobic threshold. XXIInd World congress on sports medicine. Int J Sports Med 3: 11 (abstract)
6. Bunc V, Heller J, Šprynarová Š, Zdanowicz R (1983) Mathematical description of changes in BE with increasing load. Physiol Bohemoslov (to be published)
7. Bunc V, Heller J, Leso J, Šprynarová Š, Kuta I, Novák J (1983) Seuil anaérobique au cours de la charge spécifique et non spécifique. J Physiol (Paris) (to be published)
8. Bunc V, Heller J, Šprynarová Š, Zdanowicz R (1983) Respiratory parameters in exercise diagnostics. I. Determination of anaerobic threshold in laboratory using the changes of some respiratory parameters. Congress SEPCR, Bratislava
9. Costill DL (1970) Metabolic responses during distance running. J Appl Physiol 28: 251–255
10. Daniels JT, Yarbrough A, Foster C (1978) Changes in $\dot{V}O_{2max}$ and running performance with training. Eur J Appl Physiol 39: 249–254
11. Hollmann W, Liesen H (1973) Über die Bewertbarkeit des Lactats in der Leistungsdiagnostik. Sportarzt 24: 175–182
12. Ivy JL, Withers RT, Van Handel RJ, Elger DH, Costill DL (1979) Muscle respiratory capacity and fiber type as determinants of the anaerobic threshold. J Appl Physiol 48: 523–527
13. Katch V, Weltman A, Sady S, Freedson P (1978) Validity of the relative percent concept for equating training intensity. Eur J Appl Physiol 39: 219–227
14. Keul J, Simon G, Berg A, Dickhuth H, Gürtler I, Kübe R (1979) Bestimmung der individuellen anaeroben Schwelle zur Leistungsbewertung und Trainingsgestaltung. Dtsch Z Sportmed 30: 212–218
15. Kindermann W, Simon G, Keul J (1979) The significance of the anaerobic-aerobic transition for the determination of work load intensities during endurance training. Eur J Appl Physiol 42: 25–34
16. Lehmann M, Keul J, Hesse A (1982) Zur aeroben und anaeroben Kapazität sowie Catecholaminexkretion von Kindern und Jugendlichen während langdauernder submaximaler Körperarbeit. Eur J Appl Physiol 48: 135–145
17. Leso J, Bunc V, Máček M, Pirič J (1982) Energy supply at the level of anaerobic threshold. Teorie a Praxe TV 30: 366–370 (in Czech)
18. Liesen H, Mader A, Heck H, Hollmann W (1977) Die Ausdauerleistungsfähigkeit bei verschiedenen Sportarten unter besonderer Berücksichtigung des Metabolismus: zur Ermittlung der optimalen Belastungsintensität im Training. Beiheft zu Leistungssport 9: 8–62
19. Liesen H, Dufaux B, Heck H, Mader A, Rost R, Lötzerich S, Hollmann W (1979) Körperliche Belastung und Training im Alter. Dtsch Z Sportmed 30: 218–226
20. Mader A, Liesen H, Heck H, Philipi H, Rost R, Schürch P, Hollmann W (1976) Zur Beurteilung der sportartspezifischen Ausdauerleistungsfähigkeit im Labor. Sportarzt 27: 80–88, 109–112
21. Pugh LGCE (1970) Oxygen intake in track and treadmill running with observation on the effect of air resistance. J Physiol (Lond) 207: 823–835
22. Reiterer W (1977) Kriterien körperlicher Leistungsfähigkeit. Wien Med Wochenschr 127 [Suppl 42]: 1–19
23. Reiterer W, Bachl N, Czitober H, Prokop L (1979) Verlaufsbeobachtung über die Dauerleistungsfähigkeit von Skilangläufern mittels rechnerunterstützter Ergospirometrie. Österr J Sportmed 9: 8–12
24. Skiner JS, Mc Lellan TH (1980) The transition from aerobic to anaerobic metabolism. Res Q Exerc Sport 51: 234–248
25. Weltman A, Katch V (1979) Relationship between onset of metabolic acidosis (anaerobic threshold) and maximal oxygen uptake. J Sports Med Phys Fitness 19: 135–142
26. Withers RT, Sherman WM, Miller JM, Costill DL (1981) Specificity of the anaerobic threshold in endurance trained cyclists and runners. Eur J Appl Physiol 47: 93–104
27. Yoshida T, Suda Y, Takeuchi N (1982) Endurance training regimen based upon arterial blood lactate: effects on anaerobic threshold. Eur J Appl Physiol 49: 223–230

Appendix
Proposals for Quality Control in Ergometry

The following proposals comprise recommendations which have been suggested and elaborated at the 5th International Seminar on Ergometry in Titisee/Freiburg in 1983. This list contains recommendations made by the American Heart Association (Table 1), a checklist for quality control (Table 2), and a glossary of terms currently in use in ergometry (Table 3) (for references see the contribution by H. Löllgen).

Table 1. Checklist for operation of an exercise laboratory from the AHA

...	Physician present for exercise test
...	Laboratory well ventilated with comfortable temperature and humidity range
...	Calibrated bicycle (date) ergometer or treadmill or step ergometer
...	Oscilloscope visible at working distance
...	ECG machine meeting AHA standards
...	Defibrillator present and tested before each day of use
...	Procedures outlined for management of cardiopulmonary complication
...	Medical kit fully stocked with drugs which are within the date for use
...	Procedure for accepting patients
...	Pre-exercise ECG
...	Pre-exercise procedures for history taking and examination by physician
...	Exercise protocol
...	Initial and subsequent loads identified
...	End point identified
...	Indications to stop test clearly understood by operator of test
...	ECG interpreted by physician
...	Exercise test interpretation adequate
...	Work capacity recorded in patient's chart
...	ST changes recorded in patient's chart
...	Dysrhythmias recorded in patient's chart
...	Format for report to referring physician

Date: Physician:

Table 2. Checklist for quality control in ergometry

Ergometer: Construction following the standardization guidelines as suggested by the ICSSPE 1983 [23]. (This concerns width of base, visible indications for speed and power etc.)
Calibration before the purchase by the vendor or independent institutions, later on every two years. Detailed information on ergometers (construction, calibration) by the Bundesanstalt für Technische Überwachung, Braunschweig, Germany.
Treadmill: Construction as recommended by the American Heart Association. Regular calibration of speed and elevation; service after 1000 h of use.
Recorders: ECG: see recommendations of the American Heart Association.
Direct writers: Control of stability of *gain* and *baseline*.
Response time should be adequate for the signal to be recorded. Chart speed control: accuracy of paper speed better than 99.5%. Definite frequency limits.
Receptors of air flow: For gases, see Smidt et al. [17]; for pressures, see recommendations of the American Heart Association
Environment: Regular analysis of temperature and humidity (regular recording!)
Emergency equipment: Checked every 4 weeks (defibrillator).
Drugs: checked every 4 weeks.
Procedure: training every 3 months.
Regular Checks for the following points: medical history, physical examination, ECG, medication, instructions to the patient, informed consent *available before the exercise test*
Laboratory staff: Continuing training, education, emergency training, testing procedure known to the operator (protocol, endpoint, indications and contraindications).
Documentation: Regular documentation of the checklist items.
Self-certification: Special checklist to be filled out every six months and to be maintained in the laboratory files.

Table 3. Glossary of terms

Simplicity	Quality of a test with a complement of brevity for the operator (although it sometimes requires complex apparatus)
Acceptability	Lack of (a) the need for active cooperation and motivation on the part of the subject, (b) complexity of test, and (c) unpleasantness for the subject.
Objectivity	Ability of a test to provide results independent of the instruments used, of the subjects' emotional participation, and of the operator's personality (lack of subjective influences).
Reproducibility	Lack of variation of test results after repeated measurements; reproducibility and variation are measured in terms of confidential interval, standard deviation, and coefficient of variation. Error is due to instrumental or observer "bias" or to random biological variability. Bias is avoided or decreased by multiple or at least duplicate measurements. Variation should be small in relation to its absolute magnitude. Coefficient of variation should not exceed 8% to 10%.
Reliability	Overall variability of a test, including *intra-* and *inter*individual deviations. Physical reliability includes the manufacturing tolerances of an apparatus and the ease of maintenance and availability of spares.
Discrimination	Ability of a test to describe a functional attitude which may become impaired during the course of a particular disease, by means of distinctly demonstrating the difference between healthy and diseased subjects. There should be little overlap between the "normal" and "abnormal" values, meaning that the scatter of results is small.
Validity	Physiological appropriateness. The appropriateness of the test for detecting or for proving the impaired function or disease.
Sensitivity	Probability of a positive test result in patients with disease.
Specificity	Probability of a negative test result in patients without disease.

Table 3 (continued)

False positive	Positive test result when disease is not present.
False negative	Negative test result when disease is present.
Predictive accuracy	Percentage of positive results in patients with disease (true positive results); gives the predictive value of a positive test.
Pretest likelihood	Probability of disease in a patient to be tested:
	Number of patients with disease in the test population
	Total number of patients in the test population
Posttest likelihood	Number of patients with disease showing a given test result
	Total number of patients showing the test result
Likelihood ratio	Ratio of the true positive ratio to the false positive ratio.
Prevalence	Number of patients with disease in the number of patients in the study group (or in a population of 100000 subjects).
Risk ratio	Percentage of subjects with a positive test who manifest a disease in relation to the percentage.
Incidence	The number of cases which become manifest in a period of time of subjects with negative test who manifest a disease.
Summary	(Abbreviations: TP: true positive
	TN: true negative
	FN: false negative
	FP: false positive)

$$\text{Sensitivity (\%)} = \frac{TP}{TP+FN} \times 100$$

$$\text{Specificity (\%)} = \frac{TN}{TN+FP} \times 100$$

$$\text{False positives (\%)} = \frac{FP}{TP+FP} \times 100$$

$$\text{False negatives (\%)} = \frac{FN}{TN+FN} \times 100$$

$$\text{Predictive value (\%)} = \frac{TP}{TP+FP} \times 100$$

$$\text{Risk ratio} = \frac{TP}{TP+FP} \Big/ \frac{FN}{FN+TN}$$

Subject Index

Arterial Hypertension

Pathogenesis, Diagnosis, and Therapy

Editor: **J. Rosenthal**
With contributions by numerous experts
Foreword by I. H. Page
1982. 209 figures. XV, 529 pages
ISBN 3-540-90611-8
Distribution rights for Japan: Igaku Shoin Ltd., Tokyo

Atherosclerosis VI

**Proceedings of the Sixth International
Symposium**

Editors: **F. G. Schettler, A. M. Gotto, G. Middelhoff,
A. J. R. Habenicht, K. R. Jurutka**
1983. 264 figures, 214 tables.
XXVIII, 982 pages
ISBN 3-540-11450-5

Beta-Blockers
in the Elderly

Editors: **E. Lang, F. Sörgel, L. Blaha**
With contributions by numerous experts
1982. 29 figures, 17 tables. X, 107 pages
ISBN 3-540-11682-6

Springer-Verlag
Berlin
Heidelberg
New York
Tokyo

Coronary Artery
Diseases

Diagnostic and Therapeutic Imaging Approaches

By **M. Amiel, A. Maseri, H. Petitier, N. Vasile**
With contributions by numerous experts
1984. 175 figures, 15 tables. Approx. 250 pages
ISBN 3-540-13209-0

Fluid Dynamics as a Localizing Factor for Atherosclerosis

The Proceedings of a Symposium Held at Heidelberg, FRG, June 18–20, 1982

Editors: G. Schettler, R. M. Nerem, H. Schmid-Schönbein, H. Mörl, C. Diehm
1983. 90 figures. X, 230 pages
ISBN 3-540-12393-8

Hypertrophic Cardiomyopathy

The Therapeutic Role of Calcium Antagonists

Editors: M. Kaltenbach, S. E. Epstein
1982. 172 figures. XIV, 334 pages
ISBN 3-540-11065-8

F. Z. Meerson
Adaptation, Stress, and Prophylaxis

Translated from the Russian by J. Shapiro
1984. 87 figures, 61 tables. X, 329 pages
ISBN 3-540-12363-6

Microcirculation of the Heart

Theoretical and Clinical Problems

Editors: H. Tillmanns, W. Kübler, H. Zebe
1982. 177 figures. XIV, 353 pages
ISBN 3-540-11346-0

Myocardial Infarction at Young Age

International Symposium Held in Bad Krozingen, January 30 and 31, 1981

Editor: H. Roskamm
1981. 83 figures. XII, 228 pages
ISBN 3-540-11090-9

Silent Myocardial Ischemia

Editors: W. Rutishauser, H. Roskamm
1984. 71 figures, 49 tables. XVIII, 206 pages
ISBN 3-540-13193-0

Viral Heart Disease

Editor: H.-D. Bolte
1984. 119 figures, 45 tables. XV, 248 pages
ISBN 3-540-13112-4

Springer-Verlag
Berlin
Heidelberg
New York
Tokyo